# COMPUTER VISION FOR ASSISTIVE HEALTHCARE

# Computer Vision and Pattern Recognition Series

## Series Editors

**Horst Bischof**  Institute for Computer Graphics and Vision, Graz University of Technology, Austria

**Kyoung Mu**  Department of Electrical and Computer Engineering, Seoul National University, Republic of Korea

**Sudeep Sarkar**  Department of Computer Science and Engineering, University of South Florida, Tampa, United States

Also in the Series:

Lin and Zhang, Low-Rank Models in Visual Analysis: Theories, Algorithms and Applications, 2017, ISBN: 9780128127315

Zheng et al., Statistical Shape and Deformation Analysis: Methods, Implementation and Applications, 2017, ISBN: 9780128104934

De Marsico et al., Human Recognition in Unconstrained Environments: Using Computer Vision, Pattern Recognition and Machine Learning Methods for Biometrics, 2017, ISBN: 9780081007051

Saha et al., Skeletonization: Theory, Methods and Applications, 2017, ISBN: 9780081012918

Murino, Group and Crowd Behavior for Computer Vision, ISBN: 9780128092767

# COMPUTER VISION FOR ASSISTIVE HEALTHCARE

Edited by

**MARCO LEO**

National Research Council of Italy,
Institute of Applied Sciences and Intelligent Systems,
Ecotekne Campus, Via Monteroni, 73100 Lecce (Italy)

**GIOVANNI MARIA FARINELLA**

University of Catania,
Department of Mathematics and Computer Science,
Viale A. Doria 6, 95125 Catania (Italy)

**ACADEMIC PRESS**
An imprint of Elsevier

Academic Press is an imprint of Elsevier
125 London Wall, London EC2Y 5AS, United Kingdom
525 B Street, Suite 1800, San Diego, CA 92101-4495, United States
50 Hampshire Street, 5th Floor, Cambridge, MA 02139, United States
The Boulevard, Langford Lane, Kidlington, Oxford OX5 1GB, United Kingdom

**Notices**

Knowledge and best practice in this field are constantly changing. As new research and experience
broaden our understanding, changes in research methods, professional practices, or medical
treatment may become necessary.

Practitioners and researchers must always rely on their own experience and knowledge in evaluating
and using any information, methods, compounds, or experiments described herein. In using such
information or methods they should be mindful of their own safety and the safety of others,
including parties for whom they have a professional responsibility.

To the fullest extent of the law, neither the Publisher nor the authors, contributors, or editors,
assume any liability for any injury and/or damage to persons or property as a matter of products
liability, negligence or otherwise, or from any use or operation of any methods, products,
instructions, or ideas contained in the material herein.

**Library of Congress Cataloging-in-Publication Data**
A catalog record for this book is available from the Library of Congress

**British Library Cataloguing-in-Publication Data**
A catalogue record for this book is available from the British Library

ISBN: 978-0-12-813445-0

For information on all Academic Press publications
visit our website at https://www.elsevier.com/books-and-journals

Working together
to grow libraries in
developing countries

www.elsevier.com • www.bookaid.org

*Publisher:* Mara Conner
*Acquisition Editor:* Tim Pitts
*Editorial Project Manager:* Thomas Van Der Ploeg
*Production Project Manager:* Sruthi Satheesh
*Designer:* Matthew Limbert

Typeset by VTeX

# CONTENTS

## 3.  Real-Time 3D Tracker in Robot-Based Neurorehabilitation 75

Fabio Stroppa, Mine Saraç Stroppa, Simone Marcheschi, Claudio Loconsole, Edoardo Sotgiu, Massimiliano Solazzi, Domenico Buongiorno, Antonio Frisoli

## 4.  Computer Vision and Machine Learning for Surgical Instrument Tracking 105

Nicola Rieke, Federico Tombari, Nassir Navab

**5. Computer Vision for Human–Machine Interaction**    **127**

Qiuhong Ke, Jun Liu, Mohammed Bennamoun, Senjian An, Ferdous Sohel,
Farid Boussaid

**6. Computer Vision for Ambient Assisted Living**    **147**

Sara Colantonio, Giuseppe Coppini, Daniela Giorgi, Maria-Aurora Morales,
Maria A. Pascali

## 9.  Computer Vision for Lifelogging    249
Peng Wang, Lifeng Sun, Alan F. Smeaton, Cathal Gurrin, Shiqiang Yang

## 10.  Computational Analysis of Affect, Personality, and Engagement in Human–Robot Interactions    283
Oya Celiktutan, Evangelos Sariyanidi, Hatice Gunes

## 11.  On Modeling and Analyzing Crowds From Videos    319
Nicola Conci, Niccoló Bisagno, Andrea Cavallaro

# CONTRIBUTORS

**Senjian An**
The University of Western Australia, Perth, WA, Australia

**Mohammed Bennamoun**
The University of Western Australia, Perth, WA, Australia

**Niccoló Bisagno**
University of Trento, Trento, Italy

**Farid Boussaid**
The University of Western Australia, Perth, WA, Australia

**Domenico Buongiorno**
Scuola Superiore Sant'Anna, Pisa, Italy

**Andrea Cavallaro**
Centre for Intelligent Sensing, Queen Mary University of London, UK

**Oya Celiktutan**
Imperial College London, Electrical Engineering, Personal Robotics Lab, United Kingdom

**Shayok Chakraborty**
Center for Cognitive Ubiquitous Computing (CUbiC), Arizona State University, AZ, United States

**Manuela Chessa**
DIBRIS, University of Genoa, Italy

**Sara Colantonio**
National Research Council, Institute of Information Science and Technologies, Pisa, Italy

**Nicola Conci**
University of Trento, Trento, Italy

**Giuseppe Coppini**
National Research Council, Institute of Clinical Physiology, Pisa, Italy

**Mariella Dimiccoli**
University of Barcelona and Computer Vision Center, Barcelona, Spain

**Corneliu Florea**
University Politehnica of Bucharest, Image Processing and Analysis Laboratory, Bucharest, Romania

**Laura Florea**
University Politehnica of Bucharest, Image Processing and Analysis Laboratory, Bucharest, Romania

**Antonio Frisoli**
Scuola Superiore Sant'Anna, Pisa, Italy

**Daniela Giorgi**
National Research Council, Institute of Information Science and Technologies, Pisa, Italy

**Hatice Gunes**
University of Cambridge, Computer Laboratory, United Kingdom

**Cathal Gurrin**
Insight Centre for Data Analytics, Dublin City University, Dublin, Ireland

**Feng Hu**
City University of New York (CUNY), The Graduate Center, New York, NY, United States

**Qiuhong Ke**
The University of Western Australia, Perth, WA, Australia

**Jun Liu**
Nanyang Technological University, Singapore

**Claudio Loconsole**
Polytechnic University of Bari, Bari, Italy

**Simone Marcheschi**
Scuola Superiore Sant'Anna, Pisa, Italy

**Chiara Martini**
DIBRIS, University of Genoa, Italy

**Troy McDaniel**

Center for Cognitive Ubiquitous Computing (CUbiC), Arizona State University, AZ, United States

**Meredith Moore**

Center for Cognitive Ubiquitous Computing (CUbiC), Arizona State University, AZ, United States

**Maria-Aurora Morales**

National Research Council, Institute of Clinical Physiology, Pisa, Italy

**Nassir Navab**

Computer Aided Medical Procedures, Technical University of Munich, Garching, Germany

Johns Hopkins University, Baltimore, MD, United States of America

**Nicoletta Noceti**

DIBRIS, University of Genoa, Italy

**Francesca Odone**

DIBRIS, University of Genoa, Italy

**Sethuraman Panchanathan**

Center for Cognitive Ubiquitous Computing (CUbiC), Arizona State University, AZ, United States

**Maria A. Pascali**

National Research Council, Institute of Information Science and Technologies, Pisa, Italy

**Nicola Rieke**

Computer Aided Medical Procedures, Technical University of Munich, Garching, Germany

**Mine Saraç Stroppa**

Scuola Superiore Sant'Anna, Pisa, Italy

**Evangelos Sariyanidi**

Centre for Autism Research, Philadelphia, PA, United States

**Alan F. Smeaton**

Insight Centre for Data Analytics, Dublin City University, Dublin, Ireland

**Ferdous Sohel**
Murdoch University, Murdoch, WA, Australia
The University of Western Australia, Perth, WA, Australia

**Fabio Solari**
DIBRIS, University of Genoa, Italy

**Massimiliano Solazzi**
Scuola Superiore Sant'Anna, Pisa, Italy

**Edoardo Sotgiu**
INL International Iberian Nanotechnology Laboratory, Braga, Portugal

**Fabio Stroppa**
Scuola Superiore Sant'Anna, Pisa, Italy

**Lifeng Sun**
National Laboratory for Information Science and Technology, Tsinghua University, China

**Hao Tang**
CUNY, Borough of Manhattan Community College, New York, NY, United States

**Federico Tombari**
Computer Aided Medical Procedures, Technical University of Munich, Garching, Germany

**Aleksandr Tsema**
CUNY, The City College, New York, NY, United States

**Hemanth Venkateswara**
Center for Cognitive Ubiquitous Computing (CUbiC), Arizona State University, AZ, United States

**Constantin Vertan**
University Politehnica of Bucharest, Image Processing and Analysis Laboratory, Bucharest, Romania

**Peng Wang**
National Laboratory for Information Science and Technology, Tsinghua University, China

**Shiqiang Yang**

National Laboratory for Information Science and Technology, Tsinghua University, China

**Zhigang Zhu**

CUNY, The City College, New York, NY, United States

City University of New York (CUNY), The Graduate Center, New York, NY, United States

# ABOUT THE EDITORS

**Marco Leo** is a Researcher at the Institute of Applied Sciences and Intelligent Systems, which is part of the National Research Council of Italy. He received an Honours Laurea Degree in Computer Engineering from the University of Salento (Italy) in 2001. In 2004 he was a visiting researcher at the Massachusetts Institute of Technology where he conducted research on understanding and predicting human interactions under the guidance of Prof. Alex (Sandy) Pentland. In 2007 he was a visiting researcher at the University of Central Florida where he performed research on human activity recognition under the guidance of Prof. Mubarak Shah. In 2009 he was a visiting researcher at the Imperial College of London where he performed research on non-invasive inspection of composite materials under the guidance of Prof. Danilo Mandic. His main research interests are in the fields of computer vision and pattern recognition. He participated in a number of national and international research projects focusing on assistive technologies, automatic video surveillance of indoor and outdoor environments, human attention monitoring, real-time event detection in sport contexts, and non-destructive inspection of aircraft components. He is the author of more than 100 papers in national and international journals and in conference proceedings. He is also a coauthor of three international patents on visual systems for event detection in sport contexts. He is a Co-organizer of the International Workshop on Assistive Computer Vision and Robotics (ACVR), held annually in conjunction with either the European Conference on Computer Vision (ECCV) or the International Conference on Computer Vision (ICCV). He was guest editor of the Special Issue on Assistive Computer Vision of the Elsevier Journal on Computer Vision and Image Understanding Journal (CVIU). He served as Technical Program Chair at the 14th IEEE International Conference on Advanced Video and Signal based Surveillance (AVSS 2017), as Workshop Chair at the 19th International Conference on Image Analysis and Processing (ICIAP 2017), as local Organizer of the International Conference on Advanced Concepts for Intelligent Vision Systems (ACIVS 2016), as Local Organizer of the 5th International Workshop on Computer Vision (IWCV 2016), as Co-organizer of the Special Session on Computational Intelligence for safe and secure environments and transport at the IEEE International Symposium on Industrial Electronics (ISIE 2010), as Co-organizer of the 1st Workshop on Activity monitoring by multiple distributed sensing in conjunction with the 10th IEEE International Conference on Advanced Video and Signal based Surveillance (AVSS 2013), Co-organizer of the 3rd Workshop on Activity monitoring by multi-camera surveillance systems (AMMCSS) in conjunction with the 8th IEEE International Conference on Advanced and Signal based Surveillance (AVSS 2011), Co-organizer of the 2nd Workshop on Activity monitoring by multi-camera surveillance systems (AMMCSS) in conjunction with the 7th IEEE International Conference on Advanced and Signal based Surveillance (AVSS 2010). He is a member of many committees of international conferences and workshops. More information is available at people.isasi.cnr.it/~m.leo.

**Giovanni Maria Farinella** is an Assistant Professor at the Department of Mathematics and Computer Science, University of Catania, Italy. He received a Master of Science degree in Computer Science (egregia cum laude) from the University of Catania in April 2004. He was awarded a PhD in Computer Science from the University of Catania in October 2008. Since 2008, he has served as Professor of Computer Science for undergraduate courses at the University of Catania. From 2004 to 2016 he was Adjunct Professor at the School of the Art of Catania in the field of Computer Vision for Artists and Designers. Since 2007, he has been a board research member of the Joint Laboratory STMicroelectronics–University of Catania, Italy. His research interests lie in the field of Computer Vision, Pattern Recognition, and Machine Learning. He is author of one book (monograph); editor of five international volumes; guest editor of five international journals (CVIU, IEEE J-BHI, PRL, JOGO, JIVP); author of more than 100 papers in international book chapters, international journals, and international conference proceedings; and of 18 papers in Italian book chapters, Italian journals, and Italian conference proceedings. He is co-inventor of four patents involving industrial partners. Dr. Farinella is a reviewer and serves on the board program committee for major international journals (IEEE: TIP, TIFS, TMI, TMM, TCSVT; ELSEVIER: PR, PRL, CVIU, CBM; SPRINGER: JMIV, MTA, PAA) and international conferences (CVPR, ICCV, ECCV, BMVC). He has been Area Chair for the International Conference on Computer Vision 2017, Video Proceedings Chair for the International Conferences ECCV 2012 and ACM MM 2013, General Chair of the International Workshop on Assistive Computer Vision and Robotics (ACVR; held in conjunction with ECCV 2014,

ICCV 2015, ECCV 2016, ICCV 2017), and General Chair of the International Workshop on Multimedia Assisted Dietary Management (MADiMa) 2015/16/17. He has been an invited speaker at international events, as well as invited lecturer at industrial institutions. In 2006 Giovanni Maria Farinella founded, and currently directs, the International Computer Vision Summer School (ICVSS). In 2014 he founded, and currently directs, the Medical Imaging Summer School (MISS). Dr. Farinella is an IEEE Senior Member and a CVF/IAPR/GIRPR/AIxIA/BMVA member. More information is available at www.dmi.unict.it/farinella.

# PREFACE

This book reflects the importance of considering the "real world" social impact of technology alongside the fundamental goals of basic R& D. Computer vision is an area of scientific and technological development that will continue to have a profound impact on society. It will redefine the way that information technology intersects and interfaces with medicine and other disciplines, and it will play a key role in the care of an aging population and in improving the quality of life in our modern society.

The main aim of this book is to present the state of the art in the context of Computer Vision for Assistive Healthcare. The different chapters present the latest progress in this domain and discuss novel ideas in the field. In addition to the technologies used, emphasis is given to the definition of the problems, the available benchmark databases, the evaluation protocols, and procedures.

Chapter 1, by Zhigang Zhu et al., presents a vision-based assistive indoor localization approach with various techniques in different stages for helping the visually impaired to localize in and navigate through indoor environments. Unlike other computer vision research, whose problems are already well-defined and formalized by the community and whose major tasks are to apply their developed algorithms to standard datasets by tuning the parameter of models and evaluating the performance, this work studies the navigation needs of the visually impaired, and then helps us develop techniques in data collection, model building, localization, and user interfaces in both pre-journey planning and real-time assistance.

Chapter 2, by Corneliu Florea et al., approaches computer vision solutions to the diagnostic aid of several cognitive-affective psychiatric disorders. It reviews contributions that investigate cognitive impairments that appear at all stages of human development: from childhood and cognitive accumulation (autism, dyslexia), through adulthood and trauma-related cognitive degradations (such as phobias and PTSD), and ending with the ultimate degenerative cognitive degradations induced by dementia.

Chapter 3, by Antonio Frisoli et al., describes a computer vision-based robot-assisted system used in neurorehabilitation of post-stroke patients, which allows the subjects to reach and grasp objects in a defined workspace. In particular, a novel RGB-D-based algorithm used to track generic unknown objects in real time is proposed. The novelty of the proposed tracking algorithm comes from combining different features to achieve object recognition and tracking

Chapter 4, by Nassir Navab et al., outlines how computer vision can support the surgeon during an intervention using the example of surgical instrument tracking in retinal microsurgery, which incorporates challenges and requirements that are common when using this technique in various medical applications. In particular, how to derive algorithms for simultaneous tool tracking and pose estimation based on random forests and how to increase robustness to problems associated with retinal microsurgery images, such as strong illumination variations and high noise levels, is shown.

Chapter 5, by Qiuhong Ke et al., focuses on the gesture recognition task for HMI and introduces current deep learning methods that have been used for human motion analysis and RGB-D-based gesture recognition. More specifically, it briefly introduces the convolutional neural networks (CNNs), and then presents several deep learning frameworks based on CNNs that have been used for gesture recognition by using RGB, depth, and skeleton sequences.

Chapter 6, by Sara Colantonio et al., offers a brief survey of existing, vision-based monitoring solutions for personalized health care and wellness, and introduces the Wize Mirror, a multisensory platform featuring advanced algorithms for cardio-metabolic risk prevention and quality-of-life improvement.

Chapter 7, by Mariella Dimiccoli, focuses on those aspects of egocentric vision that can be directly exploited to develop platforms for ubiquitous context-aware personal assistance and health monitoring, also highlighting potential applications and further research opportunities in the context of assistive technologies.

Chapter 8, by Sethuraman Panchanathan et al., presents the computer vision research contributions of the Center for Cognitive Ubiquitous Computing (CUbiC) at Arizona State University in the design and development of a Social Interaction Assistant (SIA), which is an Augmentative and Alternative Communication (AAC) technology that can enrich the communication experience of individuals with visual impairment. The proposed solutions place emphasis on understanding the individual user's needs, expectations, and adaptations towards designing, developing, and deploying effective multimedia solutions. Empirical results demonstrate the significant potential in using person-centered AAC technology to enrich the communication experience of individuals with visual impairments.

Chapter 9, by Peng Wang et al., focuses on the most recent research methods in understanding visual lifelogs, including semantic annotations of visual concepts, utilization of contextual semantics, recognition of activities, and visualization of activities. Some research challenges which indicate potential directions for future research are also discussed.

Chapter 10, by Oya Celiktutan et al., focuses on recent advances in social robots that are capable of sensing their users, and support their users through social interactions, with the ultimate goal of fostering their cognitive and socio-emotional well-being. This chapter sets out to explore automatic analyses of social phenomena that are commonly studied in the fields of affective computing and social signal processing, together with an overview of recent vision-based approaches used by social robots. The chapter then describes two case studies that demonstrate how emotions and personality, which are two key phenomena for enabling effective and engaging interactions with robots, can be automatically predicted from visual cues during human–robot interactions. The chapter concludes by summarizing the open problems in the field and discussing potential future directions.

Chapter 11, by Andrea Cavallaro et al., covers models and algorithms for the analysis of crowds captured in videos that can facilitate personal mobility, safety, and security and can enable assistive robotics in public spaces. The main challenges and solutions for the analysis of collective behavior in public spaces are discussed; these challenges include understanding how people interact and their constantly changing interpersonal relations under clutter and frequent visual occlusions.

Finally, Chapter 12, by Manuela Chessa et al., considers assistive environments and discusses the possible benefits for an aging population. As a study case the current state of research on a protected discharge model adopted by Galliera hospital (Genova, Italy) to assist elderly users after they have been dismissed from the hospital and before they are ready to go back home, with the perspective of coaching them towards a healthy lifestyle, are discussed. The chapter focuses in particular on the vision-based modules designed to automatically estimate a frailty index of the patient, which allows a physician to assess the patient's health status and state of mind.

Marco Leo
Giovanni Maria Farinella

# CHAPTER 1

# Computer Vision for Sight
## Computer Vision Techniques to Assist Visually Impaired People to Navigate in an Indoor Environment

**Feng Hu\*, Hao Tang[†], Aleksandr Tsema[‡], Zhigang Zhu[‡],\***
\*City University of New York (CUNY), The Graduate Center, New York, NY, United States
[†]CUNY, Borough of Manhattan Community College, New York, NY, United States
[‡]CUNY, The City College, New York, NY, United States

## Contents

Computer Vision for Assistive Healthcare.
DOI: https://doi.org/10.1016/B978-0-12-813445-0.00001-0
Copyright © 2018 Elsevier Ltd. All rights reserved.

## Abstract

This chapter focuses on computer vision techniques to assist visually impaired people to navigate in an indoor environment. First, the problem is defined in terms of tasks, sensors, devices, and the performance requirements (real-time, accuracy, and robustness). Then, a recommended paradigm is proposed to build these systems for real-world applications, which include three important components: environment modeling, localization algorithms, and user interfaces. A broad review of the recent research achievements is provided in two categories: omnidirectional image-based and three-dimensional (3D) model-based approaches. As an example, an omnidirectional-vision-based indoor localization solution is described with algorithms and corresponding implementations in maximizing the use of the visual information surrounding a user. The system includes multiple components: floor plan parsing and path planning for scene modeling, deep learning for place recognition, image indexing for initial localization, 3D vision for position refinement, and portable user interfaces. Finally, we summarize the work and present some discussions for future work.

## Keywords

Indoor navigation, Omnidirectional vision, Environmental modeling, Localization, Path planning, Portable devices, User interfaces, Structure from motion, Place recognition

## 1.1 INTRODUCTION

Computer vision has been used to build assistive technology (AT) supporting many aspects of human needs, including mental functions, personal mobility, sensory functions, daily living activities, communication and skills training, and recreation and sports, as well as housing, work, and environmental improvement, as indicated in a recent survey [1] on computer vision for assistive technologies. This comprehensive survey [1] concentrates on a set of cross-application computer vision tasks that are set as the pivots to establish a categorization of the AT already used to assist some of the user's needs. These vision tasks include localization/mapping, object detection/tracking, human activity recognition, biometrics, head pose and gaze estimation, image retrieval, and Optical Character Recognition (OCR). In this chapter, we will mainly focus on using computer vision techniques for indoor localization.

An indoor localization system is of significant importance to the visually impaired in their daily lives if it can help them localize themselves and further navigate unfamiliar indoor environments. There are 285 million visually impaired people (VIPs) in the world according to the World Health Organization (as of March 2017), among whom 39 million are blind.[1] Compared to sighted people, it is much harder for VIPs to navigate indoor environments. Nowadays, too many buildings are unfortunately mainly designed and built for sighted people; therefore, navigational tasks and functions that sighted people take for granted could be huge problems to VIPs. Although a large amount of research has been carried out for robot navigation by the robotics community, and several assistive systems have been designed for blind people, efficient and effective portable solutions for VIPs are not yet available. In this chapter, after a general survey of the approaches using computer vision to assist VIPs, an omnidirectional-vision-based localization and navigation solution is provided using portable devices, e.g. GoPano lens[2] and iPhone.

## 1.1.1 Problem Statement

When it comes to the task of indoor localization, normally sighted people usually localize themselves by looking at the surrounding visual information to identify their current locations, e.g. on a floor plan, and then determine the appropriate paths to their desired locations. Due to the lack of visual input, VIPs face great challenges in the localization and navigation tasks when entering a new indoor environment.

With the advances in image processing and computer vision techniques, and the emergence of ubiquitous portable imaging systems and mobile computing devices, we formalize our problem statement in this chapter as follows:

*How can we make full use of the already existing visual information around a visually impaired person, and of the nonintrusive portable devices available to them, to provide a real-time, accurate, and robust indoor localization service?*

## 1.1.2 Important Considerations

In the following, a number of important issues under this problem statement will be elaborated in detail by describing the concepts, designs, and implementations that could lead to the solutions to this problem.

[1] http://www.who.int/mediacentre/factsheets/fs282/en/.
[2] http://eyesee360.com.

(1) Indoor localization

Indoor localization entails using one or more sensors such as cameras, magnetic sensors, inertial measurement units (IMUs), and RGB-D sensors (with both R, G, B colors and depth D) to automatically determine the location of a robot or a person in real-time in an indoor environment. This chapter mainly focuses on the indoor localization solutions with visual sensors, including normal camera, camera with omnidirectional lens, or 3D sensors, for assisting VIPs with performing indoor navigation tasks.

A successful assistive indoor localization system will have great social and economic impact on the visually impaired community and the society. The usual way for VIPs to navigate indoor environments is by memorization, which means by remembering the already traveled route and recalling it when they come to the same place the second time [2]. If we review this psychological process, we find a lot of effort is involved in the mental task. For example, they have to precisely remember turn-by-turn information and all the important points of interest (i.e. landmarks) along the routes. Without these landmarks, they lose the waypoints to help themselves judge whether they are on the correct path or not. Therefore, it would be very meaningful to create and build assistive indoor navigation systems for the visually impaired to relieve them from such memory burdens, and free their minds to deal with other important issues in their daily lives.

(2) Why visual information?

Visual information is predominantly used by human beings, and by many other species, to perceive their environments and carry out daily tasks. As an important perspective of artificial intelligence, visual information processing is the visual reasoning skill that enables us to process and interpret meaning from visual information that we gain through our eyesight.

Visual perception plays a big role in daily life, but probably because of the ease with which we rely on the perception we tend to overlook the complexity behind it. Understanding how we interpret what we see can help us design and organize our visual information and inspire many useful applications for the real world.

Visual-based localization is one such application mainly used to achieve real-time and accurate localization; it is now attracting more and more interest in the computer vision community and in the robotics community [3–6].

Visual information encodes very rich knowledge about the indoor environment. First, it includes all the visible physical objects existing in the environment, for example, doors, windows, notice boards. Even though the general problem of object recognition has not been fully solved yet, there are sufficient techniques that we can use to explore this visual information, for example, by extracting primitive geometric elements like vertical lines to differentiate environmental landmarks for an accurate localization. Second, visual information can provide structural information, for example a 3D model, of an environment. This structural information can be used as a global matching base for a later localization query from a user providing just an image, an image sequence, or a local 3D model.

(3) Nonintrusive portable devices

Some researchers have designed special robots or other systems for helping VIPs [7]; however, according to previous research and discussion with many visually impaired individuals, the majority of them do not want to carry extra devices [8]. There are a number of reasons for this. First, they might already have several devices to hold every day (e.g. white canes and Braille notetakers). Second, most of them do not want to use systems that attract extra attention and distinguish them from normally sighted people, especially in public spaces. Third, a usability study on assistive technology shows that new functionality provided on existing devices, such as smart phones, or hands-free portable devices, such as wearable glasses, are preferable [9].

The most common portable nonintrusive devices include iOS and Android smart phones, tablets, and wearable glasses. These devices are used daily not only by the visually impaired, but also by normally sighted people. They are also becoming more and more powerful both in terms of computation and storage, and in providing high-quality visual sensors such as onboard cameras for sensing the environment.

Three popular and potentially promising devices are GoPano plus iPhone, Google Glass, and the Project Tango tablet for the unique functionality they provide. GoPano and iPhone combined together provide a full 360 degree field of view (FOV) images; Google Glass provides a very natural and convenient user interface; and the Project Tango device has fast direct 3D sensing units. Even though it seems that we need multiple devices to make use of their unique functionalities, this will likely not be necessary in future devices as we see that more and more companies are trying to embed all these features together into their latest products. For example, the iPhone 7 plus[3] is integrating a wide FOV camera with the built-in camera, and the Lenovo Phab phone[4] integrates 3D sensors into their smart phone. In the future, there will be devices that have all the features built in (portable, wide FOV, 3D sensing enable, natural user interface, etc.), and we will be able to make use of them for an integrated, accurate, and convenient localization solution for VIPs.

(4) Real-time performance

For localization applications, especially assistive localization for the blind, real-time performance is one of the most critical requirements as the users will usually lose patience if the system is too slow. There are many challenges in designing and implementing a real-time system. First, since the actual devices used by the visually impaired are portable devices, they have limited computation power compared to desktops or computer clusters. Second, as the environment area size scales up, devices such as smart phones do not have sufficient memory space to store such large environmental databases.

To ensure a real-time performance, two strategies can be used in the design and implementation process. First, a client–server architecture can be adopted to transfer the storage and computation load from the client's smart phone, which has limited resources, to the server, which theoretically has unlimited storage and computation power. Second, GPU acceleration mechanisms can be used to further improve the time performance especially when the database is large.

(5) Levels of accuracy

Accuracy is one of the most important indicators for a useful localization system. Localization result is critical for any further navigation instructions, so an inaccurate localization will make the entire navigation system dysfunctional. However, indoor localization accuracy standards have not been clearly defined by researchers or industries yet. Even though extremely high accuracy, e.g. up to millimeter level, is not necessary for localizing a person for navigation purpose, the higher the accuracy, the better. A suggested industrial-level accuracy is 2 meters,[5] a distance within which a user

---

[3] http://www.apple.com/iphone-7/specs/.

[4] http://shop.lenovo.com/us/en/tango/.

[5] http://spreo.co/.

can differentiate between the doors of two nearby offices, detect on which floor of a building the user is located, or find a car in a parking lot.

There are also a few challenges in the visual-based indexing-and-retrieving indoor localization for achieving high accuracy. First, repetition of the scenes. If multiple scenes are alike, it is hard for the system to distinguish one location from the other based on just two still images obtained at these locations. Second, limited FOV of the cameras. Even though scenes are different from each other, if only a portion of the entire scene is used to extract scene representation features, the representation power of these features will be damaged. Third, trade-off between storage consumption and accuracy. To reduce the storage and computation load, dimensions of the scene features are usually reduced, for example, using Principle Component Analysis (PCA) [10]; different features may therefore appear to be similar, and thus influence the accuracy of the localization system.

(6) Robustness

The system should be able to withstand or overcome many kinds of challenges and adversity. Note that all the information used in vision-based localization is originally from the reflections of environmental light that are then accumulated by the imaging system. If this imaging procedure is interfered with by illumination change, occlusion, distortion, etc., the same scene will generate different visual representations, which makes the system unstable.

There are multiple ways to increase the robustness of a system, and some uniqueness of the assistive indoor localization problem can be used as the assumptions. For example, VIPs usually walk slowly and smoothly without dramatic location changes. If there is a localization result significantly far away from previously obtained result, e.g. the result obtained 5 seconds ago, it could be treated as an unstable result and can be discarded.

## 1.2 A RECOMMENDED PARADIGM

Depending on the forms of localization results needed, using visual information for assistive localization can be divided into two categories: metric localization and topological localization. Metric localization provides detailed coordinates of the user (or the sensor) in a given pre-defined global two-dimensional (2D) or three-dimensional (3D) coordinate system. Topological localization does not provide absolute measurements of coordinates within any coordinate system, but divides the location results into discrete classes, and therefore provides a discrete location class name for any input visual information. In this chapter, we will focus on metric localization since it provides more accurate results and is needed in the daily life of the visually impaired, even though in some cases place categorization as a topological localization approach will be used as a first step for metric localization.

There are many different methods that can be used for visual-based indoor localization, but if we extract their common ideas and analyze them to a high level, locating any user (or sensor) using visual information conforms to a paradigm of registering newly obtained visual information with the pre-built surrounding visual model.

In this section, instead of discussing any specific methods, we in general discuss the paradigm. In this paradigm, a vision based indoor localization task is divided into three components: environmental modeling, localization algorithms, and assistive user interfaces.

## 1.2.1 Environmental Modeling

Representing the environment is one of the fundamental components of visual-based indoor localization systems. The physical space we live in is 3D, therefore we can use 3D representations, for example, a 3D point cloud, to represent the environment. We can, however, also represent the 3D environment by projecting it into 2D space, and use 2D representations, for example floor plan images, to organize environment scenes.

In this section, we will discuss two approaches in environmental modeling: omnidirectional imaging-based environmental modeling and 3D structure-based environmental modeling. The first approach belongs to the 2D category, but utilizes wide FOV images, and thus is more powerful than normal FOV representation. We divide the second approach into two types: sparse 3D points-based reconstruction using Structure from Motion technique, and dense 3D points-based reconstruction using direct sensing.

(1) 2D omnidirectional imaging-based environmental modeling

Two-dimensional image-based environmental modeling registers 2D image information—usually in terms of unique 2D features or feature combinations—and the physical coordinates of the space together, then when another occurrence of these unique features appears, we infer the corresponding location by searching the previous feature-coordinate mapping pairs.

This modeling problem can therefore be converted into a feature-coordinate indexing or mapping problem. In general, it includes two steps: (1) acquiring useful images or features, and (2) relating them to the physical locations in the space. To reduce the database scale without losing much accuracy, key frame extraction techniques are used to minimize the memory and computation needed by the features in the modeling process.

The indexing procedure is conducted in the modeling stage before the system is delivered to real users, which means that the system developer has to construct the mapping for the area they want to cover. The modeling procedure is, however, only needed once and is done offline.

A common way to relate images and the physical world is to use floor plans. For example, we can use the corner of one floor of a building as the origin of the world coordinate system, and all the other image features are registered in this coordinate system afterwards in indoor localization applications.

One advantage of using a global coordinate system is that it can utilize the coordinates of some known position, e.g. the entrance of a building, as the initial location. By setting up an initial position value when a user arrives, instead of a global search of the entire database of an area to get the initial position, we greatly reduce the computation load.

Note that what we are actually making use of are the features, or feature spatial distributions, to decide whether we are at some specific location or not; we do not have to recognize specific objects, doors, signs, etc., or utilize the geometric relationship among the objects in the environment in this 2D omnidirectional imaging-based environmental modeling.

(2) 3D structure-based environmental modeling

Real physical space is 3D, thus it is natural to model the world with 3D sensors or via 3D reconstruction techniques, and then use these 3D models to achieve reliable indoor localization. There are two major methods used to construct 3D environmental models: Structure from Motion (SfM) and direct 3D sensing.

Structure from Motion is the process of estimating 3D structure from 2D image sequences [11]. Depending on the sources of the 2D images, SfM-based modeling approaches can be classified into two categories: crowd-sourced image modeling and user-captured image modeling. In indoor localization tasks for VIPs or the blind, the areas we want to localize are not popular places of interest for the general public. Therefore, the images obtained from the Web are neither sufficient nor dense enough to model such 3D environments. So crowd-sourced image-based 3D reconstruction is not appropriate for our purpose. In real applications, we have to collect images/video data by ourselves and use this visual information for 3D reconstruction.

The creation of 3D models using SfM typically involves three parts [11]: Graphics Processing Unit (GPU) Scale Invariant Feature Transform (SIFT) feature detection, pair-wise feature matching, and sparse and dense 3D reconstruction.

The 3D environment information can also be obtained directly with portable and real-time 3D sensors, such as Google's Project Tango,[6] Microsoft Kinect,[7] ASUS Xtion[8], Structure io,[9] etc. The 3D environment modeling methods using the above-mentioned sensors are denoted direct 3D sensing methods, differentiating them from SfM methods.

One issue of the 3D data obtained by these devices is that the FOV of a single scan is limited. Sensors can only scan a very small local area in one snapshot, within a short range (0.8 to 4.0 meters).[10] So we need to align these discrete and unregistered local 3D scans together for a complete environmental model.

A common way to deal with this problem is to combine and register all the local 3D data captured at different locations/times together and construct a globally consistent 3D model under some assumptions, e.g. all global physical space is static and will not change during the localization and navigation procedure.

## 1.2.2 Localization Algorithms

Once the environmental models have been built, the newly captured input visual information can therefore be used to compute the location where this information is acquired. The methods for aligning new visual input with global models are denoted localization algorithms.

Depending on the modeling methods used to build the environmental models, as discussed in Section 1.2.1, the algorithms for localizing new input visual information are correspondingly different.

---

[6] http://get.google.com/tango/.
[7] https://developer.microsoft.com/en-us/windows/kinect.
[8] https://www.asus.com/3D-Sensor/Xtion_PRO/.
[9] http://structure.io.
[10] http://get.google.com/tango/.

Two-dimensional visual-based modeling uses image features as landmark representations, and the localization task can essentially be modeled as an information retrieval process [4,12]. Three-dimensional SfM modeling, however, relies on geometric calculation for localizing new visual information, for example using 2D-to-3D matching or a 3D-to-3D matching strategy. We will briefly introduce these two approaches in the following paragraphs.

The 2D retrieval process includes the following steps: (1) capturing one or more new images and extracting representation features; (2) searching for features against the database of a scene; (3) finding the most likely matched ones; and (4) finally outputting the localization results aligned with these features under certain matching criteria.

For the SfM models, given an input image, the localization task is initialized by again extracting local features from the image and then finding which points in the model match with this image. After corresponding points are found, we can use the perspective N-points (pNp) algorithm [13] to determine the camera pose, including the camera location and orientation, that is equivalent to the user's location.

For direct dense sensing 3D models, for example models captured by the Google Project Tango Tablet device, a user can perform the localization task by capturing new RGB-D data (both color and depth information) at a new location by holding the tablet, and can register the new data with the pre-built global 3D model using the ICP algorithm [14,15].

## 1.2.3  Assistive User Interfaces

The assistive user interface is another important component in assistive technologies and solutions. Assistive localization systems and the corresponding interfaces described in this chapter, even though they can be used by the general population or by other special communities such as the elderly, have been designed mainly for VIPs.

There are several features that we will take into consideration when designing assistive technology according to our collaboration with the visually impaired community. First, the interface should be on devices that are both light and portable. It is better if the solution can be embedded into existing devices in daily use, such as smart phones. Visually impaired people already have several things to carry, e.g. canes or Braille printers, so light and portable objects are highly preferable. Second, the device should be non-intrusive. As little extra attention as possible should be drawn from the surrounding people so the visually impaired can feel comfortable in a normal social life. Third, the design should be low-cost. Many VIPs live with low incomes or have already spent a certain amount of money in other assistive expenditures, so affordability is a critical feature.

In addition to the overall system design, the assistive user interface is an important component of an assistive vision system. Depending on the methods to be chosen, three currently available devices have good user interfaces: the iPhone with a customized iOS app, Google Glass with voice command control, and Google Project Tango with enabled touch screen.

In the iOS app, since VIPs and especially blind people cannot see the screens and they usually use their smart phones with the lowest brightness to save power, the key idea in our design is to make it simple to use. We may design a big button in the middle of the screen, and by pressing this button, the process of image capture, feature extracting, client–server communication, matching, and candidate aggregating will be executed automatically. The result will be read out to the users via voice feedback after the localization result is returned. Also, since this method depends on a user holding the smart phone vertically to extract the correct image features, we have a component that detects the smart phone's inertial measurement unit (IMU) data, and suggests adjusting the orientation via voice if the holding tilt angle is larger than a threshold.

In the Google Glass application, the user interface can be divided into two parts: input of instructions and the resulting voice output. The localization service can be initialized either by voice command or by tapping the side of the Glass. The result is read out to the Glass wearer via the bone conduction transducer [16]. In the Google Project Tango approach, we can use a similar design to that of the iOS app by creating a clickable area on the Tango screen; by tapping this area, all the underlying processes are accomplished and the results will be read out via voice feedback [6].

## 1.3 RELATED WORK

Indoor localization solutions, depending on what kinds of sensors are used to obtain the environmental information, can be divided into two categories: non-vision-based systems and vision-based systems. In this section, we will only focus on the computer vision-based localization methods for indoor navigation.

It is more natural for human beings to prefer vision over using other modalities to perceive the environment and carry out daily tasks. Visual-based localization is also generating more and more interest in both the computer vision community and the robotics community to achieve real-time and accurate localization [17,18].

Mobile device cameras and consumer level 3D sensors are becoming very inexpensive and widely used in our daily lives. A majority of the existing vision-based localization services use 2D images or 3D RGB-D data from these devices to perform the task. However, a regular camera has a narrow FOV, and this limits its capacity to perceive as much information as possible at a given time. In particular, the narrow FOV makes aiming the camera a serious problem for blind users. Sufficient visual information plays a key role in achieving an accurate, fast, and robust localization goal, so this leads us to review how the wide FOV camera systems, especially omnidirectional vision systems, can improve the performance of a localization task.

We will first review two groups of approaches: 2D (especially the omnidirectional approach) and 3D (especially using SfM and 3D sensors). Then, we will discuss the assistive technology in general and unique features about their user interfaces.

## 1.3.1 Omnidirectional-Vision-Based Indoor Localization

A human being's FOV is about 120 degrees horizontally, and around 135 degrees vertically. The FOV of a typical smart phone is only 50 degrees horizontally and 40 degrees vertically. If images are captured with different viewing angles around the same location, the scenes covered by the images will differ considerably. Meanwhile, solving the matching problem of multiple images with motion parallax would be much easier if the coverage of the images is similar rather than between images with dramatically different coverage. This is why we utilize an omnidirectional imaging system.

Omnidirectional imaging is usually achieved by integrating a lens and a mirror, such as a catadioptric optics mirror and a perspective camera [19, 20]. There are already a number of portable omnidirectional imaging systems available for smart phones; a typical configuration, whose cost is under a hundred dollars, is to mount a special mirror on top of a smart phone camera.[11]

In general, an omnidirectional-vision-based localization system usually includes three modules: image capture with omnidirectional lens/mirrors, environment mapping or model construction, and feature extraction and

[11] http://www.gopano.com/products/gopano-micro-iphone5.

searching. Though different researchers may use varying combinations of methods for each module, almost every system has these three components.

*Omnidirectional imaging* is the process of using one or multiple omnidirectional imaging systems to move along the area where we want to provide localization services, and capture and store visual information for the purpose of model construction and/or real-time searching [21,22]. Since the system does not estimate location for only one static spot, it should be mobile and have the ability to provide continuous localization services. Currently, the most widely used omnidirectional imaging systems are wearable devices or smart phones with specially designed omnidirectional lenses or mirrors built into the original systems or mounted as accessories. To provide stereo information, two or more omnidirectional imaging systems may need to cooperate and output the 3D environment information [23]. The sensing procedure can be carried out by robots, developers, or even users depending on the methods used in the environment modeling process.

*Environment mapping or model building* is the process of constructing a one-to-one or many-to-one relationships between the 3D points in the real physical space and points in the digital space [24,4,12]. If a 2D model is applied, for example using a floor plan, we can project the 3D space into a 2D plane, and all the 3D points at the same location but at different heights will project to the same 2D point [16]. For the purpose of localization, in most cases a 2D map is sufficient and can satisfy all the localization service requirements. However, in some cases, for example multi-storey buildings, a single map is not enough. This can be solved by providing one map for each floor; however, many discrete 2D virtual spaces are involved and the positions between the floors, for example the stairs between floors, cannot be distinguished. To solve this problem, some researchers use 3D digital models and all the 3D points in the real physical space have a one-to-one corresponding point in the digital space [25,26].

*Feature extraction and searching* is the process of extracting representations of local images/3D points and using them to construct the digital space model or to find the correspondence in the pre-built digital space. Even though images, videos, or 3D points generated by computing devices themselves are already discrete representations of the 3D physical space, in most cases they are still not concise and descriptive enough for efficient matching or searching. We need to extract additional features for these images, videos, or sparse 3D points. Two-dimensional image feature extraction and matching have been studied for decades in the computer vision community and are relatively mature algorithms in visual-based localization [27,28]. Three-dimensional feature matching and searching or combined 2D/3D feature matching and searching are relatively new and have attracted a lot of attention in recent years [29,30].

Appearance-based localization and navigation have been studied extensively in the computer vision and robotics communities using a wide variety of methods and camera systems. Outdoor localization in urban environments with panoramic images captured by a multicamera system (with five side-view cameras mounted on the top of a vehicle) is proposed by Murillo et al. [5]. Another appearance approach proposed by Cummins and Newman [31] is based on Simultaneous Localization and Mapping (SLAM) for a large-scale road database, which is obtained by a car-mounted sensor array and by a GPS. Kanade et al. [32,33] localize a vehicle by first driving along a route and then comparing the image captured in the current drive with the images in the database created from the first drive. These systems deal with outdoor environment localization with complex camera systems. In our work, we focus on the indoor environment with simple but effective mobile sensing devices (smart phone + lens, or Google Glass + SfM model) to serve the visually impaired community.

Since a single image from a normal camera has a very limited FOV, and thus cannot make use of all the visual information available, creating mosaic or panorama images from discrete images is being studied by many researchers [34,35]. A visual-noun-based orientation and navigation approach for blind people was proposed by Molina et al. [36], which aligns images captured by a regular camera into panoramas, and extracts three kinds of visual-noun features (signage, visual text, and visual icons) to provide location and orientation instructions to VIPs, using visual-noun matching and PnP localization methods [13]. In their work, obtaining panoramas from images requires several capture actions and relatively heavy computation resources. Meanwhile, sign detection and text recognition procedures face a number of technical challenges in a real environment, such as illumination changes, perspective distortion, and poor image resolution. In our research work, a GoPano omnidirectional lens [37] is used to capture panorama images in real time, and only one snapshot is needed to capture the entire surroundings rather than multiple captures. No extra image alignment process is required, and no sign detection or recognition is needed.

Another related navigation method in indoor environments is proposed by Aly and Bouguet [38] as part of the Google Street View service; it uses six photos captured by professionals to construct an omnidirectional image at each viewpoint inside a room, and then estimates the camera pose and moving parameters between successive viewpoints. Since their inputs are unordered images, they construct a minimal spanning tree among the complete graph of every viewpoint to select triples for parameter estimation, which is computationally intensive. In our method, since we use sequential video frames, we do not need to find such spatial relationships between images, and the computation cost is therefore reduced for a real-time solution.

Representing a scene with extracted features for different purposes, such as image classification or place recognition, has been studied by many researchers [39–42]. An early work in representing and compressing omnidirectional images into compact rotation invariant features was proposed by Zhu et al. [43], where the Fourier transform of the radial principle components of each omnidirectional image is used to represent the image. Different from [43], and based on the observation that an indoor environment usually includes a great number of vertical line segments, we embed this vertical line segment distribution information into a 1D omnidirectional feature, and then use the Fourier transform components of these features as the representation of omnidirectional images. Another major difference is that we find a user's location and orientation from each omnidirectional image, whereas in [43], only six road types are classified using a neural network-based approach.

Another similar research area is image retrieval. Direct image retrieval entails retrieving those image(s) in a collection that satisfy a user's need, either using a keyword associated with the image, which forms the category of concept-based image retrieval, or using the content (e.g. color, shape) of the image, which forms the category of content-based image retrieval [44]. The applications of content-based image retrieval include sketch-based image searching [45], changing environment place recognition [46], etc., but very few research projects are conducted on using the content-based image retrieval for localization purpose, especially indoor localization. One major reason is that to represent a specific location via images, all the visual information around this location has to be collected and stored to make this landmark distinguishable, which requires many images if using a normal FOV camera instead of a single image. Also, the features used for retrieval from the input image should depend on the location only, independent of the camera's orientation, but this is hard for normal FOV cameras because the scene images are usually captured from different perspectives, and therefore they generate different features even though images are captured at the same location.

## 1.3.2 Other Vision-Based Indoor Localization

Utilizing vision information for the purpose of localization is currently attracting more and more attention in the industrial and the academic communities [47]. Structure from Motion (SfM), a technique used to create 3D street models in the outdoor environment and to recognize the places utilizing images from the Internet is one such technique and is being studied by a number of researchers [48–50].

We denote the points in the 2D images as features, and a point in the 3D images in the SfM model as points that can be associated with many features. A general localization problem can be stated as follows: given an input image, find which points in the model are matched with this image. If corresponding points are found, we can then determine the camera location and orientation, i.e. the camera pose [16].

Since the cameras taking the images are not calibrated, we cannot determine the final scale of the 3D model built by SfM from images alone. However, it is still meaningful to localize input images within these 3D models up to a scale. To solve the scale problem, we need extra knowledge about the environment. Kume resolves the scale problem by using odometry [3]. Another method [51,16] used to solve this problem is to relate a 2D image feature with the 3D physical world by manually labeling it, for example by manually relating the SIFT feature with a 3D point. The scale of the model can therefore be determined by using known 3D points.

Lee et al. [26,52] use Microsoft Kinect to construct a 3D voxel model for an indoor environment and analyze whether each voxel is free or occupied. Based on this information, a decision is made and then the system notifies the user what the next move should be. This system, however, requires Kinect, which is an extra device not in daily use and not easy to integrate into blind people's daily life.

Other researchers use Bag of Words (BoW) [53] or ConvNet features [54] to represent outdoor environments for localization. However, few researchers focus on the indoor scenarios, especially for assistive localization purpose.

## 1.3.3 Assistive Technology and User Interfaces

Assistive technology (AT) is an umbrella term that includes assistive, adaptive, and rehabilitative devices or systems for people with disabilities and also includes the procedures used in selecting, locating, and utilizing them. Assistive technology promotes greater independence by enabling people to perform tasks that they were previously unable to accomplish or had great difficulty accomplishing by providing enhancements to, or changing methods of interacting with, the technology required to accomplish these tasks. Many factors are important for a successful assistive technology: consideration of user opinion in selection, easy device procurement, device performance, and change in user needs or priorities [55].

In this chapter, we do not intend to review all the devices/systems used for people with disabilities in general. Instead, we focus only VIPs and the localization tasks for VIPs in their daily life.

The user interface (UI) is the place where interactions between humans and machines occur for certain functions. The purpose of this interaction, especially in the assistive technology field, is to allow effective operation and control of the machine from the human end, and at the same time allow the machine to provide feedback information that aids the operators' decision-making process.

Mobile and wearable devices are cheap and ubiquitous nowadays; this accelerates the advancement of general computer vision research and of assistive applications. For example, Farinella et al. [56] use Android phones to implement an image classification system with a DCT-GIST-based scene context classifier. Altwaijry et al. [57] apply Google Glass and develop an outdoor university campus tour guide application system by training and recognizing the images captured by the Glass camera. Paisios [2] (a blind researcher) has created a smart phone UI for a Wi-Fi based blind navigation system. Manduchi [58] proposes a sign-based way-finding system and tests the blind volunteers with smart phones to find and decode the information embedded in the color marks pasted on the indoor walls. All these research projects use mobile or wearable devices and have been designed with either touch screen or voice command (feedback) interface for communicating with VIPs. However, there is very little research work being done on designing user-friendly smart phone apps (e.g. iOS apps) for helping VIPs to localize themselves and navigate through an indoor environment.

## 1.4  AN OMNIDIRECTIONAL VISION APPROACH

Localization service is critical to a VIP's normal life, not only because it provides the current position and orientation of a user, but also because it could supply them with additional information about the environment, for example locations of doors and positions of doorplates, which are very useful for judging and making decisions. The essential idea of omnidirectional imaging-based localization is that we employ for each scene a unique and distinguishable feature, so each scene position is indexed by the omnidirectional feature. In this section, we will introduce in detail how to localize a user via indexing using our omnidirectional image features (omni-features) for VIPs.

In the following, the work of constructing a real-time assistive indoor localization system using an omnidirectional GoPano lens[12] and a smart phone (iPhone)[13] is introduced, and some experiment results are provided to demonstrate the effectiveness of the solution.

The system consists of a mobile vision front end with a portable panoramic lens mounted on the smart phone, and a remote image feature-based database of the scene on a GPU-enabled server. Compact and effective omnidirectional image features are extracted and represented in the smart phone front end, and then transmitted to the server in the cloud. Short video clips of these features are used to search the database of the indoor environment via image-based indexing to find the location of the current view within the database, which is associated with floor plans of the environment. One of the indexing approaches is a 2D multiframe aggregation strategy based on a candidate's distribution densities for multi-path environmental modeling to provide more robust location estimation. To deal with the high computational cost in searching a large database for a realistic application, a place recognition step is used to reduce the search space, and GPUs are involved in an implementation [59,60]. In our work, data parallelism and task parallelism properties are identified in the database indexing process, and computation is also accelerated by using multi-core CPUs and GPUs. After an initial location estimation is obtained, a 3D vision-based approach is used to refine the position estimation. User-friendly UIs particularly for the visually impaired are designed and implemented on an iPhone, which also supports system configurations and scene modeling for new environments. Experiments on a database of an eight-storey building are carried out to demonstrate the capacity of the proposed system, with real-time response (14 fps) and robust localization results.

In summary, the system includes five major components: (1) a mobile app with a user-friendly interface; (2) floor plan parsing and path planning for scene modeling; (3) deep learning-based image classification for place recognition; (4) image indexing for initial localization; and (5) 3D vision for position refinement (see Fig. 1.1 for the system diagram). Map generation and path planning is used not only in scene modeling, but also in

[12] http://eyesee360.com.
[13] https://support.apple.com/kb/sp587?locale=en_US.

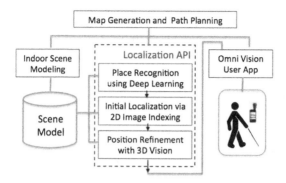

**Figure 1.1** System diagram for the omnidirectional-vision-based indoor navigation approach.

localization steps and user navigation. The three localization steps obtain user locations with a coarse-to-fine strategy (from place/zone recognition to initial localization to location refinement) to improve the efficiency and robustness of the system.

## 1.4.1 User Interfaces and System Consideration

There are two possible ways of using this system. The first is the hands-free mode, as shown in Fig. 1.2A. The camera is mounted on the top of a bicycle helmet worn by a blind user. The second is the hand-held mode, as shown in Fig. 1.2B. During a usability study, we received a lot of feedback from the blind community for the project. Originally, we wanted to mount the smart phone onto a helmet, so users would not need to hold their phones by themselves. However, their feedback suggested that the majority of blind people prefer to hold their phones in their hands to avoid unnecessary attentions. Also, since the users only need to hold their smart phones for a few seconds when a localization service is needed, they do not have to hold them the entire way while walking. As a result, in our experiments, we mainly focus on the second working mode.

Like any other image-based localization method, occlusion by moving persons and scene changes due to varying illuminations will influence the scene representation and thus challenge the effectiveness of the localization system. In extreme cases, if the scene representations are identical to each other (e.g. empty rooms with uniformly colored walls) or there are large self-occlusions (e.g. the camera is blocked by the user's hand), using an appearance-based image localization system alone will have challenging problems. Our system has two advantages over other approaches: (1) the

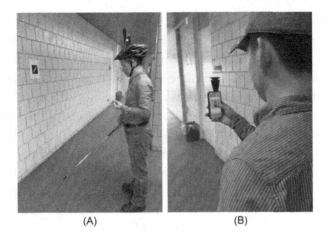

(A)                                    (B)

**Figure 1.2** Two system working modes.

omni-lens has a $360 \times 82$ degrees FOV, which captures a wider area of visual information and is less vulnerable to local occlusions or scene changes if they only occupy a small part of an omnidirectional image; (2) short video sequences are captured while the user is walking rather than a single image, so even though one of the frames might have significant occlusions or changes, the other frames can help to provide stable visual information of the environment for an effective localization.

The modeling procedure is space intensive both during the processing (with intermediate data and the final results) and the database storage. The querying procedure, on the other hand, could be very time consuming, particularly with multiframe querying. The basic idea of multiframe indexing is to use a sequence of newly captured video frames to query a pre-built video frame database to increase the success rate. Even with the pre-processing step (in the front end of an iPhone) to reduce the size of the input image data, and a place recognition step to narrow down the areas of search, the computational cost of multiframe query in the database would be high if the database were too large. As a result, both modeling and querying procedures are designed to be manipulated in the cloud server, where parallel processing can be applied for acceleration.

One parallel strategy is to partition the input video into individual frames and query the database with each frame in parallel. Then, for every input frame, we compare it one-by-one with all the frames in the database. After all the queries return their matching candidates, for example the top five matches, an aggregation step can be obtained by assuming that both the input and the matching sequences are temporal sequences. In this way, the querying process could have four levels of parallelisms. First, we divide the search database into multiple subspaces based on the place recognition results; for example, the data of each storey is counted as one subspace, and the selected subspaces can be searched in parallel with CUDA streams. Second, within each subspace, we process all the frames of an input query video clip in parallel. Third, we use multiple threads to compare each input frame with multiple database frames simultaneously rather than comparing them one-by-one. Fourth, some of the operations in obtaining rotation-invariant projection curves, such as the Fourier transform algorithm, can take advantage of the parallel processing. To do this, all of the original omni-projection curves will be sent from the front end to the server, which still communicates very efficiently because of the low-dimensional data.

## 1.4.2 Path Planning for Scene Modeling

In order to localize a user in an indoor environment, the first question we should ask is, how can we effectively and efficiently select paths to model an environment using our omnidirectional imaging system? We need a path planning algorithm to determine the paths for capturing omnidirectional images in order to model an environment.

This path planning work, even though intended for convenient environmental modeling, is also valuable to the visually impaired themselves. When VIPs enter an indoor environment for the first time, for example a multi-storey building, they need knowledge about the structure of the building or layouts of each floor to successfully navigate within it to reach the desired destination, such as a room. This knowledge can be acquired in multiple ways: self-exploring multiple times and memorizing the layout; asking other people within the building; asking other people who are not within the building but remotely connected, for example via video [61]; or independently learning the architecture information before leaving. The first three methods can sometimes be effective, but at the cost of losing the users' independence, which is one of the core spirits we need to consider when designing an assistive technology. Learning the architecture maps of a building pre-journey, before physically entering the building, is an effective way for them to navigate inside the building, as is using existing localization and navigation services [62–64].

### 1.4.2.1 Map Parsing and Path Planning

In this section, we present the use of architectural drawings to achieve automatic path planning for both pre-journey planning by VIPs and environment modeling by developers [65,66]. An architectural drawing is a technical drawing of a building, usually using Computer Aided Design (CAD) software (e.g. AutoCAD[14]) and is available for buildings constructed in the past three decades. They are widely used by architects and others for a number of purposes: to develop a design idea into a coherent proposal, to communicate ideas and concepts, to convince the clients of the merits of a design, or to make a record of a completed construction project.

Using the same path planning algorithm for both environment modeling and pre-journey planning also makes the modeling and localization stages more consistent. Once the traversability between any two places, e.g. the entrance and a restroom, is calculated, single-path or multi-path based omnidirectional imaging modeling can be performed to provide a robust indoor localization service. Then in the localization stage, when the user follows a path very similar to the path that has been planned by the model, the image-based localization would be much more effective.

The system of traversability calculation includes the following four steps [65]: (1) the system reads an architectural floor plan (such as an Auto-CAD file), extracts the information of each entity (room, corridor, etc.) and layer, and then stores them in a database; (2) an image-based analytics method is applied to extract each room's layout (entity polygon); (3) the system identifies the geometric relations between neighboring rooms and corridors, which allows a topological graph of the entire building to be computed; (4) a 3D floor map and a traversable map are finally generated. The system diagram is shown in Fig. 1.3.

Fig. 1.4 shows an example of an AutoCAD architectural floor plan of a complex campus building. The previous four steps are carried out one by one with this map: we first parse the AutoCAD file, and then extract useful layer and semantic information, which are stored in a database. We then render a new floor image from the database that only includes wall structures. The region growing method is applied to the new floor image and entities are identified from the image. In Fig. 1.5, the green regions represent rooms and blue regions represent corridors. Once the entities are extracted, the geometric relations among entities are calculated and a topological map (Fig. 1.6) is successfully built. In addition, the contours of the

[14] http://www.autodesk.com/products/autocad/overview.

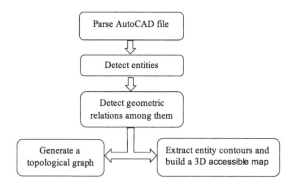

**Figure 1.3** Automatic path planning for data collection.

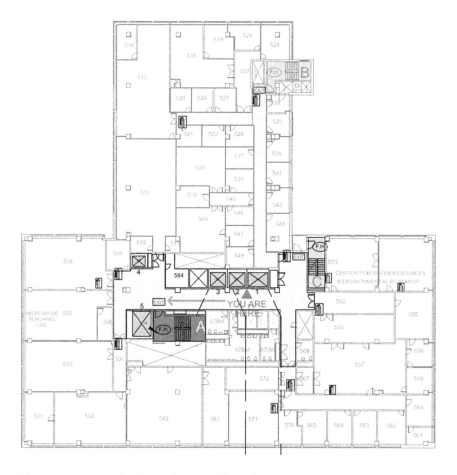

**Figure 1.4** An example of an architectural floor plan.

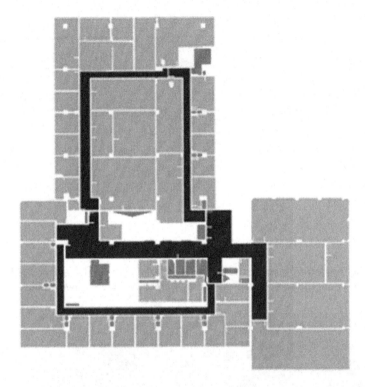

**Figure 1.5** A visualization of the lookup table after region growing is performed on the input AutoCAD map in Fig. 1.4. (For interpretation of the colors in this figure, the reader is referred to the web version of this chapter.)

entity polygons in the floor image are extracted and saved in a JSON file. A 3D traversable floor map is built, as shown in Fig. 1.7, and a turn-by-turn navigation direction can be calculated using the $A^*$ algorithm [67].

We manually verify the accuracy of the topological and the traversable maps. For the topological map, we compare it with the original AutoCAD map. We derive a topological map from the AutoCAD map and compare it with the one generated by our proposed algorithm, and they match correctly. For the 3D traversable map, we project it into a 2D space and it aligns with the original AutoCAD map correctly as expected.

To test the performance of the topological and the 3D traversable maps, each time we randomly select two entities from the AutoCAD map and we manually calculate the shortest path between two entities. We then compute the navigation summary, using the Dijkstra algorithm [68] from the topological map, and the turn-by-turn directions, using the $A^*$ algo-

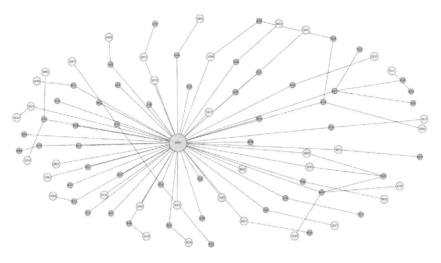

**Figure 1.6** The topological map for the input AutoCAD map in Fig. 1.4, each node is represented by a note and an edge represent the connectivity between two entities.

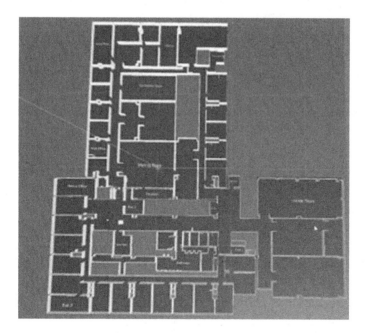

**Figure 1.7** The rendered 3D traversable map for the input AutoCAD map in Fig. 1.4.

rithm [67] from the 3D traversable map. We compare the three paths and they also match correctly.

In summary, given any building available with AutoCAD documentations, we can detect the traversability between any two entities A and B (e.g. offices and restrooms). Also, we can generate turn-by-turn navigation guidance from A to B, even though the user—either a VIP or environment modeling staff—has never been to this building before. Therefore, the proposed system can allow our administrative staff to use the omnidirectional imaging system and model the indoor environment and our visually impaired users to navigate the same environment more efficiently and effectively.

### 1.4.2.2 Scene Modeling

In the modeling stage, the system developer carries a smart phone and moves along the corridors at an even normal walking pace (0.5 m/s to 2 m/s), covering and recording the video into a database. Geotags (e.g. physical locations of current frames) are manually labeled for associated key frames. Motion estimation algorithms can be used in the future to ease the constraint of linear motion.

For the modeling stage, we use all the frames of the video to compose the database in order to make full use of the visual information and obtain the highest possible localization resolution for each scene. During the labeling procedure, we select associated key frames where the landmarks (e.g. door plates, corners) are located while these frames are captured, and then label the rest frames in between by interpolation in the defined floor plan coordinate system. In the testing stage, we sample the testing videos by selecting one from every 5–10 frames to reduce the query computation while still making use of the multiframe query advantage.

Using all the pixels in the images of even a short video clip to represent a scene is too expensive for both communication and computation, and the data are not invariant to environment variables such as illumination or user heading orientation. Therefore, we propose a concise and effective representation for the omnidirectional images by using a number of 1D omnidirectional projections (omni-projections) for each panoramic scene. It is observed that an indoor environment often has many vertical lines (door edges, pillar edges, notice boards, windows, etc.), and the distribution of these lines around the user provides unique landmarks for localization. However, for computational efficiency and robustness, we do not extract these line features directly; instead, we embed them inside of the proposed omni-projection representations, even though these line features can be

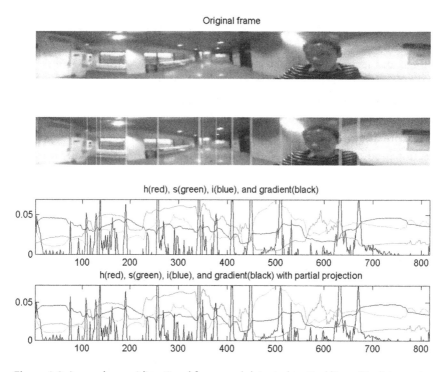

**Figure 1.8** A sample omnidirectional frame and detected vertical lines. (For interpretation of the colors in this figure, the reader is referred to the web version of this chapter.)

utilized in the future with a correlation-based matching method to estimate a viewer's location.

The omni-feature extraction algorithm [12] is summarized below (Fig. 1.8). After an original image is warped into its cylindrical representation (Fig. 1.8 row 1) it is converted from the RGB (red, green and blue) color space to the HSI (hue, saturation and intensity) space, and a gradient map is generated for the intensity channel. Then each of the hues, saturations, intensities, and gradient channels is projected vertically to generate an omnidirectional profile vector (Fig. 1.8 rows 3 and 4; the difference between the results of the two rows is the projection for the gradient map: in row 3, the whole gradient map is used, whereas in row 4, the top and bottom of the images are excluded from the project thus creating more useful peaks for the detection of vertical lines). Finally, the omni-vectors are normalized and transformed into Fourier domain using Fast Fourier Transform (FFT), and the magnitudes of their FFT coefficients (which are rotation-invariant) are saved as the omni-feature vectors for the images.

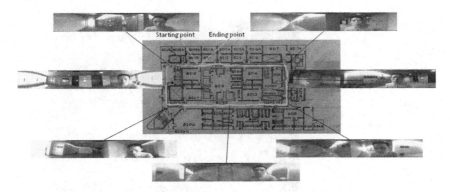

**Figure 1.9** One of the eight-storey testing environments and some sample omnidirectional images. The red line in the map is the modeled path. The blue and green lines are the testing paths. (For interpretation of the colors in this figure, the reader is referred to the web version of this chapter.)

Note that our work does not directly use individual vertical lines to represent or match scene images, even though vertical lines can be easily detected from the projection of the gradient map, as shown in Fig. 1.8, row 2.

To provide localization service in large areas, e.g. large rooms or wide corridors, single-path modeling is not sufficient when the testing path is far away from the modeling path. Therefore, in one of our experiments, we model each floor of an eight-storey building with multiple paths (five parallel paths). We first capture video sequences of the entire floor along these five paths as the training databases, one of which is shown by the red line in Fig. 1.9. We then capture two other video sequences as the testing datasets, as shown by the blue and green lines. Some sample omnidirectional images used in the modeling process, and their geolocations attached to the floor map, are also shown around the map in Fig. 1.9.

### 1.4.3 Machine Learning for Place Recognition

Deep learning has been used to solve a wide spectrum of different problems, including wide variety of problems in the field of computer vision. The most popular deep learning models for image classification tasks are convolutional neural networks (CNNs) [69,70]. The main difference from regular artificial neural networks like multilayer perceptron is that CNN takes advantage of the fact that inputs are raw images and the architecture is adjusted to work with them efficiently. Thus, CNNs can classify high-resolution raw images quickly and accurately.

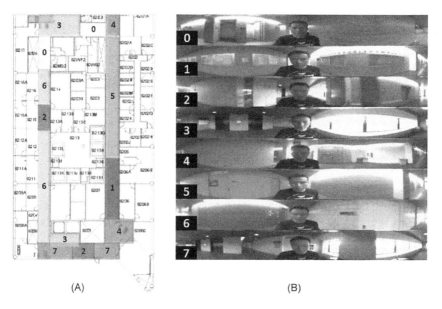

(A)                                             (B)

**Figure 1.10** Data labeling for place recognition: (A) an architectural floor plan and (B) images in the eight zones.

A CNN model consists of many stacked layers. The key layers are the convolutional layers, which perform convolution operations on given input data. Very often a max pooling layer follows a convolutional layer and performs downsampling of the inputs. Last layers are fully connected layers that perform classification. In this work, we use a CNN model to categorize an input omnidirectional image into one of several predefined categories of an indoor scene. For example, we divide the corridor of a floor into zones, as shown in Fig. 1.10. All data in the database will be represented in nonlinear data structure like a graph or tree. So, the task of searching an input omnidirectional image in the database will be divided into two steps: (1) recognizing the class of the input image (place recognition); and (2) finding the best match in the database under this class (Section 1.4.4). Thus, the search time complexity can be reduced by the factor of number of classes we have, assuming that each category has the same number of images. In the following we will report the data, the architecture, the learning process, and the experimental results of place recognition using CNN on omnidirectional images.

**Table 1.1** Classes in the dataset and the corresponding number of images

| Label | Class | # of images | Notes |
|---|---|---|---|
| 0 | Bathrooms | 1228 | Large opened doors, black bathroom sign |
| 1 | Corridor with message boards | 2207 | Message boards on the walls |
| 2 | Doors | 2000 | Large orange doors along the way |
| 3 | Elevators | 3947 | Black elevator doors |
| 4 | Exit | 754 | Wide gray doors with red exit sign |
| 5 | Long corridor | 4378 | Walls with pattern, doors |
| 6 | Short corridor | 4227 | Walls with pattern |
| 7 | Windows | 3759 | Large bright windows |

### 1.4.3.1 Dataset

In this work, we generated a dataset from the videos captured on one of the floors of our building. We have 22,500 manually labeled images divided into eight classes (see Table 1.1). Here the manual labeling was simply the division of the video clips into these eight zones; therefore, the labeling process is quite manageable.

To prevent overfitting and to monitor generalization ability of the model we randomly divided the dataset into three distinct subsets: training, validation, and testing sets with proportions of 0.7, 0.2, and 0.1, respectively. All images were resized from the initial $256 * 2149 * 3$ pixels to $32 * 268 * 3$ pixels to reduce computational cost.

### 1.4.3.2 Architecture

We tested several CNN topologies [69,70], some of which have won international image classification competitions. Looking for a quasi-optimal model, we relied on the principal of Occam's razor [71] and simplified complicated models iteratively. The best results were produced by the architecture represented in Fig. 1.11.

This model contains 10 layers: 4 convolutional, 4 max pooling, and 2 fully connected layers with Leaky RELU activation functions. In total, there are six layers with optimizable weights. The first convolutional layer takes raw image as an input and filters it with 64 kernels of size $3 * 3 * 3$ with

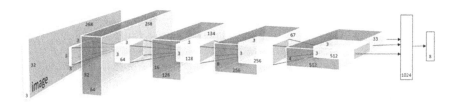

**Figure 1.11** The CNN structure for omnidirectional image classification.

a stride of 1 pixel. Then we apply max pooling of size $2*2$ with stride 2. The next three convolutional and pooling layers do the same, but each convolutional layer doubles the number of kernels of the previous convolutional layer. Finally, the pooled output of the fourth convolutional layer goes into the first fully connected layer with 1024 neurons. The second and last fully connected layer with eight neurons performs classification.

### 1.4.3.3 Learning Process

Along the research process we tried many different global parameter variations and used some heuristics to tune the model, but this discussion is beyond the scope of this book, so we will go directly to the description of the quasi–optimal parameters. We trained our CNN with randomly shuffled batches of 128 examples per batch. As a loss function optimizer, we used an Adam optimizer with exponentially decreasing learning rate, and trained and validated the model along 10,000 steps, which is around 70 epochs. We set the target loss value as 0.07, so the training process stops when the loss function reaches this bound. It took around 7000 steps to complete the training.

To better understand the insights of the learning process we visualized the outputs of all the convolutional layers and the 64 kernels from the first convolutional layer, as shown in Fig. 1.12.

### 1.4.3.4 Results

We tested a trained model with the generated test dataset, and the result was quite promising, with an accuracy of up to 89% using just the top candidate. Table 1.2 shows the confusion matrix of place recognition on the test data, and Fig. 1.13 shows two examples. In the left example, the corridor with message boards was confidently recognized, whereas in the right example the recognition of the scene was about equal for window and door classes since it includes both doors (in the middle) and windows

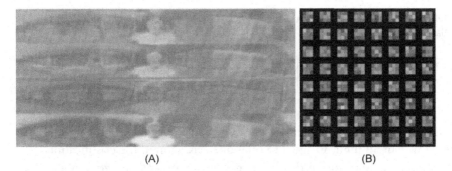

(A)                                    (B)

**Figure 1.12** Visualization of the place recognition CNN: (A) four randomly picked activations and (B) the 64 kernels, both of the first convolutional layer.

**Table 1.2** Confusion matrix of place recognition for the eight classes. The first column of the table lists the correct class labels and the first row shows the predicted class labels; starting from the second column, each row shows the number of samples predicted in classes 0–7

|   | 0 | 1 | 2 | 3 | 4 | 5 | 6 | 7 |
|---|---|---|---|---|---|---|---|---|
| **0** | 130 | 0 | 0 | 0 | 0 | 0 | 0 | 0 |
| **1** | 0 | 159 | 0 | 0 | 0 | 2 | 0 | 3 |
| **2** | 0 | 0 | 63 | 0 | 0 | 0 | 0 | 0 |
| **3** | 0 | 0 | 0 | 399 | 0 | 0 | 0 | 0 |
| **4** | 0 | 0 | 0 | 0 | 215 | 0 | 25 | 0 |
| **5** | 0 | 0 | 0 | 0 | 0 | 277 | 0 | 134 |
| **6** | 0 | 0 | 0 | 0 | 12 | 0 | 409 | 0 |
| **7** | 0 | 15 | 0 | 0 | 0 | 161 | 0 | 230 |

Short Corridor
Long Corridor
Corridor With Boards

Short Corridor
Windows
Doors

**Figure 1.13** Successful class prediction of two samples from the testing set.

(at the two far ends). This means that our CNN has a good generalization ability and can theoretically produce good results in real-life work. The feed forward propagation is extremely fast and classification of one single image can be easily done in real time.

Since the purpose of the place recognition step is to reduce the number of matches for the following steps, we also evaluated the accuracy of the recognition when we consider the correct answer to be in the top two

**Table 1.3** Accuracy (in percent) of place recognition for the top two candidate classes

|       | 0   | 1   | 2   | 3   | 4   | 5   | 6   | 7   | Average |
|-------|-----|-----|-----|-----|-----|-----|-----|-----|---------|
| Top 1 | 100 | 97  | 100 | 100 | 90  | 67  | 97  | 57  | 89      |
| Top 2 | 100 | 100 | 100 | 100 | 100 | 100 | 100 | 100 | 100     |

candidate classes. Table 1.3 shows the results. We have two observations. First, if we put an input image into only one class (zone), the classification results are very good for all the classes except for two classes, zone 5 and zone 7; as we can see in Table 1.2, many of the images in these two zones are classified as the other, 33% and 43%, respectively. Second, if we put the input image into the top two candidate classes, the average recognition rate increases to 100%, which means that the correct class label of each input test image is one of the top two predicted classes. For a large indoor scene that has more than eight classes, putting an input image into two class categories correctly can still reduce the computation time of the following indexing step considerably.

### 1.4.3.5 Discussions

The current trained convolutional neural network is not optimal and it will take many hours to find the optimal architecture and global parameters if we only use CPUs. To speed things up, we will train our model on a GPU cluster in a cloud, such as the AWS, Azure, or Google Cloud Platform.

In addition, data collection and manual labeling is still tedious and time consuming, so we will explore an efficient strategy to automate this process.

Another important thing is that we work with a video stream, which means that captured frames have one more crucial dimension: time. This means that each image depends on the previous images and has an influence on the following ones. Thus, for the time series data it is common to use Recurrent Neural Networks, and we will try to combine RNN with our existing approach. Using temporal information also applies to the following steps.

## 1.4.4  Initial Localization Using Image Retrieval

In the querying stage, a visually impaired user can walk into the area covered in the above modeling stage and take a short video clip. The smart phone extracts video features and sends them to the server via a wireless connection. The server receives the query, searches for the image candidates in the database concurrently, and returns the localization and orientation result to the user.

Newly extracted features of an input image are used as query keys to localize and navigate in the environment represented by the database. Because the omni-projection presentation will be different even at the same location if the smart phone is not held vertically, we developed a function in the database acquisition and testing procedure by extracting the smart phone's built-in gyroscope values, and give the developer or the user a hint if the phone is not held properly.

### 1.4.4.1 Two-Dimensional Multiframe Aggregation Based on Candidates' Densities

For an environment with wide corridors or large rooms, modeling with a single traverse is not sufficient and we need multiple traversals (paths) to be able to cover the complete area and provide accurate results. Also, for a multiframe-based localization query, the query frame is sent to the server together with a number of frames before and after it. After receiving a large number of position candidates for each query frame, a more sophisticated aggregation algorithm is needed to integrate all these candidates together before we deliver a finalized location coordinate to the user.

We demonstrate the application of the system for the areas such as corridors or rooms, which are large, by increasing the densities of the environmental modeling with multiple paths of video capture. Instead of capturing a video on only one path, we use multiple parallel paths to build the environment model. After that, we correlate the paths with the physical space coordinates by performing associated frame labeling and interpolation. Then, the final location is obtained by aggregating the possible candidates in the 2D coordinate plane.

If there is no noise for any input testing frame, in an ideal case all the returned candidates from the server should be at the exact same position where the testing frame was captured since the testing frame and the modeling frames (which generate the candidates) should capture identical environmental information. However, there are various reasons (noise, scene similarity, etc.) why some of the candidates may be disturbed and are far away from the ground truth position, while the majority of the candidates cluster around the ground truth area. In this work, according to the above observation, we have designed and implemented a 2D multiframe aggregation algorithm by utilizing the densities of the candidates' distribution, and calculate the most likely finalized location estimation.

The 2D multiframe aggregation algorithm is detailed in [12] and can be summarized as follows. In the initialization step, the candidates' indexes in the modeling frames are mapped to their actual floor plan 2D coordinates by checking the geo-referenced mapping table pre-generated while modeling the environment. In our experiments, we discretize the floor plan coordinates as 30 cm by 30 cm tiles.

Then in Step 2, we count the number of candidates in each tile by a 2D binning operation, and obtain the query frame's location distribution array, *locationDistriArray*. We sort this array in descending order in Step 3, and obtain the first *TopC* number (currently *TopC* = 10) of tiles in Step 4.

In a perfect case, the top tile of these *TopC* candidates should be the best estimated result since it has the largest number of candidates dropped in. This is not always the case, however, because the tile with the most candidates may be a false positive estimation, and its nearby tiles may have very few candidates.

To increase the robustness of the estimation, in Step 5 we set a tolerant circle around each tile of a given radius (in our experiment, this radius is 3 m). Afterwards, we count the total number of candidates within this circle. If the number is greater than a threshold percent, *tolerPer* (e.g. *tolerPer* = 20%), of the total number of candidates, the corresponding tile is estimated as the final position, and is returned to the user via the sockets.

### 1.4.4.2 Localization in Wide Areas: Experimental Results

To provide localization service in large areas, e.g. large rooms or wide corridors, single-path modeling is not sufficient when the testing path is far away from the modeling path. In this experiment, we model a whole floor with multiple paths (five parallel paths), and test our 2D aggregation algorithm. Some sample images, one of the training paths, and the two test paths are shown in Fig. 1.9.

For each testing frame, we query the entire training database and find the most similar stored features with the smallest Euclidean distance. Many scenes in the environment may be very similar, for example the two images in Fig. 1.14, so the returned feature with the smallest distance may not necessarily be the real match. We then select the top $N$ ($N = 15$ in this experiment) features as the candidates. Also, to solve the problem of scene similarity for each querying frame, we use the scene itself as the query key and then we also use $M - 1$ ($M = 11$) other frames—$(M - 1)/2$ before and $(M - 1)/2$ after this frame—as the query keys to generate candidates. We test all these frames within all the $P$ ($P = 5$) paths, the frames of which

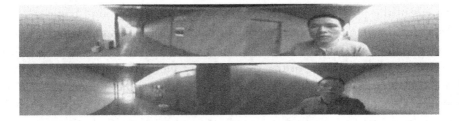

**Figure 1.14** Two very similar scenes.

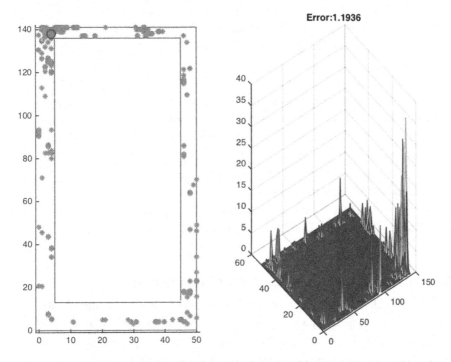

**Figure 1.15** Matching results of a frame and its neighbors against a multi-path database. (For interpretation of the colors in this figure, the reader is referred to the web version of this chapter.)

are all labeled on the floor plan coordinate system, and retrieve all the candidates. Consequently, there are a total of $N \times M \times P$ candidates.

In Fig. 1.15, the red stars are the 825 location candidates of a query frame and its related 10 frames. Note that several frames may return the same location, so each red star may represent more than one candidate. In the left image of Fig. 1.15, the coordinate system has the same scale as the floor plan. Each unit in both horizontal and vertical directions corresponds to 30 cm in the physical space. The green star shows the ground truth location and the blue circle shows the final estimation of the testing frame. The distribution of the candidates is also illustrated in the right plot.

The $x$ and $y$ coordinates of the right panel of Fig. 1.15 are the floor plan coordinates, and the height of each point represents the number of candidates falling into this position. Since all the modeling frames captured near the testing frame have similar content and thus have similar features, the majority of the candidates drop near the location where the testing frame is captured.

In this example (Fig. 1.15) the testing frame's ground truth location is in the top left corner of the floor plan (the green star); as we can see, the majority of the candidates drop in the top corner area. The rightmost corner in the right panel of Fig. 1.15 also shows that the number of candidates (representing by the height) near the ground truth position is much larger than in the other places. There are some false positives distributed around the whole floor, as shown in both the left and right panel. A short video clip can be found here[15] to visualize the robustness of the 2D aggregation algorithm with more queries. The number in the top right of Fig. 1.15 shows the estimated error in meters.

After the aggregation of all the candidates is completed, we obtain the final candidate (blue circle in Fig. 1.15), and use voice feedback to notify the user.

## 1.4.5 Localization Refinement With 3D Estimation

Omnidirectional imaging based indoor localization has the advantage of utilizing all the available visual information surrounding a visually impaired person for localizing themselves in the pre-built environmental models. However, in some cases, for example in high similarity scenes or where higher accuracy results are demanded, extra localization refinement is needed.

In this section, we propose a multiview omnidirectional geometry approach for localization refinement. The overall strategy is that the omnidirectional image-based localization approach narrow the user's location to a small area, and then the 3D based approach can be more effectively used in terms of both the computation time needed and the robustness of matches. We discuss the basic algorithm and present some preliminary results. Future

---

[15] https://youtu.be/gPGGcAfFsvY.

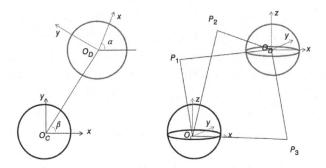

**Figure 1.16** Top view of current and destination locations (left) and 3D view of current and destination locations (right).

work is needed to fully integrate the refinement algorithms into a workable system for assistive navigation.

### 1.4.5.1 Geometric Constraints-Based Localization

We assume that after the 2D aggregation algorithm for initial localization, we find the most similar omnidirectional images in the database that were modeled and mapped to the physical space beforehand, using for example floor plan. Because of the modeling accuracy and environment noise, these query images will have very similar visual information as the database images, meaning we can find corresponding matching features between them, but they are not in the same locations. In this section, we will show how to find the relative location of the query omnidirectional image against the matched omnidirectional image in the database, whose position was labeled in the modeling process.

We model the omnidirectional imaging system with the spherical camera models [72,38] instead of the traditional pinhole camera model. Without losing generality, let us assume that there are two spherical camera coordinate systems, one for the query omnidirectional image whose position is unknown, and the other for the reference omnidirectional image in the database whose position is already recorded. The geometric relationship is illustrated in Fig. 1.16, where the bottom left is a query coordinate system, and the top right is a reference coordinate system. We align the origin and the three axes of the world coordinate system with the corresponding origin and axes of the query spherical camera coordinate system. We parameterize the motion from the query spherical camera coordinate system to the reference spherical camera coordinate system with two arguments: $\alpha$ and $\beta$, where $\alpha$ stands for the angles of the query coordinate system that will rotate counterclockwise to be able to align with the reference coordinate system, and $\beta$ stands for the direction the query coordinate system will translate to in order to align itself with the reference coordinate. In other words, the query spherical coordinate system can align with the reference spherical coordinate system by first a translation (represented by a translating direction angle $\beta$) and then a rotation (representing by a heading angle $\alpha$).

Note that the motion of the camera is a 2D motion in the floor plane, with two parameters, a rotation angle $\alpha$ and a heading direction $\beta$. This is because we are assuming that the users, i.e. VIPs, are standing on the floor, so the $z$-axis—the vertical direction—can be ignored.

After the problem is formulated, we can estimate the two parameters by utilizing the relationship between these two images. We detect and match local SIFT features between the two omnidirectional images. Before we move to the relationship formulation, we need to find the geometric relationship between the pixels in the omnidirectional image $I(u_{omni}, v_{omni})$ and our unit spherical models $(x_{sph}, y_{sph}, z_{sph})$, which is formalized as

$$\begin{cases} x_{sph} = \cos\left((H - v_{omni}) * D\right) * \cos(u_{omni}/W * 2\pi) \\ y_{sph} = \cos\left((H - v_{omni}) * D\right) * \sin(u_{omni}/W * 2\pi) \\ z_{sph} = \sin\left((H - v_{omni}) * D\right) \end{cases}, \qquad (1.1)$$

where $W$ is the width of the cylindrical image, $H$ is the height (in pixels) of the center of the projection of the camera from the ground in the cylindrical representation (estimated by the calibration procedure), and $D$ is degree per pixel in the vertical direction. With Eq. (1.1), we can calculate the corresponding camera coordinates $(x_{sph}, y_{sph}, z_{sph})$ for any image point $I(u_{omni}, v_{omni})$ in the un-warped cylindrical image space.

After we detect, describe, and match SIFT features for the query omnidirectional image and the reference omnidirectional images, we can obtain corresponding image feature pairs. Given a pair of image feature points, we can use Eq. (1.1) to calculate its corresponding unit sphere point pairs. These point pairs satisfy the constraints expressed in Eq. (1.2)

$$p_r^T \times E \times p_l = 0, \qquad (1.2)$$

where $p_r$ and $p_l$ are the 3D spherical points before and after the motion, and $E$ is the essential matrix between the two views [38]:

$$E = \begin{bmatrix} 0, 0, \cos\alpha\sin\beta - \sin\alpha\cos\beta \\ 0, 0, -\sin\alpha\sin\beta - \cos\alpha\cos\beta \\ -\sin\beta, \cos\beta, 1 \end{bmatrix}. \quad (1.3)$$

Finally, using Eq. (1.2) we can estimate the two angles $\alpha$ and $\beta$ by using at least three pairs of points in two images. Note that by using only two omnidirectional images, we cannot determine the translation vector amplitudes, and so we need to involve a third omnidirectional image to determine the scale factor [38]. The third image with known location can be easily obtained by finding the second closest match of the current image in the database.

### 1.4.5.2 Experiment Using Multiview Omnidirectional Vision

In this section, we show the results of utilizing synthetic data and real omnidirectional image data for localization. Since there is unavoidable noise in SIFT feature detection and matching, we will first examine how the detection/matching error will influence the parameter estimation using synthetic data. Then we will show our real data experiment as well as robustness improvement via the RANSAC algorithm [73].

(1) Error analysis on simulated data

As we can see from Eq. (1.1), because of feature detection noise there will be errors between the real 3D space position and calculated unit sphere position. This error propagates and influences the accuracy of the estimated pose parameters when we are calculating the parameters with Eq. (1.2). We will first examine how the noise in the feature detection process will influence the angle estimation accuracy.

In our experiments, we evenly distribute random noise with interval $[-n, n]$, where $n = 2, 4, 8$ pixels, which is added to the original point image coordinates. The estimated angles with and without the noise are shown in Table 1.4.

Two sets of results are obtained here for estimating angle and using Eq. (1.2). When estimating and from the essential matrix $E = [e_{ij}]_{3\times3}$ in Eq. (1.3), we can determinate the scale based on $e_{13}$ and $e_{23}$ (i.e. the third column of $E$, leading to solution $S_1$ in Table 1.4) or $e_{31}$ and $e_{32}$ (i.e. the

**Table 1.4** Error analysis results in pose estimation

| Parameters | Ground truth | $S_1$ | | | $S_2$ | | |
|---|---|---|---|---|---|---|---|
| pixel error | 0 | 2 | 4 | 8 | 2 | 4 | 8 |
| $\alpha$ | 60 | 61.9 | 61.3 | 64.4 | 59.7 | 57.6 | 61.2 |
| $\beta$ | 45 | 47.3 | 46.3 | 50.4 | 45.6 | 43.4 | 48.3 |

third row of $E$, leading to solution $S_2$). As we can see from Table 1.4, errors in feature point localization in omnidirectional images only cause minor errors in the final estimation angles, less than 5 degrees on average even if there is an 8-pixel offset of the ground truth feature location. Since we only need approximate direction instructions for VIPs, this amount of error is within the acceptance limits of angle errors. Solutions $S1$ and $S2$ are comparable for our simulated experimental results, but $S1$ gives slightly more stable results. So we choose solution $S1$ for our real data experiments.

(2) Real data experiment with RANSAC

RANdom SAmple Consensus (RANSAC) [73] is an iterative method used to estimate parameters of a mathematical model from a set of observed data which possibly contains outliers. Given $N$ pairs of corresponding points, three point pairs are randomly selected to estimate a set of angle pairs $(\alpha, \beta)$. Then the average error of the estimation is calculated, as shown in Eq. (1.4),

$$D(\alpha, \beta) = \sum_{i=1}^{N} p_{i_r}^T \times E \times p_{i_l}, \qquad (1.4)$$

where $p_{i_r}$ and $p_{i_l}$ are corresponding points of the $i$th $(i = 1, 2, ..., N)$ selected points pairs from the matched points candidates. Fig. 1.17 shows a real localization refinement example, where the top figure is the reference omnidirectional image from the modeling database whose position is known, and the bottom is a new query omnidirectional image whose rough location is provided, but is waiting for refinement.

Ten pairs of points are selected as the testing point pairs from these two images for RANSAC based parameter estimation. Every time, three of the ten points are used to calculate a set of pose parameters and all ten points are used to evaluate the result during RANSAC.

Fig. 1.18 shows the estimated results of the 120 ($C_{10}^3$) trials. The horizontal axis stands for the trial index of the experiments, while the vertical axis stands for the estimated $\alpha$ and $\beta$ results from the triples, denoted

**Figure 1.17** An example of a reference omnidirectional image (top) and a new query omnidirectional image (bottom).

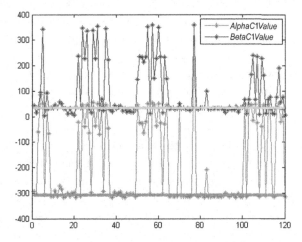

**Figure 1.18** Estimated pose parameter results (x-axis stands for the trial indexes, and y-axis stands for the estimated values). (For interpretation of the colors in this figure, the reader is referred to the web version of this chapter.)

*AlphaC1Value* and *BetaC1Value*. The yellow line indicates the best estimated $\beta$ value of all the trials, and the pink line indicates the best estimated $\alpha$ value. We can see that even though there are a few noisy results whose values are away from the final $\alpha$ and $\beta$ result (the red and blue impulse values), by utilizing RANSAC, we can eliminate these errors and integrate the correct results. Using RANSAC can also improve the accuracy of the estimation compared with using just mean values of the 120 estimation results, as shown in Table 1.5. The errors between the estimated and the ground truth are also listed in the table, while each error value is calculated as the ratio of difference between an estimation and the ground truth, and the full range of the angles (360 degrees).

**Table 1.5** Estimated results with real data

| Parameters | Ground truth | Mean value and error | | RANSAC value and error | |
|:---:|:---:|:---:|:---:|:---:|:---:|
| $\alpha$ | 45 | 58.1 | 3.6% | 52.1 | 2.0% |
| $\beta$ | 27 | 34.0 | 1.9% | 32.8 | 1.6% |

## 1.5 CONCLUSIONS AND DISCUSSIONS

This chapter presents a vision-based assistive indoor localization approach with various techniques in different stages for helping VIPs to localize in and navigate through indoor environments. Different from many other computer vision research, whose research problems are already well defined and formalized by the community, and whose major tasks are to apply their developed algorithms on standard datasets by tuning the parameter of models and evaluating the performance, this work studies the navigation need of VIPs, and then develop techniques in data collection, model building, localization, and user interfaces in both pre-journey planning and real-time assistance.

In summary, we use a smart phone with a panoramic camera and a high-performance server architecture to ensure portability and the mobility of the user and to take advantage of the huge storage and the high computation power of the server. An image indexing mechanism is used to find the location of an input image (or a short sequences of images/multiple images). To improve the query speed and ensure real-time performance, we use a place recognition pre-step and many-core GPUs to parallelize the query procedure.

For ensuring automatic environmental modeling using the previous omnidirectional imaging system, we use the building AutoCAD files for traversability, checking beforehand to find the optimal necessary paths in navigation from point A to point B in an indoor environment. A coarse-to-fine approach is used to obtain an initial localization efficiently and then to further improve the localization accuracy in three steps. First, we used a machine learning approach to classify images into categories so that one or more input images can be assigned to a category of the scene, therefore reducing the search space in the second step. Second, with a small clip of input video, the current location of the user is searched in the database of the scene considering the place recognition results. Finally, using pre-calibrated omnidirectional images and a new input omnidirectional image, as well as their geometric relationship to refine the new image's location and orientation. While the place recognition work and the refinement work are still in their early stages, they hold promise for a more efficient and accurate localization. We foresee that vision-based indoor localization, and its potential applications such as assistive navigation, will attract more research and will be widely integrated into daily life, especially that of VIPs.

## GLOSSARY

**Assistive technology (AT)**   assistive, adaptive, and rehabilitative devices or systems for people with disabilities and the procedures used in selecting, comparing, and using these devices.

**Convolutional Neural Network (CNN)**   a deep learning model consisting of many stacked layers; its key layers are called convolutional layers that perform convolution operations on the given input.

**Environmental modeling**   a process to build physical world 3D representations (e.g. a CAD model) or its 2D projections (e.g. a floor plan) to organize environment scenes and build mappings between the visual input (e.g. images) and their physical world locations.

**Field Of View (FOV)**   the extent of the area that a human eye or an imaging device can cover, usually in terms of degrees in both the horizontal or vertical directions.

**Indoor localization**   using one or more sensors such as cameras, magnetic sensors, IMUs, RGBD sensors, etc., to automatically determine the location of a robot or a person in real time in an indoor environment.

**Iterative Closest Point (ICP)**   an algorithm used to align two sets of 3D point cloud of the same scene.

**Omnidirectional imaging**   an imaging process to capture 360 degree field of view visual information, usually in the horizontal direction, or an imaging process to cover a visual field of (approximately) the entire sphere.

**Path planning**   the algorithm to find the shortest or other optimal path between two given points in a 2D or 3D space.

**Perspective N-point (pNp) problem**   the problem of estimating the pose (position and orientation) of a calibrated camera given a set of 3D points in the world and their corresponding 2D projections in an image.

**RANdom SAmple Consensus (RANSAC)**   an iterative method to estimate parameters of a mathematical model from a set of observed data that may contain outliers.

**Structure from Motion (SfM)**   a process of estimating 3D structure from a 2D image sequence.

**User Interface (UI)**   the place where interactions between humans and machines occur to allow effective operation and control of the machine and at the same time receive feedback from the machine to aid in the operators' decision-making processes.

## ACKNOWLEDGMENTS

This work is supported by the US National Science Foundation (NSF) through Award # EFRI-1137172 and # CBET-1160046, VentureWell (formerly NCIIA) through Award 10087-12, and the CUNY Graduate Center Science Fellowship (2012–2014). This work has also been supported by CUNY Graduate Center Doctoral Student Research Grant, PSC-CUNY research grant (2014–2016) and BMCC faculty development grant

(2016–2017). We would like to thank Ms. Barbara Campbell and her colleagues of the New York State Commission for the Blind (NYSCB) for their comments and suggestions in user interface designs.

## REFERENCES

[1] M. Leo, G. Medioni, M. Trivedic, T. Kanade, G. Farinella, Computer vision for assistive technologies, Computer Vision and Image Understanding 154 (2017) 1–15.
[2] N. Paisios, Mobile Accessibility Tools for the Visually Impaired, Ph.D. thesis, New York University, 2012.
[3] H. Kume, A. Suppé, T. Kanade, Vehicle localization along a previously driven route using image database, in: MVA, Citeseer, 2013, pp. 177–180.
[4] F. Hu, Z. Zhu, J. Zhang, Mobile panoramic vision for assisting the blind via indexing and localization, in: Computer Vision—ECCV 2014 Workshops, Springer, 2014, pp. 600–614.
[5] A.C. Murillo, G. Singh, J. Kosecka, J.J. Guerrero, Localization in urban environments using a panoramic gist descriptor, IEEE Transactions on Robotics 29 (1) (2013) 146–160.
[6] F. Hu, N. Tsering, H. Tang, Z. Zhu, Indoor localization for the visually impaired using a 3D sensor, Journal on Technology and Persons with Disabilities 4 (2016) 192–203.
[7] J. Xiao, S.L. Joseph, X. Zhang, B. Li, X. Li, J. Zhang, An assistive navigation framework for the visually impaired, IEEE Transactions on Human-Machine Systems 45 (5) (2015) 635–640.
[8] W.L. Khoo, Z. Zhu, Multimodal and alternative perception for the visually impaired: a survey, Journal of Assistive Technologies 10 (1) (2016) 11–26.
[9] W. Lun Khoo, J. Knapp, F. Palmer, T. Ro, Z. Zhu, Designing and testing wearable range-vibrotactile devices, Journal of Assistive Technologies 7 (2) (2013) 102–117.
[10] I. Jolliffe, Principal Component Analysis, Wiley Online Library, 2002.
[11] J.L. Carrivick, M.W. Smith, D.J. Quincey, Structure from Motion in the Geosciences, John Wiley & Sons, 2016.
[12] F. Hu, Vision-Based Assistive Indoor Localization, Ph.D. thesis, Department of Computer Science, The CUNY Graduate Center, 2016.
[13] V. Lepetit, F. Moreno-Noguer, P. Fua, EP$n$P: an accurate $O(n)$ solution to the P$n$P problem, International Journal of Computer Vision 81 (2) (2009) 155–166.
[14] Z. Zhang, Iterative point matching for registration of free-form curves and surfaces, International Journal of Computer Vision 13 (2) (1994) 119–152.
[15] F. Hu, N. Tsering, H. Tang, Z. Zhu, RGB-D sensor based indoor localization for the visually impaired, in: 31st Annual International Technology and Persons with Disabilities Conference, San Diego, CA, USA, March 2016.
[16] F. Hu, K. Yamamoto, Z. Zhu, 3D assistive indoor localization with a Google Glass, in: CIE-USA/GNYC 2016 Annual Convention, 2016, p. 50.
[17] W.B. Thompson, T.C. Henderson, T.L. Colvin, L.B. Dick, C.M. Valiquette, Vision-based localization, in: DARPA Image Understanding Workshop, 1993, pp. 491–498.
[18] S. Se, D. Lowe, J. Little, Vision-based mobile robot localization and mapping using scale-invariant features, in: Proceedings, 2001 ICRA, vol. 2, IEEE International Conference on Robotics and Automation, 2001, IEEE, 2001, pp. 2051–2058.
[19] Z.J. Geng, Method and apparatus for omnidirectional imaging, 2001, US Patent 6,304,285.

[20] S.K. Nayar, Omnidirectional imaging apparatus, 1998, US Patent 5,760,826.

[21] Z. Zhu, K.D. Rajasekar, E.M. Riseman, A.R. Hanson, Panoramic virtual stereo vision of cooperative mobile robots for localizing 3D moving objects, in: Proceedings, IEEE Workshop on Omnidirectional Vision, 2000, IEEE, 2000, pp. 29–36.

[22] A. Nishimura, T. Okuyama, Y. Yagi, Omnidirectional imaging system, 2016, US Patent 9,244,258.

[23] M. Drulea, I. Szakats, A. Vatavu, S. Nedevschi, Omnidirectional stereo vision using fisheye lenses, in: IEEE International Conference on Intelligent Computer Communication and Processing (ICCP), 2014, IEEE, 2014, pp. 251–258.

[24] A. Chapoulie, P. Rives, D. Filliat, Appearance-based segmentation of indoors/outdoors sequences of spherical views, in: IEEE/RSJ International Conference on Intelligent Robots and Systems, 2013, IEEE, 2013, pp. 1946–1951.

[25] F. Hu, N. Tsering, H. Tang, Z. Zhu, Indoor localization for the visually impaired using a 3D sensor, Journal on Technology and Persons with Disabilities 4 (November 2016) 192–203.

[26] Y.H. Lee, G. Medioni, Wearable RGBD indoor navigation system for the blind, in: Workshop on Assistive Computer Vision and Robotics, ECCV2014, IEEE, 2014.

[27] A.I. Awad, M. Hassaballah, Image Feature Detectors and Descriptors: Foundations and Applications, vol. 630, Springer, 2016.

[28] T. Sattler, B. Leibe, L. Kobbelt, Improving image-based localization by active correspondence search, in: European Conference on Computer Vision, Springer, 2012, pp. 752–765.

[29] T. Sattler, B. Leibe, L. Kobbelt, Efficient & effective prioritized matching for large-scale image-based localization, IEEE Transactions on Pattern Analysis and Machine Intelligence (2016).

[30] T. Sattler, B. Leibe, L. Kobbelt, Fast image-based localization using direct 2D-to-3D matching, in: International Conference on Computer Vision, 2011, IEEE, 2011, pp. 667–674.

[31] M. Cummins, P. Newman, Appearance-only slam at large scale with FAB-MAP 2.0, The International Journal of Robotics Research 30 (9) (2011) 1100–1123.

[32] H. Kume, A. Suppé, T. Kanade, Vehicle localization along a previously driven route using an image database, in: IAPR International Conference on Machine Vision Applications, Kyoto, Japan, 2013.

[33] H. Badino, D. Huber, T. Kanade, Real-time topometric localization, in: IEEE International Conference on Robotics and Automation (ICRA), 2012, IEEE, 2012, pp. 1635–1642.

[34] F. Hu, T. Li, Z. Geng, Constraints-based graph embedding optimal surveillance-video mosaicing, in: The First Asian Conference on Pattern Recognition, IEEE, 2011, pp. 311–315.

[35] E. Molina, Z. Zhu, Y. Tian, Visual nouns for indoor/outdoor navigation, in: 13th International Conference, ICCHP 2012, vol. 7383, 2012, pp. 33–40.

[36] E. Molina, Z. Zhu, Visual noun navigation framework for the blind, Journal of Assistive Technologies 7 (2) (2013) 118–130.

[37] GoPano, GoPano micro camera adapter, 2015, http://www.gopano.com/products/gopano-micro.

[38] M. Aly, J.Y. Bouguet, Street view goes indoors: automatic pose estimation from uncalibrated unordered spherical panoramas, in: IEEE Workshop on Applications of Computer Vision (WACV), 2012, IEEE, 2012, pp. 1–8.

[39] A. Oliva, A. Torralba, Modeling the shape of the scene: a holistic representation of the spatial envelope, International Journal of Computer Vision 42 (3) (2001) 145–175.

[40] G. Farinella, S. Battiato, Scene classification in compressed and constrained domain, Computer Vision, IET 5 (5) (2011) 320–334.

[41] J. Xiao, K.A. Ehinger, A. Oliva, A. Torralba, Recognizing scene viewpoint using panoramic place representation, in: IEEE Conference on Computer Vision and Pattern Recognition (CVPR), 2012, IEEE, 2012, pp. 2695–2702.

[42] Y. Li, Video Representation, Springer, Boston, MA, USA, ISBN 978-0-387-39940-9, 2009, pp. 3296–3303.

[43] Z. Zhu, S. Yang, G. Xu, X. Lin, D. Shi, Fast road classification and orientation estimation using omni-view images and neural networks, IEEE Transactions on Image Processing 7 (8) (1998) 1182–1197.

[44] A. Kovashka, D. Parikh, K. Grauman, Whittlesearch: image search with relative attribute feedback, in: IEEE Conference on Computer Vision and Pattern Recognition (CVPR), 2012, IEEE, 2012, pp. 2973–2980.

[45] C. Jin, Z. Wang, T. Zhang, Q. Zhu, Y. Zhang, A novel visual-region-descriptor-based approach to sketch-based image retrieval, in: Proceedings of the 5th ACM on International Conference on Multimedia Retrieval, ACM, 2015, pp. 267–274.

[46] D. Mishkin, M. Perdoch, J. Matas, Place recognition with WxBS retrieval, in: CVPR 2015 Workshop on Visual Place Recognition in Changing Environments, 2015.

[47] J. Fuentes-Pacheco, J. Ruiz-Ascencio, J.M. Rendón-Mancha, Visual simultaneous localization and mapping: a survey, Artificial Intelligence Review 43 (1) (2015) 55–81.

[48] A. Torii, R. Arandjelovic, J. Sivic, M. Okutomi, T. Pajdla, 24/7 place recognition by view synthesis, in: Proceedings of the IEEE Conference on Computer Vision and Pattern Recognition, 2015, pp. 1808–1817.

[49] B. Zeisl, T. Sattler, M. Pollefeys, Camera pose voting for large-scale image-based localization, in: Proceedings of the IEEE International Conference on Computer Vision, 2015, pp. 2704–2712.

[50] T. Sattler, M. Havlena, F. Radenovic, K. Schindler, M. Pollefeys, Hyperpoints and fine vocabularies for large-scale location recognition, in: Proceedings of the IEEE International Conference on Computer Vision, 2015, pp. 2102–2110.

[51] D. Zhang, D.J. Lee, B. Taylor, Seeing eye phone: a smart phone-based indoor localization and guidance system for the visually impaired, Machine Vision and Applications 25 (3) (2014) 811–822.

[52] Y.H. Lee, G. Medioni, A RGB-D camera based navigation for the visually impaired, in: RSS 2011 Workshop on RGB-D Cameras, 2010, pp. 1–6.

[53] J. Cao, T. Chen, J. Fan, Landmark recognition with compact BoW histogram and ensemble ELM, Multimedia Tools and Applications (2015) 1–19.

[54] N. Sunderhauf, S. Shirazi, F. Dayoub, B. Upcroft, M. Milford, On the performance of ConvNet features for place recognition, in: IEEE/RSJ International Conference on Intelligent Robots and Systems (IROS), 2015, IEEE, 2015, pp. 4297–4304.

[55] B. Phillips, H. Zhao, Predictors of assistive technology abandonment, Assistive Technology 5 (1) (1993) 36–45.

[56] G. Farinella, D. Ravì, V. Tomaselli, M. Guarnera, S. Battiato, Representing scenes for real-time context classification on mobile devices, Pattern Recognition 48 (4) (2015) 1086–1100.

[57] H. Altwaijry, M. Moghimi, S. Belongie, Recognizing locations with Google Glass: a case study, in: IEEE Winter Conference on Applications of Computer Vision (WACV), Steamboat Springs, CO, USA, 2014.

[58] R. Manduchi, Mobile vision as assistive technology for the blind: an experimental study, in: 13th International Conference on Computers Helping People of Special Needs, ICCHP 2012, Springer, 2012.

[59] J. Zhang, S. You, L. Gruenwald, Parallel online spatial and temporal aggregations on multi-core CPUs and many-core GPUs, Information Systems 44 (2014) 134–154.

[60] S. You, J. Zhang, L. Gruenwald, Large-scale spatial join query processing in cloud, in: 31st IEEE International Conference on Data Engineering Workshops (ICDEW), 2015, IEEE, 2015, pp. 34–41.

[61] G. Olmschenk, C. Yang, Z. Zhu, H. Tong, W.H. Seiple, Mobile crowd assisted navigation for the visually impaired, in: IEEE 12th International Conference on Ubiquitous Intelligence and Computing, 2015, pp. 324–327.

[62] H. Petrie, V. Johnson, T. Strothotte, A. Raab, S. Fritz, R. Michel, MOBIC: designing a travel aid for blind and elderly people, Journal of Navigation 49 (1) (1996) 45–52.

[63] T. Völkel, G. Weber, RouteCheckr: personalized multicriteria routing for mobility impaired pedestrians, in: Proceedings of the 10th International ACM SIGACCESS Conference on Computers and Accessibility, ACM, 2008, pp. 185–192.

[64] R. Ichikari, T. Yanagimachi, T. Kurata, Augmented reality tactile map with hand gesture recognition, in: International Conference on Computers Helping People with Special Needs, Springer, 2016, pp. 123–130.

[65] H. Tang, N. Tsering, F. Hu, Z. Zhu, Automatic pre-journey indoor map generation using autocad floor plan, Journal on Technology and Persons with Disabilities 4 (2016) 176–191.

[66] H. Tang, N. Tsering, F. Hu, Automatic pre-journey indoor map generation using autocad floor plan, in: 31st Annual International Technology and Persons with Disabilities Conference, San Diego, CA, USA, March 2016.

[67] P.E. Hart, N.J. Nilsson, B. Raphael, A formal basis for the heuristic determination of minimum cost paths, IEEE Transactions on Systems Science and Cybernetics 4 (2) (1968) 100–107.

[68] E.W. Dijkstra, A note on two problems in connexion with graphs, Numerische Mathematik 1 (1) (1959) 269–271.

[69] A. Krizhevsky, I. Sutskever, G.E. Hinton, ImageNet classification with deep convolutional neural networks, in: ImageNet ILSVRC Challenge, Advances in Neural Information Processing Systems 25, NIPS 2012, 2012.

[70] K. Simonyan, A. Zisserman, Very Deep Convolutional Networks for Large-Scale Image Recognition, Technical report, 2014.

[71] C. Rasmussen, Z. Ghahramani, Occam's razor, in: T. Leen, T.G. Dietterich, V. Tresp (Eds.), Neural Information Processing Systems, vol. 13, The MIT Press, 2001, pp. 294–300.

[72] A. Torii, A. Imiya, N. Ohnishi, Two- and three-view geometry for spherical cameras, in: Proceedings of the Sixth Workshop on Omnidirectional Vision, Camera Networks and Non-classical Cameras, Citeseer, 2005.

[73] M.A. Fischler, R.C. Bolles, Random sample consensus: a paradigm for model fitting with applications to image analysis and automated cartography, Communications of the ACM 24 (6) (1981) 381–395.

# CHAPTER 2

# Computer Vision for Cognition
## An Eye Focused Perspective

**Corneliu Florea, Laura Florea, Constantin Vertan**
University Politehnica of Bucharest, Image Processing and Analysis Laboratory, Bucharest, Romania

## Contents

## Abstract

In this chapter we review computer vision techniques related to gaze tracking and eye area analysis in general that may help the detection and investigation of cognitive impairments. In the first part, we summarize the normal anatomical and psychological traits of the human eye and gaze to establish the middle line that can act as a reference for abnormal behavior. Next, we discuss the main approaches to eye analysis and gaze tracking and their potential use in various psychological and neurological experiments. The second part presents actual contributions to the early detection of impairments. Mainly, these are related to children's autism, dyslexia, and anxiety disorders, which all strongly feature abnormal gaze behavior. We end by discussing various perspectives.

Computer Vision for Assistive Healthcare.
DOI: https://doi.org/10.1016/B978-0-12-813445-0.00002-2

## Keywords

Automatic gaze tracking, Early detection of mental impairment, Abnormal gaze pattern, Eye and gaze anatomy

## Chapter Points

- We discuss the reasons behind considering eyes as the "gates to the soul" and what their "normal behavior" is.
- We review the main anatomical and psychological features of the eye and gaze in the context of nonverbal communication.
- Analysis of gaze movement provides valuable cues for the early detection of various conditions.
- Autism or, more precisely, Asperger Syndrome is associated with avoiding eye contact in nonverbal interactions and different patterns while performing face recognition.
- Those with dyslexia or reading disability features show different patterns of fixations and saccades with respect to normal people while attempting to read.
- Post Traumatic Stress Disorder is associated with hypervigilance, which is an abnormal pattern of fixations and saccades separable in specific scenarios, thus allowing early identification.
- Computer vision methods related to eye and gaze tracking facilitate early detection of several cognitive impairments and help separation and identification of various symptoms.
- The research is still limited to clinical laboratory scenarios and tests. Further development is necessary to ease diagnosis by using in-the-wild environment investigation and for milder cases.

## 2.1  WHY EYES ARE IMPORTANT FOR HUMAN COMMUNICATION

The human face is probably the most striking visual stimulus as it is the main interface for nonverbal communication [1]. The human visual and cognitive system is tuned towards the processing of facial information (as demonstrated for instance by the Thatcher illusion [2]) as it is normally present during interactions, and specific processes have been developed for face differentiation and understanding. The eye region of the face is a focal point in this communication, due to the extensive amount of information that can be extracted from it. Kobayashi and Kohshima [3] noted that among the primates only humans have a dark iris against a white sclera through the palpebral fissure (the aperture between the eyelids), thus providing a number of potential cues to gaze direction; they also suggested that the combination of a dark iris with light sclera evolved to provide a cue for gaze direction (an advantage in a highly social and cooperative species such as humans), in contrast to the darker sclera of other primates that probably evolved for the opposite purpose, to make gaze direction obscure, such that predators cannot tell that they have been spotted.

## 2.1.1 Eyes in Nonverbal Communication

With respect to other facial features, the eye region is fundamentally implicated in all aspects of nonverbal communication such as the expression of emotion, direction of attention, and identity. To give a glimpse, let us recall that Frischen et al. [4] evaluate the social impact of gaze direction and conclude that "people's eyes convey a wealth of information about their direction of attention and their emotional and mental states". They further note that "eyes and their highly expressive surrounding region can communicate complex mental states such as emotions, beliefs, and desires" and "observing another person's behavior allows the observer to decode and make inferences about a whole range of mental states such as intentions, beliefs, and emotions". In recent years there has been tremendous activity in the fields of Cognitive and Behavioral Neuroscience while investigating the processing of the eye region and gaze direction in various tasks and social situations; in parallel, the underlying neural systems associated with these processes are far from being understood. The process of understanding these aspects favors early discovery of various social impairments or physical disabilities.

The high mobility of the eyeball (bulbus oculi) in the human head makes it a highly effective message transmitter. Thus, interfaces based on the human gaze are able to replace standard verbal communication for impaired people.

Regarding the importance of eye regions in communication, Itier and Batty [1] conclude in their review that ocular movements during face perception have shown that people spend more time on internal features (eyes, nose, mouth) than external ones (hair, face contour, forehead, ears). On this path, multiple studies [5,6] concluded that the eye region is the most attended of the facial features and the primary source of information regardless of the task, whether it focuses on gaze, head orientation, identity, gender, facial expression, or age.

In a similar manner to faces, eyes vary greatly from one individual to another and the eye region may in fact be the facial zone that varies most between people. An illustration of this aspect may be found for instance in the recent Multi-view Gaze dataset [7]. Eye color, shape, interocular or interpupillary distances are specific to each individual and can be used as soft biometrics. Other elements of the eye region such as eyebrows, eyelids, eyelashes, and their respective distances constitute elements necessary for the recognition of face identity [8]. Psychologists have shown that in the absence of eye features the accuracy of face recognition drops considerably; in contrast, masking the nose or the mouth has little or no effect [9]. Similarly, face detection in humans drops when the eye region is occluded compared to the nose, mouth, or forehead cases [10]. Also, it is the principal element that people use to decide whether a face is male or female. The eye region is, thus, a key element of face recognition. In addition to its important role in processing identity, the eye region carries information necessary for emotion recognition [11]. In conclusion, normal people (i.e. those without psychological impairments) rely heavily on eye regions in interactions and nonverbal communication.

## 2.1.2 Eye Movements

The purpose of this section is to recall, from an anatomical point of view, the mobility capabilities of the eye. Typical description of eye movements in the computer vision literature [12] divides the movements only between *saccades* and *fixations*. However the medical and psychological literature identifies more categories. Furthermore, the classification is approached from several perspectives, such as spatial movements and temporal movements.

### Spatial Movements

From a medical point of view, eye movement is accomplished by a selective increase/decrease in the activity of the muscles surrounding the eye globe, which can be divided into two categories [13]:

- *intrinsic* ocular muscles, which control the lens and pupil;
- *extraocular* muscles, which are designed to stabilize and move the eyes.

As our limited spatial resolution impedes us from identifying and interacting based on lens and pupil movements unless we are very close, the main nonverbal interaction is related only to the movements of the eyeball.

The list of extraocular movements[1] includes the following [13]:

- adduction—the pupil directed toward the nose;
- abduction—the pupil directed laterally;
- elevation—the pupil directed up;
- depression—the pupil directed down;
- intorsion—the top of the eye moving toward the nose;
- extorsion—the superior aspect of the eye moving away from the nose.

From the medical point of view, horizontal eye movements are rather simple as they require the activation of only one muscle (either the lateral rectus or the medial rectus); instead, movements above or below the horizontal plane are complicated as they require, at the very least, the activation of a pair of muscles. The explanation lies in the fact that the orbit is not directed straight forward in the head and there is no single muscle positioned to direct the eye straight up or down without the simultaneous occurrence of unwanted movements. Consequently, such eye movements are less fluid and generate potential difficulties while tracking.

## Temporal Movements

While various authors distinguish multiple classification possibilities, the most popular classification of eye movement assumes four basic types [14]: saccades, smooth pursuit movements, vergence movements, and vestibulo-ocular movements.

**Saccades** are fast, ballistic movements of the eyes that assume an abrupt change of the fixation point. Their amplitude range goes from small (e.g. while reading) to significant (e.g. while gazing a distant scene). Saccades can be voluntarily produced or they can occur involuntarily, even if the gaze is fixed onto a target. The latter category is related, for instance, to a controversial hypotheses from the Neuro-Linguistic Programming theory [15].

Regarding temporal development, after the onset of a target, a saccade takes about 200 ms to begin [16]. During this delay, the position of the target with respect to the fovea is determined by the brain and the difference with respect to the intended position is converted into a neurological command that activates the extraocular muscles to correct the distance and the direction. Saccadic movements are often described as "ballistic" because the saccade generating system cannot respond to subsequent changes in the

---

[1] For more details and video illustration of movements we refer the reader to the following Internet pages: https://www.dartmouth.edu/~dons/part_1/chapter_4.html or https://cyhsanatomy2.wikispaces.com/Bones+and+muscles+of+the+orbitals.

position of the target during the course of the eye movement. If the target moves again during this time (which is in the range of 15–100 ms), the saccade will miss the target, and a second saccade will be needed to correct the error.

Saccadic eye movements are a focal point in eye related investigation as many applications are based upon them, such as the detection of fatigue or drowsiness, studies concerning the mechanisms of human vision, assistance in diagnosing neurological disorders, or in sleep studies [17].

Between saccades, the gaze is held in **fixations**. Previous works [18] range their duration from 100 to 500 ms. While reading, a fixation averages approximately 250 ms. A 2 degree saccade usually takes about 30 ms in a typical reading task, while a 5 degree saccade (typical while doing scene perception) lasts about 40–50 ms.

Fixations are often analyzed in vision science, neuroscience, and psychological studies to determine a person's focus and level of attention.

Recalling the Donder law, Hansen and Ji [12] stressed that gaze direction determines eye orientation uniquely, and that the orientation is independent with respect to the previous eye positions. For stationary heads and far reaching gazes (i.e. pointing towards the horizon line), the Listing law describes the three-dimensional valid subset of eye positions and eye axes of rotation as those that can be reached from the "primary" position by a single rotation about an axis perpendicular to the gaze direction [19].

**Smooth pursuit movements** are used to stabilize the image of a moving object of interest on the fovea, facilitating continuous high-acuity scrutiny of the target [20]. These movements are voluntarily controlled.

**Vergence movements** are disconjugate movements (i.e. the eyes can move separately) that aim to align the fovea of each eye with targets located at different distances.

**Vestibulo-ocular movements** stabilize the eyes with respect to the outer world by compensating for head movements. They prevent visual images to be shaky on the retinal surface as the head shakes.

## 2.2 GAZE DIRECTION RECOGNITION AND TRACKING

The information carried by eye movements has fascinated scientists and, over the years, a plethora of methods to track the motion of the eye have been proposed, forming what is known as *oculography*. There are three typical categories:

1. **ElectroOculography** is based on electrical sensors that are attached to the skin around the eyes and measure the changes in the electric field produced by eye rotation. This process is based on the existence of a permanent potential difference between the cornea and the eye fundus of approximately 1 mV [21]. Small voltage differences appear around the eye as the eye position varies. Using carefully located electrodes the horizontal and vertical movements can be recorded. The disadvantage is that signal changes may also appear when there is no eye movement. ElectroOculography is considered to be less reliable for quantitative measurement, particularly for medium and large saccades. The advantage of the method, which is rather noninvasive, is that it is able to record large eye movements. The method is also able to record eye movements when the eyelid is closed.

2. **Scleral Search Coils** are still considered the golden standard for measuring eye movements [22]. A small coil of wire is placed on the subject's eye embedded in a contact lens. When a coil of wire moves in a magnetic field, the field induces a voltage in the coil, which is then read and processed, being related in the end with the position of the eye. Typically, the procedure requires anesthesia and a surgical intervention (being an invasive method) and thus is restricted to laboratory medical and psychological research.

3. **Image/Video Infrared Oculography** is based in the illumination of the eye by a light source that is further reflected by the sclera. The difference between the amounts of infrared light reflected back is recorded by an IR image sensor and decoded as it carries the information about the eye position modifications. The tracker may use the corneal reflection (which forms the first Purkinje image) or the reflection from the back of the lens of the eye (the fourth Purkinje image). The main disadvantage of this method is that it can measure eye movements only for a limited horizontal and vertical range. In many embodiments, the light source is infrared, as it has the advantage of being invisible to the subject's eye, and thus is nonintrusive.

**Noninvasive Pattern Recognition Gaze Tracking Solutions.** Alternatively, direct image/video analysis methods based on standard pattern recognition systems may be employed to determine the gaze direction or eye movements.

The first step in determining eyes and gaze is the identification of the face within the recorded image/image stream. In the last decade the dominant face detection method was based on Haar boosted cascades [23], but recent advances have suggested methods based on the deformable part model (DPM) [24] or Deep Convolutional Networks [25]. Next, the eyes are searched for in the upper part of the detected face by exploiting their particular characteristics. Initial approaches explored the contrast between the darker iris within the brighter sclera either directly [26], or indirectly by means of integral projections [27,28]. The high contrast between the iris and the sclera produces strong gradients, identifiable directly [29] or by means of machine learning [30]. The circular shape of the eye can be aggregated by means of isophotes [31,32]. More recently, gaze tracking systems have also benefited from the advance of deep convolutional neural networks (DCNN) [33].

Although the noninvasive pattern recognition gaze tracking methods have greatly evolved in recent years, they still provide limited accuracy [12] due to image sensor resolution and inherent pattern recognition limitations. For instance, the DCNN method [33] reaches an average error of 18 mm in terms of fixation localization on a mobile phone; this corresponds to approximately 3.4 degrees compared to less than 0.5 degrees for a fixed, display-mounted IR-based eye tracker. The image/video-based solutions, however, have a significantly lower price and are easier to build.

**Head-Mounted vs Remote.** Another perspective of categorizing methods refers to the relative position of the tracker with respect to the eyes and head:

1. Head mounted devices (e.g. glasses or head mounted camera);
2. Stationary and/or remote devices.

Because they are closer to the eye, head mounted devices have access to higher resolution images and a more controlled illumination, thus yielding better precision. They do exhibit some inherent shortcomings. First, the price is one order of magnitude higher than the remote systems, restricting the area of usability. Second, they must be worn, which is a distinct indicator that the user is subject to exploration and investigation by nontraditional means.

## 2.2.1 Eye Tracking Metrics

In the development of any practical application, there is a need for quantitative feedback in order to determine the weak areas and to improve upon them; in other words, there is a need for metrics related to eye tracking [34]. In the context of assistive technology based on eye tracking, practical applications require specific measurements. Some of the most frequently used metrics include the following:

- Fixation duration: typically the time taken by the fovea to process an image is measured;
- Time interval between two fixations, saccades: the time taken by the fovea to switch from one image to another is measured;
- Gaze duration: the cumulative duration and average spatial location of a series of consecutive fixations within an area of interest. Gaze duration is usually composed of several fixations and sometimes a small amount of time for the short saccades between these fixations;
- Area of interest: the area of the viewed scene that is of interest for the current application; it is application defined;
- Scan path: the spatial arrangement of a series of fixations.

Additionally, other metrics have been used in practice, such as the number of fixations for each area of interest, the fixation duration mean on each area of interest, fixation duration total for each area of interest, number of areas of interest fixated, scan path length, scan path direction, etc.

## 2.3 EYE TRACKING AND COGNITIVE IMPAIRMENTS

Following the creation and development of eye tracking technology, many studies and investigations of cognitive related impairments based on eye and gaze behavior appeared. In their initial stages the studies looked for atypical behavior in previously diagnosed people in order to search for symptoms. In later years, the gaze and eyes became part of the diagnosis. A summary of various cognitive impairments and the way that tracking technology impacted it is given in Table 2.1.

## 2.4 COMPUTER VISION SUPPORT FOR DIAGNOSIS OF AUTISM SPECTRUM DISORDERS

Autism spectrum disorder (ASD) is a highly variable neurodevelopmental disorder characterized by impaired social interaction, verbal and nonverbal communication, and restricted and repetitive behavior [35,36]. Researchers are still investigating and debating whether autism is a set of many different disorders akin to intellectual disability or a few disorders sharing common aberrant pathways in the brain circuitry.

One standing characteristic of a person suffering from autism is difficulty in extracting cues from the face of the interlocutor [1]. A more serious deviation into this direction is Asperger Syndrome (AS) [35], which is a developmental disorder characterized by a significant amount of restricted

**Table 2.1** Cognitive impairments and the eye tracking technologies used to study them

| Impairment | Aim | Technology | Requirements |
|---|---|---|---|
| Autism | Early diagnosis in children and infants | Temporal and spatial gaze tracking | High spatial accuracy |
| Autism | Symptom analysis | Spatial gaze tracking | High spatial accuracy |
| Autism | Symptom analysis | Combined gaze and multimodal analysis tracking | Nonintrusive |
| Dyslexia | Symptom analysis | Temporal gaze tracking | High sampling rate |
| Phobia | Phobia subject | Point of gaze | Low cost |
| PTSD | Symptom analysis | Spatial gaze tracking | Noninvasive |
| Dementia | Symptom analysis | Electro-oculographic | Closed eyelids |

social interaction and deficiency in nonverbal communication, along with restricted and repetitive patterns of behavior.

Currently, the main guidelines in diagnosing autism refer to a specific protocol named the Autism Diagnostic Observation Schedule (ADOS) [37], which requires specific clinical evaluation by trained professionals and restricts the number of cases approached. With respect to these aspects, the average age of ASD diagnosis in the United States is at approximately 5 years of age [38], while many children affected by autism start to exhibit specific characteristics as early as 2 years of age [39].

Given this motivation, multiple solutions based on recent developments in the field of computer vision have been proposed to assist in the process of early identification of ASD in general or AS in particular [1]. The particular aim is to identify and to measure the amount of deviation from the normal behavior.

Abnormal behavior in an autistic person is related to the lower likelihood of engaging in movements and procedures for nonverbal communication [1]. More precisely, when told explicitly to pay attention to gaze, autistic individuals gather general knowledge about the eyes and the information transmitted; autistic persons are aware that eyes are necessary for sight and can discriminate gaze direction [40]. A point of abnormality lies in the specific strategy used to determine gaze orientation: instead of treating the eyes as social cues (according to the *social reading hypothesis*), ASD individuals rely on low-level information such as pupil direction and the contrast between iris and white sclera for spatiality by assuming the so-called feature correspondence hypothesis [41].

Furthermore, it has been rather recently shown that early signs of autism may be identified in infant gaze [42]. In this study, the authors followed 110 infants aged from 2–6 months to 36 months, where the final diagnosis was established. They concluded that the children with autism experienced different growth curves for social visual engagement. These results established that early detection is feasible.

## 2.4.1 Methods and Solutions

Typical diagnosis procedures take place in a clinical laboratory setup. Two main directions are followed: (1) based on the saliency map or analyzing "where the subject looks" and (2) based on the behavior of the subject or "how the subjects looks".

### Using Saliency Models

Let us recall that *visual saliency* means that given a scene, the majority of people's attention will be drawn by the same areas, called saliency regions [43,44]. The human visual fixation behavior is driven by sensorial bottom-up mechanisms; visual saliency refers to the physical, bottom-up distinctness of image details [45].

In the typical visual saliency setup, a user is asked to look at a displayed image while a gaze tracker identifies the point of gaze. The eye tracker is noninvasive, remote, often table-mounted and video-based [46]. It has a sampling rate of 50 to 500 Hz and uses corneal reflection to estimate the gaze location. Using a prior geometric calibration that projects the space of locations onto the display plane the tracker returns a pair of Cartesian screen coordinates that represents the area of interest. The calibration procedure assumes a set of fixations in a specific location of the screen (the point of gaze).

However, when the scenario involves small children or infants, there are concerns regarding the accuracy of gaze estimation that arise because children and infants have more significant head movements and are less responsive to calibration [47]. Counteractions are related to off-line calibration [48] and computation of the cumulative time spent looking at predefined areas of interest (AOI) [47].

A variation on the saliency approach is to extend the unique display into a pair. This approach is called *paired visual preference paradigm*. According to this paradigm, the two visual displays should be presented side-by-side and the information (or visual stimuli) displayed should be slightly different [49]. The logic of this approach is related to the ease of linking the looking time to a specific type of information. Thus, the fewer stimuli that are different, the easier it will be to interpret the results [50].

A third alternative that has been proposed recently is based on a controlled environment where the display (the source of artificial stimuli) is replaced by real stimuli. In such a study, the environment is scanned by a Kinect-3D camera [51]. In this case, the point of gaze is determined by consecutive face detection, followed by active appearance models tracking; the head pose and eye locations are determined via eye fiducial point localization, and the final determined point of gaze is correlated with pre-learned object locations.

Saliency-based studies show that people with ASD tend to ignore socially relevant stimuli (such as faces) during visual searches [52]. However, if forced to look at a face, they will tend to focus on the mouth and will ignore the eyes [53].

Taken together, the findings suggest that visual saliency when associated with ASD is atypical. More thorough examinations have shown that people with ASD have higher saliency weights for low-level properties of images, but lower weights for object- and semantic-based properties [54]. In more detail, people with ASD fixated more on the center of the images, fixated on fewer objects when multiple similar objects were present in the image, and seemed to have atypical preferences for particular objects in natural images.

## Using Behavioral Models

In the second case, the procedure refers to recording the interaction, known as *behavioral imaging*. In such a scenario, the interviewer wears glasses with built-in eye tracking technology [55]. An example of this scenario is used for the acquisition of the Multimodal Dyadic Behavior (MMDB) dataset [56], taking place within a child-friendly large space, equipped with a variety of unobtrusive sensing capabilities, including high-speed cameras, a Kinect camera, etc. In addition, the trained examiner records, via face detection and eye tracking, the point of gaze of the infant. By this development, during post-processing, the examiner is able to determine whether eye contact was made with the subject child. Furthermore, multimodal data

**Table 2.2** Studies investigating autism related subjects and the gaze related information used

| Authors | Approach | Subjects | Conclusions |
|---|---|---|---|
| Klin et al. [52] | Saliency | Children | ⋄ Children with ASD look less at faces |
| Jones et al. [53] | Saliency | Toddlers | ⋄ Toddlers with ASD look less at eyes and more at mouth |
| Sasson et al. [57] | Saliency | Children | ⋄ Children with ASD explore fewer images and make more fixations per image |
| von Hofsten et al. [58] | Behavior | Infants | ⋄ Infants with ASD fail more often to predict the interlocutor change in a dialogue |
| Klin et al. [59] | Saliency | Toddlers | ⋄ Toddlers with ASD look less at biological motion |
| Pierce et al. [60] | Saliency | Toddlers | ⋄ Toddlers with ASD look more at dynamic geometric; ⋄ 60% with ASD spend as much time as typically developing watching dynamic social images |
| Bedford et al. [61] | Behavior | Infants | ⋄ Gaze following might not be impaired |
| Rehg et al. [56] | Behavior | Infants | ⋄ Children with autism make less eye contact |
| Falck-Ytter et al. [50] | Behavior | Infants | ⋄ Gaze following capacity is correlated with communication ability |
| Wang et al. [54] | Saliency | Adults | ⋄ People with ASD fixate more on low-level features and less on objects |

also allows the construction and validation of additional hypotheses during post-processing regarding normal and abnormal behavior.

## 2.4.2 Results

Using such scenarios numerous studies regarding the correlation between abnormal gaze behavior and autism have been performed. A summary of some of the most relevant studies can be found in Table 2.2.

## 2.5 COMPUTER VISION SUPPORT FOR THE IDENTIFICATION OF DYSLEXIA

Dyslexia is a less aggressive cognitive impairment. It is a neurodevelopmental reading disability estimated to affect up to 10% of the population. It affects the speed and accuracy of word recognition and, as a consequence, impedes reading fluency and text comprehension. Since reading ability is a skill that falls along a continuum, dyslexia is best considered a difficulty along this continuum with no clear-cut or absolute limit [62].

While there is yet no full understanding of the cause of dyslexia, or agreement on its precise definition, it is certain that many individuals suffer persistent problems in learning to read for no apparent reason [63]. Although it is generally agreed that early intervention is the best form of support for children with dyslexia, there is still a lack of efficient and objective means to help identify those at risk during the early years of school. Recently it has been shown that the use of machine learning (feature selection and classification) and advanced statistics can differentiate, based on the analysis of eye movements during reading tasks, high-risk individuals with significant accuracy [63].

Distinct application of gaze tracking is the analysis of the reading process with the aim of understanding the process of learning to read or how attention correlates with understanding. For instance, Joseph et al. [64] investigated, by means of gaze tracking, insights into how children learn to read and the impact of word frequency in sentence reading. Godfroid et al. [65] used eye tracking measurements to test hypotheses concerning word complexity, attention persistence, and short- and long-term memory. Along the same lines, Rayner et al. [66] monitored subjects' eye movements while they read sentences containing high- or low-predictable target words; their findings showed word predictability (due to contextual constraints) and word length have strong and independent influences on word skipping and fixation durations. Possible applications of eye movement control in teaching are discussed, for instance, in [67].

Many previous studies showed that dyslexic individuals exhibit abnormal gaze behavior when tracking; they have longer duration of fixations and shorter saccades leading to more fixations in reading than normally developing readers of the same chronological age [68]. This behavior has been reported in the context of different languages [69]. There is debate among psychologist whether dyslexia is caused by abnormal eye movements or the reading impairment itself leads to abnormal movements [70].

**Normal Reading.** Previous tracking-based evidence showed that readers do not only process the fixated word; for instance, it has been shown that the perceptual span of alphabetic languages (that read from left to right) usually extends from 3 to 4 spaces to the left of the fixation point to about 14–15 spaces to the right; this rightward asymmetry seems to be linked to the direction of reading [71]. The harder the text is to read, the smaller the perceptual span becomes. Within this "perceptual span", the information acquired refers to (1) readers who identify words in the area closest to the fixation point and (2) readers who obtain the gross information such as the initial letters of words, letter features, and word length beyond that region. Saccades determine the position of the next point of gaze, while the fixation duration determines when this movement will be made [71].

**Dyslexia Reading.** There are many studies indicating various visual and oculomotor deficits in children diagnosed with dyslexia [71]. Examples in this direction could include the finding that dyslexic readers have an abnormally longer latency for saccades and vergence [72]. The same authors also showed that dyslexic children had poor binocular coordination of saccades and fixations when looking at paintings, suggesting an oculomotor deficit. However in conditions other than reading, it has been found that dyslexic children have the same capabilities regarding the oculomotor as normal children.

**Eye Tracking Systems.** In particular, to study the process of reading only professional, built-in eye trackers with high sampling rates have been used. To our best knowledge, there has been no successful attempt to study the process of reading and to further investigate and diagnose dyslexic people by analyzing images acquired with a normal camera illuminated by visible spectrum light. The main explanation for this feature is that the process of reading is dynamic and studying the related eye movements requires high resolution in both space and time. Eye trackers based on visible spectrum light do not have the sampling rate necessary to separate abnormal ocular movements from normal movements while the subjects are reading. Furthermore, as the differences between normal reading and dyslexic reading are small, and even when studied with the most accurate eye trackers some of these differences are still surrounded by hot debate among psychologists, there is no place yet in dyslexia studies for cheaper systems.

## 2.6 COMPUTER VISION SUPPORT FOR IDENTIFICATION OF ANXIETY DISORDERS

Anxiety is a human emotion characterized by an "unpleasant state of inner turmoil, often accompanied by nervous behavior, such as pacing back and forth, somatic complaints, and rumination" [62]. It is often confused with fear and even psychologists debate its boundaries [73]. The main difference with respect to fear is that the fear focuses on a real object or entity, while anxiety is the expectation of future threat.

According to Russell [74], anxiety has a negative valence and mild intensity. It is not included in the standard set of emotions as defined by Ekman and Friesen [11] and only recently has the facial expression been determined. Perkins et al. [73] showed that, in the case of anxiety, eye darts and head swivels are the most common characteristics of anxiety. While mild anxiety (often appearing as various phobias) has no dramatic impact on the quality of life, one of its most severe forms, Post Traumatic Stress Disorder (PTSD), does.

### 2.6.1 Assessing Phobias

A significant set of studies used eye tracking systems to study various phobias. Again gaze tracking technology is used in two basic applications: saliency oriented and behavior oriented.

The saliency oriented approach is more related to assessing phobias. In this approach, in a clinical laboratory setup, a gaze tracker calibrated in conjunction with a display is used to asses points of fixation. Further analysis associates these points of fixation with specific objects. Particular arrangements of objects, where the object assumed to ignite the phobia is not in the normal visual saliency regions, is particularly informative in establishing the source of the phobia.

In this direction, arachnophobia (fear of spiders) was investigated in a number of studies. For instance, several studies [75,76] suggested that spider phobics initially fixate faster on spider stimuli and then avoid them. Other work has shown that spider phobics may be distracted by spider pictures even when the pictures are irrelevant to the task [77].

In a similar manner, social phobias have been investigated by means of eye tracking. In these studies, the subjects look at faces with various emotional content and differences in visual scanpaths are analyzed based on a psychiatric estimation of phobic/nonphobic. It appears that social phobics carry an initial bias for the processing of threat cues, a bias that is amplified when the subject is stressed [78,79].

It is the authors' opinion that, although they do not exist, noninvasive remote visible spectrum cameras are of use in the study of phobias. In particular, weak phobias can be kept under control. A scenario with a large enough area containing various objects or stimuli is well suited for the identification of phobias, and the rather high spatial resolution is reachable by visible spectrum eye trackers.

## 2.6.2 Studying PTSD

The main characteristic of PTSD is *hypervigilance*, which is an increase in attention to threatening, potentially threatening, or trauma-relevant stimuli [35]. The list of symptoms often includes constant visual scanning for suspicious behavior in public places, being alert to unusual sounds, noting where entrances and exits are in enclosed places, constant checking of locks inside the home, or investigating circumstances that seem out of the ordinary.

Previous PTSD studies have produced substantial evidence that anxiety and depression are characterized by attentional biases [80]. The well-known hypothesis of diverting attention was initially studied by triggering an event and measuring how long it took the subject to press a key. The advances in eye tracking technology have allowed eye tracking approaches that are much closer to *covert attention* than manual responses [80]. The high sampling rates available in modern trackers enable virtually continuous recording of attention and thus provide a remarkable increase in the efficiency of measuring attention.

However, when switching to more intense forms of anxiety, such as PTSD, the situation changes. First, the main eye related feature of this type of anxiety is hypervigilance and this aspect is approached by behavioral solutions. In these cases high resolution, both spatial and temporal, are required.

For instance, while studying the gaze behavior of Iraqi veterans, Kimble et al. [81] found that not only the point of gaze may be of interest, but also the size of the pupil. More precisely, they concluded that veterans reporting higher levels of PTSD symptoms had larger pupils when shown negatively valenced pictures and that they "spent more time looking at them than did veterans lower in PTSD symptoms". Also, by studying the point of gaze and the fixation duration, Kimble et al. [81] concluded that post-traumatic pathology is associated with vigilance rather than avoidance when visually processing negatively valenced and trauma relevant stimuli.

Eye tracking studies have suggested that the behavior of PTSD patients is significantly different from people with social phobias, and thus the treatment should differ [81]. Social phobics, after the identification of the threatening stimulus, quickly switch to avoiding it; instead, PTSD sufferers continuously observe the stimulus (pointing to a lack of will to disengage) in a manner that places them closer to depressed or to obsessive–compulsive patients.

## 2.7 COMPUTER VISION SUPPORT FOR IDENTIFICATION OF DEPRESSION AND DEMENTIA

A particular troubling class of cognition-affecting disorders are the degenerative conditions, such as Parkinson's, dementia, and some of their possible precursors, such as depression. A substantial number of studies established decades ago that there is a substantial link between the onset and development of depression and dementia and sleep disorders [82–84]. Insomnia and the reduction in or lack of rapid eye movement (REM) sleeping phase (ranging from reduced REM latency, increased REM time, or the presence of REM sleep behavior disorder). This type of measurements is commonly done using standard polysomnographic recordings, which emphasizes using EEG recordings over the direct recording of the electro-oculographic eye movements and, under these circumstances, the use of computer vision is exaggerated.

Still, there is room for clear computer vision approaches in eye movement measurements, as reported in [85], where the use of saccades was tested for a differential diagnosis between Parkinson's disease dementia (PDD) and dementia with Lewy bodies (DLB). The tests recorded reflexive (gap, overlap) and complex saccades (prediction, decision, and antisaccade) via electro-oculography for both affected patients and a control group. Impaired saccade execution in reflexive tasks allowed discrimination between DLB and Alzheimer's disease, and between PDD and Parkinson's disease.

## 2.8 CONCLUSIONS AND DISCUSSION

In this chapter we approached computer vision solutions to the diagnostic aid of several cognitive-affecting psychiatric diseases. We reviewed reported contributions that investigate cognitive impairments that appear at all stages of human development from childhood and the cognitive accumulation (autism, dyslexia), through adulthood and trauma-related cognitive degradations (phobias and PTSD), and ending with the ultimate degenerative cognitive degradations induced by dementia. From a medical and social prevention point of view, it is natural that more weight should be given to the early diagnosis of conditions that affect cognitive growth, which explains the unbalanced coverage and documentation that favors autism-related research over research in the dementia-related areas.

At the end of this chapter several ideas stand out. On the psychological and neurological side, mental impairments are disorders that are still in the process of investigation; in many cases, the causes and extent are not really understood. Moreover, these disorders are defined by some outlying traits that in the popular imagination replaces the disorder.

Computer vision techniques can be placed at several stages in the process of diagnosis. On the one hand several specific scenarios have been identified where the gaze behavior is a distinct cue contributing to diagnosis. In these cases a professional eye tracker that exhibits high accuracy is part of the typical instrumentation used to assess abnormal behavior. Without the impressive accuracy of eye trackers, many aspects of autism, dyslexia, or PTSD would not be understood or have even been identified.

On the other hand, visual computational models are expected to be used in an "in-the-wild" scenario, using normal cameras, mobile phone implementation, non-laboratory setups, etc. However, while such features are promising goals and may be of actual use in some specific scenarios such as identifying the source of phobias, significantly more progress is needed before they actually become part of scenarios for the diagnosis of severe impairments. For instance, Bone et al. [86] criticized the too early introduction of machine learning support for the detection of autism; while computer tools may indeed help, without very rigorously established and highly accurate measures, their use is risky. In the case of early diagnosis of autistic children, once the diagnosis is set, the life of the child is changed; if there is a mistake in the diagnosis, it is nearly impossible to correct it. Furthermore, taking into consideration that even human experts do not fully understand many aspects regarding impairments, significant progress where computer vision scientists collaborate with neurologists and psychologists is expected before wide-scale use. Close collaboration between domains may create new tools that could provide additional insights into cognitive impairments.

## ACKNOWLEDGMENTS

This work has been partially funded by University Politehnica of Bucharest, through the "Excellence Research Grants" Program, UPB – GEX. Identifier: UPB-EXCELENTA-2016, Contract no. 95/26.09.2016 (aFAST).

## REFERENCES

[1] R. Itier, M. Batty, Neural bases of eye and gaze processing: the core of social cognition, Neuroscience and Biobehavioral Reviews 33 (6) (2009) 843–863.

[2] P. Thompson, Margaret Thatcher: a new illusion, Perception 9 (4) (1980) 483–484.

[3] H. Kobayashi, S. Kohshima, Unique morphology of the human eye, Nature 387 (2008) 767–768.

[4] A. Frischen, A.P. Bayliss, S.P. Tipper, Gaze cueing of attention, Psychological Bulletin 133 (4) (2007) 694–724.

[5] J.M. Henderson, C.C. Williams, R.J. Falk, Eye movements are functional during face learning, Memory & Cognition 33 (1) (2005) 98–106.

[6] R.J. Itier, C. Alain, N. Kovacevic, A.R. McIntosh, Explicit versus implicit gaze processing assessed by ERPs, Brain Research 1177 (2007) 79–89.

[7] Y. Sugano, Y. Matsushita, Y. Sato, Learning-by-synthesis for appearance-based 3D gaze estimation, in: CVPR, 2014, pp. 1821–1828.

[8] S. Dakin, R. Watt, Biological bar codes in human faces, Journal of Vision 9 (4) (2009) 1–10.

[9] P. Sinha, B. Balas, Y. Ostrovsky, R. Russell, Face recognition by humans: 19 results all computer vision researchers should know about, Proceedings of the IEEE 94 (11) (2006) 1948–1962.

[10] M.B. Lewis, A.J. Edmonds, Face detection: mapping human performance, Perception 32 (8) (2003) 903–920.

[11] P. Ekman, W. Friesen, The Facial Action Coding System: A Technique for The Measurement of Facial Movement, Consulting Psychologists Press, San Francisco, 1978.

[12] D.W. Hansen, Q. Ji, In the eye of the beholder: a survey of models for eyes and gaze, IEEE Transactions on Pattern Analysis and Machine Intelligence 32 (3) (2010) 478–500.

[13] A.G. Reeves, R.S. Swenson, Disorders of the Nervous System, Online version developed at: Dartmouth Medical School, 2009.

[14] D. Purves, G.J. Augustine, D. Fitzpatrick, et al. (Eds.), Neuroscience, Sinauer Associates, 2001.

[15] R. Bandler, J. Grinder, Frogs into Princes: Neuro Linguistic Programming, Real People Press, Moab, UT, USA, 1979.

[16] R.H.S. Carpenter, Movements of the Eyes, 2nd edition, Pion, London, 1988.

[17] A. Duchowski, Eye Tracking Methodology: Theory and Practice, Springer-Verlag, 2007.

[18] K. Rayner, Eye movements and attention in reading, scene perception, and visual search, Quarterly Journal of Experimental Psychology 62 (8) (2009) 1457–1506.

[19] D. Hestenes, Invariant body kinematics: I. Saccadic and compensatory eye movements, Neural Networks 7 (1) (1994) 65–77.

[20] T. Haarmeier, P. Thier, Impaired analysis of moving objects due to deficient smooth pursuit eye movements, Brain 122 (8) (1999) 1495–1505.

[21] D.A. Robinson, A method of measuring eye movement using a scleral search coil in a magnetic field, IEEE Transactions on Bio-medical Electronics (1963) 137–145.

[22] M. Shelhamer, D.C. Roberts, Magnetic scleral search coil, in: Vertigo and Imbalance: Clinical Neurophysiology of the Vestibular System, in: Handbook of Clinical Neurophysiology, vol. 9, Elsevier, 2010, pp. 80–87 (Chapter 6).

[23] P. Viola, M. Jones, Robust real-time face detection, International Journal of Computer Vision 57 (2) (2004) 137–154.

[24] M. Mathias, R. Benenson, M. Pedersoli, L.V. Gool, Face detection without bells and whistles, in: Proceedings of the European Conference on Computer Vision, vol. 8692, 2014, pp. 720–735.

[25] K. Zhang, Z. Zhang, Z. Li, Y. Qiao, Joint face detection and alignment using multi-task cascaded convolutional networks, IEEE Signal Processing Letters 23 (10) (2016) 1499–1503.

[26] J. Wu, Z.H. Zhou, Efficient face candidates selector for face detection, Pattern Recognition 36 (5) (2003) 1175–1186.

[27] Z. Zhou, X. Geng, Projection functions for eye detection, Pattern Recognition 37 (5) (2004) 1049–1056.

[28] L. Florea, C. Florea, C. Vertan, Robust eye centers localization with zero-crossing encoded image projections, Pattern Analysis & Applications 20 (1) (2017) 127–143.

[29] F. Timm, E. Barth, Accurate eye centre localisation by means of gradients, in: Proceedings of the International Conference on Computer Theory and Applications, 2011, pp. 125–130.

[30] N. Markus, M. Frljak, I.S. Pandzi, J. Ahlberg, R. Forchheimer, Eye pupil localization with an ensemble of randomized trees, Pattern Recognition 47 (2) (2014) 578–587.

[31] R. Valenti, T. Gevers, Accurate eye center location through invariant isocentric patterns, IEEE Transactions on Pattern Analysis and Machine Intelligence 34 (9) (2012) 1785–1798.

[32] M. Leo, D. Cazzato, T. De Marco, C. Distante, Unsupervised eye pupil localization through differential geometry and local self-similarity matching, PLoS ONE 9 (8) (2014) e102829.

[33] K. Krafka, A. Khosla, P. Kellnhofer, H. Kannan, S. Bhandarkar, W. Matusik, A. Torralba, Eye tracking for everyone, in: CVPR, 2016.

[34] M.L. Lai, M.J. Tsai, F.Y. Yang, C.Y. Hsu, T.C. Liu, S.W.Y. Lee, et al., A review of using eye-tracking technology in exploring learning from 2000 to 2012, Review of Educational Research 10 (2013) 90–115.

[35] AP Association, Diagnostic and Statistical Manual of Mental Disorders-Text Revision, American Psychiatric Association, 2012.

[36] D.H. Geschwind, Autism: many genes, common pathways?, Cell 135 (3) (2008) 391–395.

[37] K. Gotham, A. Pickles, C. Lord, Standardizing ADOS scores for a measure of severity in autism spectrum disorders, Journal of Autism and Developmental Disorders 39 (5) (2009) 693–705.

[38] P.T. Shattuck, M. Durkin, M. Maenner, C. Newschaffer, D.S. Mandell, L. Wiggins, et al., Timing of identification among children with an autism spectrum disorder: findings from a population-based surveillance study, Journal of the American Academy of Child and Adolescent Psychiatry 48 (5) (2009) 474–483.

[39] L. Zwaigenbaum, S. Bryson, T. Rogers, W. Roberts, J. Brian, P. Szatmari, Behavioral manifestations of autism in the first year of life, International Journal of Developmental Neuroscience 23 (2) (2005) 143–152.

[40] S. Wallace, M. Coleman, O. Pascalis, A. Bailey, A study of impaired judgment of eye-gaze direction and related face-processing deficits in autism spectrum disorders, Perception 35 (12) (2006) 1651–1664.

[41] J. Ristic, L. Mottron, C.K. Friesen, G. Iarocci, J.A. Burack, A. Kingstone, Eyes are special but not for everyone: the case of autism, Cognitive Brain Research 24 (3) (2005) 715–718.

[42] W. Jones, A. Klin, Attention to eyes is present but in decline in 2–6-month-old infants later diagnosed with autism, Nature (2013).

[43] L. Itti, C. Koch, E. Niebur, A model of saliency-based visual attention for rapid scene analysis, IEEE Transactions on Pattern Analysis and Machine Intelligence 20 (11) (1998) 1254–1259.

[44] A. Toet, Computational versus psychophysical bottom-up image saliency: a comparative evaluation study, IEEE Transactions on Pattern Analysis and Machine Intelligence 33 (11) (2011) 2131–2146.

[45] J.H. Fecteau, D.P. Munoz, Salience, relevance, and firing: a priority map for target selection, Trends in Cognitive Sciences 10 (8) (2006) 382–390.

[46] Q. Guillon, N. Hadjikhani, S. Baduel, B. Rogé, Visual social attention in autism spectrum disorder: insights from eye tracking studies, Neuroscience and Biobehavioral Reviews 42 (2014) 279–297.

[47] S.V. Wass, T.J. Smith, M.H. Johnson, Parsing eye-tracking data of variable quality to provide accurate fixation duration estimates in infants and adults, Behavior Research Methods 45 (1) (2013) 229–250.

[48] M.C. Frank, E. Vul, R. Saxe, Measuring the development of social attention using free-viewing, Infancy 17 (4) (2012) 355–375.

[49] E.S. Spelke, Preferential looking methods as tools for the study of cognition in infancy, in: G. Gottlieb, N. Krasnegor, N. Norwood (Eds.), Measurement of Audition and Vision in the First Year of Postnatal Life, Ablex, 1985, pp. 323–363.

[50] T. Falck-Ytter, S. Bolte, G. Gredeback, Eye tracking in early autism research, Journal of Neurodevelopmental Disorders 5 (28) (2013).

[51] D. Cazzato, F. Adamo, G. Palestra, G. Crifaci, P. Pennisi, G. Pioggia, et al., Non-intrusive and calibration free visual exploration analysis in children with autism spectrum disorder, in: Proceedings of 5th Eccomas Thematic Conference on Computational Vision and Medical Image Processing, VipIMAGE 2015, 2016, pp. 201–208.

[52] A. Klin, W. Jones, R. Schultz, F. Volkmar, D. Cohen, Visual fixation patterns during viewing of naturalistic social situations as predictors of social competence in individuals with autism, Archives of General Psychiatry 59 (2002) 809–816.

[53] W. Jones, K. Carr, A. Klin, Absence of preferential looking to the eyes of approaching adults predicts level of social disability in 2-year-old toddlers with autism spectrum disorder, Archives of General Psychiatry 65 (2008) 946–954.

[54] S. Wang, M. Jiang, X.M. Duchesne, E.A. Laugeson, D.P. Kennedy, R. Adolphs, Q. Zhao, Atypical visual saliency in autism spectrum disorder quantified through model-based eye tracking, Neuron 88 (2015) 604–616.

[55] J.M. Rehg, A. Rozga, G.D. Abowd, M.S. Goodwin, Behavioral imaging and autism, IEEE Pervasive Computing 13 (2) (2014) 84–87.

[56] J. Rehg, G. Abowd, A. Rozga, M. Romero, M. Clements, S. Sclaroff, et al., Decoding children's social behavior, in: Proceedings of the IEEE Conference on Computer Vision and Pattern Recognition, 2013, pp. 3414–3421.

[57] N. Sasson, L. Turner-Brown, T. Holtzclaw, K. Lam, J. Bodfish, Children with autism demonstrate circumscribed attention during passive viewing of complex social and non-social picture arrays, Autism Research: Official Journal of the International Society for Autism Research 1 (2008) 31–42.

[58] C. von Hofsten, H. Uhlig, M. Adell, O. Kochukhova, How children with autism look at events, Research in Autism Spectrum Disorders 3 (2) (2009) 556–569.

[59] A. Klin, D. Lin, P. Gorrindo, G. Ramsay, W. Jones, Two-year-olds with autism orient to non-social contingencies rather than biological motion, Nature 459 (7244) (2009) 257–261.

[60] K. Pierce, D. Conant, R. Hazin, R. Stoner, J. Desmond, Preference for geometric patterns early in life as a risk factor for autism, Archives of General Psychiatry 68 (1) (2011) 101–109.

[61] R. Bedford, M. Elsabbagh, T. Gliga, A. Pickles, A. Senju, T. Charman, M. Johnson, Precursors to social and communication difficulties in infants at-risk for autism: gaze following and attentional engagement, Journal of Autism and Developmental Disorders 42 (10) (2012) 2208–2218.

[62] M. Seligman, E. Walker, D. Rosenhan, Abnormal Psychology, 4th edition, Norton & Company, 2000.

[63] M.N. Benfatto, G.O. Seimyr, J. Ygge, T. Pansell, A. Rydberg, C. Jacobson, Screening for dyslexia using eye tracking during reading, PLoS ONE (2016), https://doi.org/10.1371/journal.pone.0165508.

[64] H. Joseph, K. Nation, S.P. Liversedge, Using eye movements to investigate word frequency effects in children's sentence reading, School Psychology Review 42 (2) (2013).

[65] A. Godfroid, F. Boers, A. Housen, An eye for words: gauging the role of attention in incidental L2 vocabulary acquisition by means of eye-tracking, Studies in Second Language Acquisition 35 (3) (2013) 483–517.

[66] K. Rayner, T.J. Slattery, D. Drieghe, S.P. Liversedge, Eye movements and word skipping during reading: effects of word length and predictability, Journal of Experimental Psychology: Human Perception and Performance 37 (2) (2011) 514–528.

[67] K. Rayner, B.R. Foorman, C.A. Perfetti, D. Pesetsky, M.S. Seidenberg, How psychological science informs the teaching of reading, Psychological Science in the Public Interest 2 (2) (2001) 31–74.

[68] C. Prado, M. Dubois, S. Valdois, The eye movements of dyslexic children during reading and visual search: impact of the visual attention span, Vision Research 47 (19) (2007) 2521–2530.

[69] F. Hutzler, H. Wimmer, Eye movements of dyslexic children when reading in a regular orthography, Brain and Language 89 (1) (2004) 235–242.

[70] G.T. Pavlidis, Do eye movements hold the key to dyslexia?, Neuropsychologia 19 (1) (1981) 57–64.

[71] S. Bellocchi, M. Muneaux, M. Bastien-Toniazzo, S. Ducrot, I can read it in your eyes: what eye movements tell us about visuo-attentional processes in developmental dyslexia, Research in Developmental Disabilities 34 (1) (2013) 452–460.

[72] M.P. Bucci, D. Brémond-Gignac, Z. Kapoula, Latency of saccades and vergence eye movements in dyslexic children, Experimental Brain Research 188 (1) (2008) 1–12.

[73] A.M. Perkins, S.L. Inchley-Mort, A.D. Pickering, P.J. Corr, A.P. Burgess, A facial expression for anxiety, Journal of Personality and Social Psychology 102 (5) (2012) 910–924.

[74] J.A. Russell, A circumplex model of affect, Journal of Personality and Social Psychology 39 (6) (1980) 1161–1178.

[75] T. Pflugshaupt, U.P. Mosimann, R. von Wartburg, W. Schmitt, T. Nyffeler, R.M. Müri, Hypervigilance–avoidance pattern in spider phobia, Journal of Anxiety Disorders 19 (1) (2005) 105–116.

[76] M. Rinck, E. Becker, Spider fearful individuals attend to threat, then quickly avoid it: evidence from eye movements, Journal of Abnormal Psychology 115 (2) (2006) 231–238.

[77] A. Gerdes, G. Alpers, P. Pauli, When spiders appear suddenly: spider-phobic patients are distracted by task-irrelevant spiders, Behaviour Research and Therapy 46 (2) (2008) 174–187.

[78] K. Mogg, P. Philippot, B. Bradley, Selective attention to angry faces in clinical social phobia, Journal of Abnormal Psychology 113 (1) (2004) 160–165.

[79] M. Garner, K. Mogg, B. Bradley, Orienting and maintenance of gaze to facial expressions in social anxiety, Journal of Abnormal Psychology 115 (4) (2006) 760–770.

[80] T. Armstrong, B.O. Olatunji, Eye tracking of attention in the affective disorders: a meta-analytic review and synthesis, Clinical Psychology Review 32 (8) (2012) 704–723.

[81] M.O. Kimble, K. Fleming, C. Bandy, J. Kim, A. Zambetti, Eye tracking and visual attention to threating stimuli in veterans of the Iraq war, Journal of Anxiety Disorders 24 (3) (2010) 293–299.

[82] N. Tsuno, A. Besset, K. Ritchie, Sleep depression, The Journal of Clinical Psychiatry 66 (10) (2005) 1254–1269.

[83] G.J. Emslie, A.J. Rush, W.A. Weinberg, J.W. Rintelmann, H.P. Roffwarg, Children with major depression show reduced rapid eye movement latencies, Archives of General Psychiatry 47 (2) (1990) 119–124.

[84] R.B. Postuma, J.A. Bertrand, J. Montplaisir, C. Desjardins, M. Vendette, S. Rios Romenets, M. Paisset, J.F. Gagnon, Rapid eye movement sleep behavior disorder and risk of dementia in Parkinson's disease: a prospective study, Movement Disorders 27 (6) (2012) 720–726.

[85] U.P. Mosimann, R.M. Müri, D.J. Burn, J. Felblinger, J.T. O'Brien, I.G. McKeith, Saccadic eye movement changes in Parkinson's disease dementia and dementia with Lewy bodies, Brain 128 (6) (2005) 1267–1276.

[86] D. Bone, M.S. Goodwin, M.P. Black, C.C. Lee, K. Audhkhasi, S. Narayanan, Applying machine learning to facilitate autism diagnostics: pitfalls and promises, Journal of Autism and Developmental Disorders 45 (5) (2015) 1121–1136.

# CHAPTER 3

# Real-Time 3D Tracker in Robot-Based Neurorehabilitation

**Fabio Stroppa\*, Mine Saraç Stroppa\*, Simone Marcheschi\*,**
**Claudio Loconsole†, Edoardo Sotgiu‡, Massimiliano Solazzi\*,**
**Domenico Buongiorno\*, Antonio Frisoli\***
\*Scuola Superiore Sant'Anna, Pisa, Italy
†Polytechnic University of Bari, Bari, Italy
‡INL International Iberian Nanotechnology Laboratory, Braga, Portugal

## Contents

## Abstract

The chapter describes a computer vision-based robot-assisted system used in neurorehabilitation of post-stroke patients that allows the subjects to reach for and grasp objects in a defined workspace.

The proposed computer vision technique is used to model objects that have not been preprocessed in a real setting, track them in real time, and provide their actual pose to the robotic device in order to accomplish grasping tasks.

Computer Vision for Assistive Healthcare.
DOI: https://doi.org/10.1016/B978-0-12-813445-0.00003-4

The robotic device is composed of three integrated modules: (i) a 4-DOF arm exoskeleton that supports the patient's impaired arm when reaching for the objects; (ii) a 3-DOF actuated wrist exoskeleton for optimizing the hand pose in the grasping task; and (iii) a 2-DOF (flexion/extension) underactuated hand exoskeleton designed to be automatically adjusted for different grasping tasks based on contact forces.

The conducted tests have demonstrated the robustness of the proposed approach, and its performance in the neurorehabilitation scenario through reaching and grasping task experiments.

## Keywords

Robotic therapy, Object recognition, 3D tracking, Grasping task, 3D point cloud, Active upper-limb exoskeletons

## 3.1 INTRODUCTION

The hemiparesis of the upper extremity represents a common impairment in post-stroke patients [1], estimated to affect approximately 795,000 people each year in the United States alone [2]. Rehabilitation robots can be successfully employed in the earlier phases of recovery from stroke [3–6] since they can overcome some of the major limitations of traditional assisted training (lack of repeatability, dependence on the availability of skilled personnel, ability to provide high intensity motor training under controlled conditions, etc.) [7].

A distinctive feature of some rehabilitation devices, such as active (Armin [8], L-Exos [9]) or passive (T-WREX [10]) exoskeletons, is that they are anthropomorphic and wearable. This allows patients to perform the movements that require more complex inter-joint coordination and gravity counterbalancing, and is then suitable for motor tasks in Activities of Daily Living (ADLs). Furthermore, robot-based rehabilitation allows the device to assist and guide the patient. As an example, reaching and grasping tasks are often performed during the ADLs: a robotic device can provide assistance to move an impaired arm toward a real object and finally grasp it.

For this purpose, the therapy setup should be endowed with a robust, smart, and fast system for the automated tracking of moving objects. In particular, the tracking system should (i) be as transparent as possible to the therapist, allowing the hands to manipulate objects without affecting the system's performance and (ii) manage different kind of objects according to the different needs of the therapist (without knowing the model of the objects). A robot-based therapy scenario proposing such a system consists of a computer-vision module providing information to an active upper limb exoskeleton for reaching for and tracking different objects within the robot workspace [11].

The computer vision module requires a hardware tool for the image acquisition that is also able to obtain depth information of the analyzed environment. Red Green Blue-Depth (RGB-D) cameras implement this hardware as a combination of a standard VGA color camera and an infrared range sensor calculating the scene depth through structured light distortion.

Once acquired, the images captured by the sensor must be properly processed to recognize the objects in the scene and track their position in time. In the literature, several methods have been proposed for real object tracking in images/video streams, most of which combine 2D tracking algorithms with 3D information using a model-based approach (known or preprocessed objects) [12–14]. Some solutions for tracking the 3D pose of a rigid object combine Scale-Invariant Feature Transform descriptors [15] with the well-known Kanade–Lucas–Tomasi (KLT) tracker [16] to estimate the three-dimensional pose of an object in real time. This is particularly appropriate for natural control of robot manipulators; however, the KLT tracking algorithm has a tendency to drift due to abrupt illumination changes, occlusions, and aperture problem.

A change in the approach for object tracking is represented by P. Alimi [17]. In this work, the author performs object recognition using 3D features. In particular, the recognition is based on a database of known objects that are compared to the objects observed in the scene. The objects are recognized using features derived from sense 3D geometry: Spin Images [18]; Viewpoint Feature Histogram (VFH) descriptors [19]; Fast Point Feature Histograms (FPFH) descriptors [20]; and Normal Aligned Radial Feature (NARF) [21]. Excellent matching performance has been demonstrated by VFH, especially when an object is viewed from the same vantage point at both training and test time, allowing the tracking to be successful when the objects are shifted, rotated, added, and removed from a scene. However, the use of only VFH descriptors may not be sufficient when the tracker is used to analyze a video stream with moving objects.

Considering the above survey of the state of the art, there are several issues that need to be solved for obtaining a robust tracking module:
- the knowledge of the objects to be used;
- the use of 2D features to lighten the performance of 3D feature-based methods;
- the drift introduced by the KLT tracker over time;
- the transparency of the computer-vision based system to the users;

- the lack of object reconstruction techniques due to occlusions;
- the need to merge different types of features to improve the robustness of the tracking.

To overcome the above issues, in this chapter we will discuss a novel and robust tracking algorithm specialized in discriminating cylindrical objects for grasping real objects. This tracking method manages unknown objects providing their 3D pose (position and orientation in space) using a different technique than the KLT tracker. Specifically, the algorithm transforms the RGB image in a three-dimensional set of points, called the "point cloud". The whole processing is executed on the PCs grabbed during the video stream, defining an object fingerprinting-based technique composed of a collection of several 3D features that also includes VFH descriptors. The comparison between objects is performed introducing a similarity distance measurement, which provides the likeness of two objects to classify them as the same object or two different ones in two different frames of the video. Moreover, the algorithm also manages object occlusions, mainly caused by the therapist's human hands performing an improved skin removal algorithm followed by object reconstruction procedures to let the system be as transparent as possible to the users.

The chapter will illustrate how the system is set up, describing both the hardware and the software. More details will be shown for the tracking module, meticulously analyzing the entire pipeline and the techniques adopted to solve all the above-mentioned issues. Finally, a preliminary experiment will show the behavior of the system when it is asked to reach for and grasp an object in the workspace. As shown in Fig. 3.1, the system is composed of an RGB-D camera facing the workspace, where one or more objects are placed on a desk. On the other side of the desk, the robotic tool is placed at a suitable distance to reach every position in the workspace.

## 3.2 TRACKING MODULE

The computer-vision module is composed of an RGB-D camera for 3D video grabbing, and two software libraries for 2D and 3D processing. In particular, the RGB-D sensor is the Microsoft Kinect[1]: it provides a $640 \times 480$ colored image and depth map at 30 Frames Per Second (FPS), which is appropriate for real-time rehabilitation system scenarios. The 2D

---

[1] http://www.microsoft.com/en-us/kinectforwindows/.

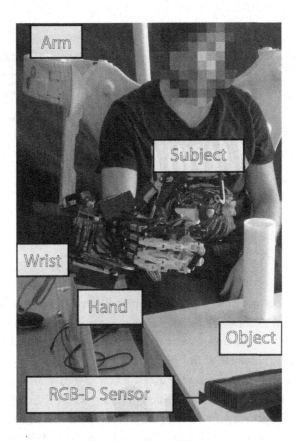

**Figure 3.1** The proposed robot-based neurorehabilitation scenario.

image preprocessing is performed with OpenCV library[2]; the processing of 3D point clouds is handled by Point Cloud Library (PCL)[3] [22], which includes filtering, feature estimation, surface reconstruction, registration, model fitting, and segmentation.

The tracking module described here is a substantial improvement on the original computer vision–based system proposed by the authors [11,23] in terms of applied techniques, devices, and potential application.

Thanks to the underactuation concept adopted by the kinematics, the hand orthosis (see Section 3.3.3) can grasp an object of any shape and size. However, even though this module is able to detect a generically shaped

---

[2] http://opencv.org/.
[3] http://pointclouds.org.

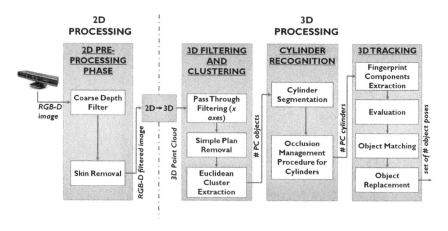

**Figure 3.2** Workflow diagram for the tracking module.

object, the overall computer vision algorithm will process and track only cylindrical ones. There are several reasons for this choice: (i) it is simple to correctly identify the orientation of a cylinder in space by providing the unit vector where its height lies; and (ii) the grasp of a cylindrical object is performed by easy and natural movements.

To correctly perform the robotic-assisted grasping with the exoskeleton, the primary aim is to provide the pose of each object in space. In particular, the pose of an object is intended as its space orientation and 3D position of the centroid. This information is provided by the computer vision-based tracking system, executing an iterative algorithm frame by frame and divided into two main parts:

- *2D Preprocessing*, in which a single 2D image is grabbed and processed with a preliminary filtering;
- *3D Processing*, in which 3D information gathered from the image is filtered and clusterized, performing the actual tracking.

Fig. 3.2 shows the flow diagram of the Tracking Module.

## 3.2.1 Two-Dimensional Preprocessing

This phase aims to speed up the entire workflow of the module, performing a set of fast and simple operations. Since the consequent 3D filtering requires higher computational costs, it is important to remove the useless data from the starting 2D image to streamline the next procedure.

At the end of this phase, the 3D point cloud is obtained from the filtered 2D image and the depth map through the intrinsic parameters of the camera model.

### 3.2.1.1 Coarse Depth Filtering

The first operation performed by most of the computer vision software solutions is to focus only on a selected subset of data identified for a particular purpose, namely the Region of Interest (ROI). In a 3D workspace, it is likely to have a ROI at the center of the environment, so everything in the background can be removed.

Hence, in this submodule, each image pixel whose corresponding cell in the depth map is not in the range of a predefined distance from the camera is discarded by applying a black patch (i.e. [0.2 m, 0.9 m]). The final 3D point cloud will be constructed only by the remaining points, which are part of the ROI.

### 3.2.1.2 Skin Removal

An important requirement of the system is that its interface should be as transparent as possible during the object handling phase. Part of the 2D preprocessing is therefore dedicated to removing those pixels that are considered skin colored, such that the therapist's hands or arms do not interfere with the object tracking. In fact, when the therapist manipulates a real object, the tracking algorithm may consider the hand and the object to be grasped as a single unit. However, by removing the skin pixels from the image, the object can result "spatially divided" from the hand, and therefore considered as two or more separate objects. This requires a proper object reconstruction procedure (see Section 3.2.2.2).

The method used to achieve this target exploits a Bayesian Skin Detector [24], which was suitably improved to provide a sufficient frame rate for real-time purposes with the introduction of a custom rapid conversion from $RGB$ to $YC_rC_b$ color space [25], obtaining higher working frequencies.

## 3.2.2 Three-Dimensional Processing

The processing of the 3D point cloud consists of three steps aiming at (i) filtering and clusterizing the point cloud to obtain the objects in the scene; (ii) detecting the cylindrical objects; and (iii) performing the actual tracking along consecutive frames.

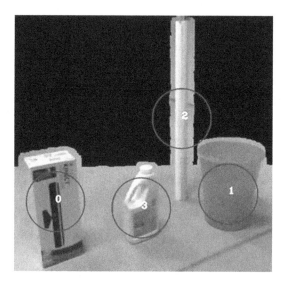

**Figure 3.3** Multiple object clustering.

### 3.2.2.1 3D Filtering and Clustering

Three-dimensional filtering and clustering involves processing the point cloud obtained from the filtered RGB image to detect the interested object within the scene. In particular, three steps are performed:

- *Pass Through Filtering*: removal of all the points out of the range [−1 m, 1 m] on the horizontal axis perpendicular to the Kinect multi-array microphone, obtaining a further focus on the ROI;
- *Simple Plan Segmentation*: based on the RANdom SAmple Consensus (RANSAC) method [26], removal of all the points that support a plane model; and
- *Euclidean Cluster Extraction*: detection of a separate point cloud for each object in the workspace, called a cluster.

As shown in Fig. 3.3, the point clouds of every object in the scene are provided at the end of the subprocedure.

### 3.2.2.2 Cylinder Recognition

The process of recognizing cylinders within the scene can be logically divided into two steps: (i) a segmentation of the point cloud to define which object is cylindrical, and (ii) a phase of point cloud reconstruction to manage those objects that are subjected to occlusions.

## Cylinder Segmentation

Each cluster resulting from the Euclidean Cluster Extraction (see Section 3.2.2.1) is processed by the Cylinder Model Segmentation method to preserve only the objects that support a cylindrical model. This segmentation method is based on RANSAC, as the Simple Plan Segmentation previously described. Furthermore, the process also provides geometrical information, such as the cylinder normal unit vector and its radius, which will be useful to the exoskeleton during the reaching and grasping procedure.

However, the Cylinder Segmentation presents a significant issue: it always looks for a cylinder regardless of the point cloud shape. For example, in the case of a parallelepiped box, the method will extract the inscribed cylinder providing a false positive. It is reasonable to test the change in the number of points within the cloud after the segmentation to overcome this issue and discard a "false" cylindrical object ($O_{fc}$) from a "true" one ($O_{tc}$). Indicating with $\|PC_O\|_b$ and $\|PC_O\|_s$ the number of points of the object $O$, respectively before and after segmentation, and with $th_c$ a threshold in $[0, 1]$:

$$\begin{cases} \textbf{if} & \frac{\|PC_O\|_s}{\|PC_O\|_b} \le th_c \quad \textbf{then } O = O_{fc} \\ \textbf{else} & O = O_{tc} \end{cases} \tag{3.1}$$

After a preliminary study on cylinder recognition, with a value of $th_c = 0.75$ it is possible to discriminate reasonably well a cylindrical object from a differently shaped object.

## Cylinder Reconstruction

Occlusions or noise might divide a unique object into two or more objects (case 1): an example is the case of the therapist's hand grasping an object. The same issue may arise due to perspective (case 2): if the camera is posed above a hollow cylinder, the Euclidean Clustering Extraction will provide the front and back sides of the object as two different cylinders.

Hence, it is necessary to check whether some of these "residual" point clouds are actually part of the same object once the cylinder recognition has been performed. The Cylinder Reconstruction method reduces the number of false positives in the scene, taking into account the normal axes of the cylinders, their centroids, and their radii to check whether they are geometrically part of the same object.

**Figure 3.4** Example of cylindrical object recognition (left). Example of skin removal keeping one of the cylinders tracked (right).

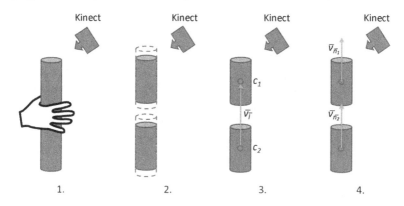

**Figure 3.5** Occlusion management for the cylinder reconstruction module.

Fig. 3.4 shows some results of the Cylinder Recognition module, providing an example of occlusion caused by skin removal.

**Occlusions.** In case 1, two "residual" point clouds are placed one on top of the other. Hence, to be part of the same cylinder, the two sub-cylinders should be parallel to each other and to the line joining their centroids. Fig. 3.5 illustrates the steps to perform the test:

1. Euclidean clustering;
2. Cylinder segmentation;
3. Centroid calculation ($c_1$ and $c_2$), and estimation of the joining line $\bar{l}$ between centroids with its unit vector $\vec{v}_{\bar{l}}$;
4. Calculation of the normals $\bar{n}_1$ and $\bar{n}_2$, and their unit vectors $\vec{v}_{\bar{n}_1}$ and $\vec{v}_{\bar{n}_2}$, finally checking if the normals are parallel to the line.

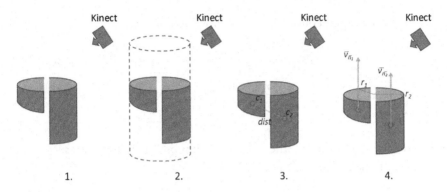

**Figure 3.6** Perspective management for the cylinder reconstruction module.

This last step is obtained via the orthogonal product between unit vectors, considering the following constraints:

$$\begin{cases} \vec{v}_{\bar{n}_1} \cdot \vec{v}_{\bar{l}} \cong 1 \\ \vec{v}_{\bar{n}_2} \cdot \vec{v}_{\bar{l}} \cong 1 \end{cases} \tag{3.2}$$

If each constraint is satisfied, the procedure imposes that the two sub-cylinders belong to the same cylinder.

**Perspective.** In case 2, two point clouds are posed one behind the other. The cylinder models provided by the segmentation should feature coinciding axes and the same radius in order to be considered part of the same cylinder. Moreover, the point cloud centroids are shifted with respect to the actual centroid of the cylinder model since the points are part of two separate objects. This leads to another check that should be performed: the distance between the two centroids must be as small as possible. Fig. 3.6 illustrates the steps to perform the test:

1. Euclidean clustering;
2. Cylinder segmentation;
3. Centroid calculation ($c_1$ and $c_2$) and estimation of the centroid–joining-line length *dist*;
4. Calculation of the normals $\bar{n}_1$ and $\bar{n}_2$ with their unit vectors $\vec{v}_{\bar{n}_1}$ and $\vec{v}_{\bar{n}_2}$, finally checking if (i) the normals are parallel to each other; (ii) the sub–cylinder radii $r_1$ and $r_2$ are equal; and (iii) the sub–cylinders are as close as possible.

This last step is obtained considering the following constraints:

$$
\begin{cases}
\vec{v}_{\bar{n}_1} \cdot \vec{v}_{\bar{n}_2} \cong 0 \\
\|r_1 - r_2\| \cong 0 \\
\|c_1 - c_2\| = dist \cong 0
\end{cases}
\tag{3.3}
$$

If each constraint is satisfied, the procedure imposes that the two sub-cylinders should belong to the same cylinder.

### 3.2.2.3 Three-Dimensional Tracking

Although the overall computer vision system is used to handle only cylindrical objects, the tracking module is designed to run with any type of object regardless of its shape. The recognition is performed by processing two consecutive frames of the video, performing real-time object matching and exploiting an object-fingerprint-recognition technique.

#### Fingerprint

The fingerprint of a 3D object is a set of data that differentiates one object from another, describing a unique item in the scene. Three-dimensional descriptors, such as VFH, are usually exploited to represent fingerprints in a three-dimensional environment. However, when an object moves among consecutive frames, using only VFH descriptors might be not sufficient for tracking. For this reason, the robustness of the algorithm has been improved by taking into account four classes of features:

- **VFH descriptors**, composed of 308 single-precision floating-point numbers having a maximum value of $3.4 \cdot 10^{38}$ (according to IEEE 754 standard);
- **average color** of the point cloud, expressed in RGB as three integers $[0 - 255]$;
- **number of points** in the cloud (size), expressed as a single integer;
- **3D position** of the point cloud centroid, expressed in meters as three single-precision floating-point numbers.

These four classes of features are considered the fingerprint components of a single object, and are stored in the memory to keep track of each object that appeared during the video stream.

## Evaluation

This tracking algorithm introduces a novel measure of likeness between two objects, determining whether they are actually the same one in two different frames.

Let us define two separate sets of objects as $X$ and $Y$. Then, for each pair $\langle x, y \rangle$, where $x \in X$ and $y \in Y$, let us define and calculate the distance $d$, which is an estimation of the likeness between two different objects:

$$d(x, y) = \frac{(f \cdot K_f) + (c \cdot K_c) + (s \cdot K_s) + (p \cdot K_p)}{K_f + K_c + K_s + K_p}. \tag{3.4}$$

Specifically, $f$, $c$, $s$, and $p$ are the Euclidean distances obtained by comparing the fingerprints in $\langle x, y \rangle$:

- $f$-distance is the Euclidean distance between two VFH descriptors;
- $c$-distance is the distance between two values of average color;
- $s$-distance is the absolute difference between two values of a number of points (expressed in percentage with respect to the size of the stored point cloud);
- $p$-distance is the Euclidean distance between two 3D centroids.

To compute the final value of $d$, each of these four components is thresholded and normalized to the range $[0, 1]$. The $K$-coefficients are used to weigh the value of each class of features, tuning the results for an optimal tracking.

The distance $d$ is named Similarity Distance. As the term "distance" suggests the lower the value, the bigger the similarity.

## Matching

The matching procedure exploits the Similarity Distance previously introduced, and it is used to assess whether two objects in two different frames (both consecutive or not) are actually the same object. Given an object $x \in X$, retrieving a similar object $y \in Y$ requires taking into account several aspects, especially to avoid false positives (not matching an object) or, worse, true negatives (matching an object with another one in the scene).

The core of the matching phase relies on a procedure called Crossed Matching. In order to illustrate this procedure, we define the Matching Relation ($\mathcal{M}$) as

$$\mathcal{M}(x, y) = \begin{cases} 1 & d(x, y) = \min_{i=1, \dots, \|Y\|} s(x, y_i) \\ 0 & \text{otherwise} \end{cases}, \tag{3.5}$$

where $\|Y\|$ indicates the number of items in $Y$.

During the algorithm execution, this procedure is performed on two defined sets: $C$, the set of object fingerprints in the current frame, and $S$, the set of object fingerprints stored in memory. Let us assume that the following is valid:

$$\forall o_C \in C, \ \exists o_S \in S : o_c = o_s. \tag{3.6}$$

The task of the matching is to find, for each object $o_c \in C$, its corresponding object $o_s \in S$. However, a new object in the scene means that the assumption in Eq. (3.6) is not always true. On the other hand, it is also not true that for every object in $S$ there is a corresponding object in $C$, as in the case of object loss from the scene.

Once the distance $d$ has been calculated for all the pairs in these two sets, a preliminary thresholding is executed to reject a priori those pairs that present high $d$ values according to a certain threshold ($th_d$). Then, the Crossed Matching is performed among the pairs passing this step. Exploiting the relation $\mathcal{M}$ (3.5), the possible outputs of the Crossed Matching are as follows:

- matching, if $\mathcal{M}(o_s, o_c) \wedge \mathcal{M}(o_c, o_s)$;
- $o_s$ is lost, if $\neg\mathcal{M}(o_s, o_c) \wedge \mathcal{M}(o_c, o_s)$;
- $o_c$ is new, if $\mathcal{M}(o_s, o_c) \wedge \neg\mathcal{M}(o_c, o_s)$.

It is important to point out that the last two outputs can occur simultaneously.

Moreover, to improve the robustness of the algorithm, two special procedures have been designed, the first for managing "false" new objects (and consequently "false" lost objects) and the second for managing "true" lost objects from memory.

In the first case, issues like noise in the video or object occlusions might lead the algorithm to fail the recognition, thus providing high values of $d$ between the same object in two different frames. This results in the recognition of a "false" new object according to $\mathcal{M}$. To fix this issue and determine whether the object is actually new, the algorithm tries to match it to the objects lost both in the previous and current frames before adding it to $S$ as new. The distance $d$ is then re-calculated considering only color and position features. In fact, in this case, the VFH and the size of an occluded object may be different between two consecutive frames.

In the second case, on the other hand, the procedure verifies whether a stored object reappears in the scene. The algorithm recalculates the distance $d$ of this object with respect to each object in $C$ without considering the

**Table 3.1** Tracker algorithm parameters

| $K_f$ | $K_c$ | $K_s$ | $K_p$ | $th_d$ |
|------|------|------|------|------|
| 3.0 | 2.0 | 1.0 | 3.0 | 0.2 |

$p$ feature. In fact, if an object leaves the field of view of the camera, it may appear again from any point in the workspace.

### Replacement

Finally, each matched old stored object fingerprint is replaced with its corresponding object fingerprint in the current frame. The new object fingerprints are added to $S$, whereas the fingerprints of lost objects can be discarded if their "age" in the dataset is less than 50 frames, or kept otherwise.

## 3.2.3 Assessment

### 3.2.3.1 Parameter Tuning

The algorithm was subjected to thirty different tests for the parameter tuning. A total of ten objects featuring different colors and textures were used. The tests investigated three different conditions by varying the algorithm parameter values for different combinations of number and object shapes in the scene:

1. all the objects are still;
2. one of the objects is moving, while the others are still;
3. one of the objects is repeatedly moved outside and inside the camera field of view, while the others are still.

The findings of the tests are shown in Table 3.1.

As the results suggest, the VFH and centroid position are the most relevant features for object recognition, assuming that the pose of an object does not change suddenly between two consecutive frames. On the other hand, the color of an object may change due to light condition, but it still remains a robust feature that should be taken into account. Finally, the less relevant feature has been found to be the number of points in the point cloud, which could drastically change in the case of occlusion. Table 3.1 also shows the best value for the thresholding discussed in Section 3.2.2.3. A value of $d = 0.1$ has been empirically chosen to correctly discriminate whether two still objects are actually the same one in two different frames. However, this value might increase when the object is moving. A threshold $th_d = 0.2$ is imposed to overcome this issue such that all the pairs achieving

**Table 3.2**  Average time (over 100 frames) required by the proposed tracking algorithm for different numbers of tracked objects

| 1 object | 3 objects | 5 objects |
|---|---|---|
| $0.085 \pm 0.015$ s | $0.088 \pm 0.024$ s | $0.137 \pm 0.0436$ s |

a value $d \leq 0.2$ pass the thresholding; the Crossed Matching procedure is then executed to avoid any mismatch.

### 3.2.3.2 Robustness

Further tests have been conducted to assess the robustness of the tracking algorithm. In particular, these tests have evaluated:

1. the mean computation time;
2. the maximum speed that an object can reach and not be lost by the tracker;
3. the robustness to change of light conditions, to object roto-translations, and to object occlusions.

Regarding the execution (test 1), the mean time registered over 200 frames has been calculated considering three different cases: tracking of a single object, three different objects, and five different objects. The tracking module was run on an Intel i7 3.20 GHz PC with 16 GB RAM and Windows 7 64-bit operating system. Table 3.2 shows the results achieved.

The maximum object speed allowed to keep the tracker working properly (test 2) was calculated using an iterative method, assessing whether the tracker recognizes a moving object between two consecutive frames. The object was initially posed at 1500 mm from the Kinect. Then, it was translated by several predefined offsets along the Kinect multi-array microphone (the horizontal direction on the camera image plane). The offset was iteratively incremented by 100 mm for each test, beginning every step at the same starting point. It was observed that the maximum distance that an object can move without been lost by the algorithm is 600 mm. With reference to the computation time of a single object in the scene shown in Table 3.2, the maximum hypothetical speed that an object can have is more than 270 km/h, which means that an object cannot be lost because of its speed with the above-mentioned setup.

The last set of experiments was focused on assessing the robustness of the tracking algorithm to different light conditions, to object roto-translation, and to occlusions (test 3). Here we present a set of figures to show the potential of the algorithm: Fig. 3.7 illustrates how the algorithm reacts to

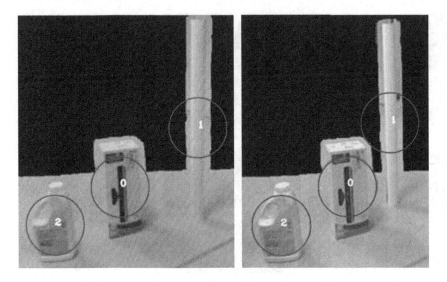

**Figure 3.7** Example of robustness to different light conditions in a video stream.

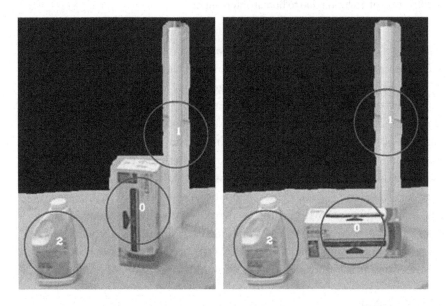

**Figure 3.8** Example of robustness to roto-translation of objects in a video stream.

light changes; Fig. 3.8 shows an example of roto–translation robustness; and Fig. 3.9 shows two different cases of occlusion. The figures demonstrate what is reported at the end of Section 3.2.3.1. In the fingerprint

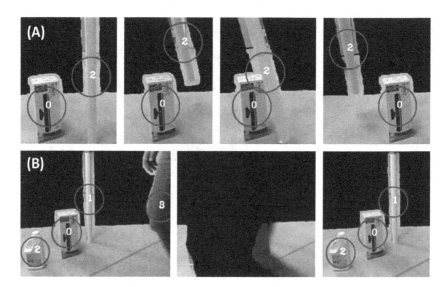

**Figure 3.9** (A) An example of occlusion caused by the overlapping of objects in the scene. (B) A case of occlusion due to human intervention.

matching procedure, the $f$- and $p$-distance are observed to be the most significant components. The potential of the VFH descriptors have already been discussed, being a robust class of fingerprints for still objects. Variation in $p$-distance can be considered of small entity, considering that even when the object is moving, the covered distance between two consecutive frames is negligible. This makes the algorithm robust to rotations or occlusions. Color is very sensitive to the variation in the light, especially in the cases where the camera is occluded for a few frames (i.e. when someone or something crosses the scene). Hence, the $c$-distance is an important feature for the matching procedure, but it is necessary to consider the color with a wider margin of error. Finally, the $s$-distance contribution is not very significant for the matching, not just because of the occlusions that may occur, but also because it is highly sensitive to noise in the point cloud grabbing.

### 3.2.3.3 Validation

The algorithm was tested on an RGB-D Object Dataset [27] to validate both the tracking method and the cylindrical object recognizer. This dataset contains image sequences of 300 objects, grouped in 51 categories. The images were captured with a Kinect-like camera at a resolution of $640 \times 480$ px. Each object was recorded from three viewing heights (30°,

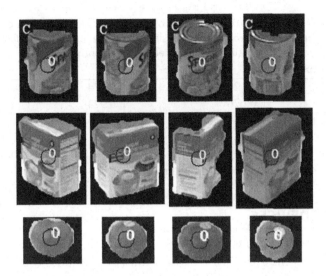

**Figure 3.10** Three examples of objects of the RGB-D dataset. The superimposed signs are the visual outputs of our algorithm. The circle indicates the object PC centroid, 0 is the object label, and C indicates that the object is a cylindrical object.

45°, and 60° above the horizon) while rotating on a turntable, resulting in approximately 150 views per object and 207,920 RGB-D images in total. The images are already cropped and segmented, obviating the need for object detection and segmentation. Fig. 3.10 shows some examples of frames included in the dataset and processed by the algorithm.

During the analysis, about 20% of the dataset was discarded from the tests due to low quality image acquisition (some portions of the object were missing). The remaining 240 objects were used as inputs to the algorithm, conducting two different series of tests to evaluate the tracking performances on the dataset and to evaluate the cylindrical object classification.

The tests performed to evaluate the tracking performance of the algorithm (test 1) presented a correctness percentage in tracking equal to 89.62%. In most of the cases, the objects that were not tracked by the algorithms actually featured a point cloud containing a very small number of points.

For cylinder classification (test 2), the Cylinder Recognition module provided an excellent results. Table 3.3 reports the confusion matrix obtained from the recognition phase (cylindrical objects were 29% of the dataset, corresponding to 87 objects). On the other hand, PCL always tries to find a cylinder. So, in order to correctly recognize real cylinders, $th_c$ is

**Table 3.3** Confusion matrix reporting the results of the test on cylindrical object classification. The rows represent the ground truth classification of the objects in the dataset, while the columns report the classification obtained by our algorithm

| Dataset/output | No cylinder | Cylinder |
|----------------|-------------|----------|
| No cylinder    | 0.925       | 0.074    |
| Cylinder       | 0.076       | 0.924    |

set in the range of 75–80% (see Section 3.2.2.2). In fact, the impact of the false positives strongly depends on the value of $th_c$.

## 3.3 ROBOTIC DEVICES

In this section we discuss the robotic devices present in the system. The main component is the Arm Light Exoskeleton (see Section 3.3.1), used to move the arm of the patient toward the desired object. The second component consists of a Wrist Exoskeleton (see Section 3.3.2) mounted on the end effector of the Arm and used to orient the patient's hand to the axis of the cylindrical object. Finally, a Hand Orthosis (see Section 3.3.3) is fastened on the Wrist and performs the actual grasping.

### 3.3.1 Arm Light Exoskeleton

The Arm Light Exoskeleton (ALEx) is a state-of-the-art mechanically compliant robotic device for the human upper limb [28]. In particular, the model exploited in the system presented here performs with two arm exoskeletons for the rehabilitation of both the right and the left limb, as shown in Fig. 3.11.

The kinematics of each arm is characterized by six serially connected rotational Degrees of Freedom (DoFs). The first four DoFs are both sensorized and actuated, while the last two are only sensorized. More details are shown in Fig. 3.12.

Further features include the localization of its motors in the fixed backpack rather than on the kinematic chain, and a patented implementation of the shoulder joint DoFs [29]. These design solutions allow the exoskeleton to

- achieve higher dynamic performance, due to a significant reduction of mass and inertia from the device's moving parts;
- cover about 90% of the natural workspace of the human arm without singularity;

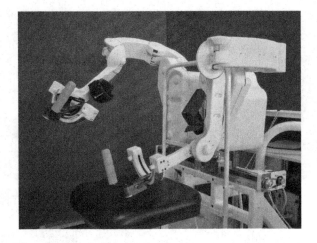

**Figure 3.11** The Arm Light Exoskeleton.

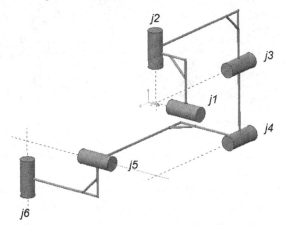

**Figure 3.12** ALEx kinematics.

- localize the power wiring in the backpack, with benefits for the Electromagnetic Compatibility and the reliability of the electric connections.

## 3.3.2 Wrist Exoskeleton

The WRist ExoSkeleton (WRES) has been designed to provide 3$DoF$ active, rotational movement to the user's forearm and wrist. In order to satisfy the requirements of the haptic technology, the high transparency is ensured by reducing the inertia, friction, and backlash of the actuators and the mechanical design. Furthermore, the device is light, easy to wear, and has an open structure, which allows it to be integrated with the upper arm

**Figure 3.13** WRES device worn by a user.

exoskeleton and the hand exoskeleton during operational tasks with no collision, as in Fig. 3.13.

In particular, the device supports pronation/supination (*PS*) to the forearm, and flexion/extension (*FE*) and radial/ulnar deviation (*RU*) to the wrist independently. This mechanism was modeled as a spherical 3*DoF* kinematics, where the range of motion of the joints is chosen with respect to the natural range of motion of a healthy user. These active joints are actuated through DC motors. Furthermore, a passive adjustment was developed in order to align the mechanical joints with the anatomical joints.

Thanks to the anodized aluminum that used for the mechanism, the device can achieve high stiffness and light weight properties simultaneously. The joints are actuated through pulleys and steel tendons to increase the force control performance with simple low-cost force sensors at the end effector of the manipulator. The actuators of *FE* and *RU* joints are located away from the joints to increase the dynamic performance of the mechanism. The *PS* joint is actuated in a direct manner, while the *FE* and *RU* joints are actuated by two independent actuators through a differential mechanism, where the transmission rates are defined by the mechanical capstan and pulley sizes.

The actuators are controlled by a digital Ethercat drive at 10 kHz, while the main control on a master computer is run at 5 kHz. The control algorithm of the device compensates the gravity and viscous friction of the actuators. The device is controlled through the actuator torques through the actuator rotations as feedback.

### 3.3.3 Hand Orthosis

The underactuated hand exoskeleton was designed to provide assistance to finger opening/closing tasks [30]. Kinematically, all of the fingers except the thumb have $3DoF$ flexion/extension and $1DoF$ abduction/adduction movements, while the thumb has $3DoF$ flexion/extension and $2DoF$ abduction/adduction movements. Nevertheless, simple grasping tasks can be assisted by controlling two joints for the fingers and the thumb using simple linear actuators for each finger component. Even though the abduction/adduction movement is left uncontrolled, it is left free for the user to move passively to increase the ease and the naturalness of the operation. The different kinematics between the thumb and other fingers lead the mechanism of the corresponding finger component to be designed differently (see Fig. 3.14) to implement the same approach.

The underactuation approach controls multiple joints with a single actuator by adjusting the grasping tasks based on the objects with various shapes and sizes in a stable manner. The proposed mechanism does not move the finger joints individually or with a predefined relation, but applies force on the phalanges to open or close the hand while the relation between the finger joints is set by the contact forces automatically with no prior knowledge of the size and shape of the grasped object. The auto-adjustability for the finger sizes might be achieved by assuming the finger as a part of the kinematic chain. Not aligning the finger and the mechanical joints increases the safety of the force transmission as well as the finger size adjustability. The device is designed to exert only normal forces to the finger phalanges during operation in order to increase the naturalness of the force feedback and the ease of connection between the user and the exoskeleton.

A Delfino[4] board has been used to control the device and read the sensors through ADC pins with 1 kHz frequency. The communication between the host computer and the control board is set by a simple USB

---

[4] http://www.ti.com/.

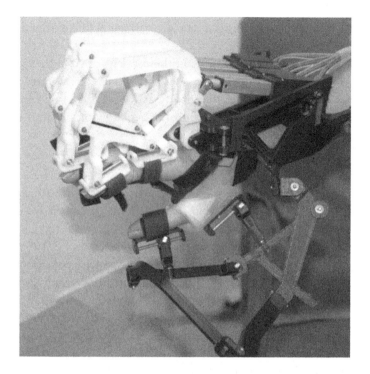

**Figure 3.14** Hand exoskeleton worn by a user.

port that limits the communication speed to 500 Hz. In addition to the control algorithm, the control board allows the additional force sensors to be read. In particular, the chosen actuators have mechanical gearboxes to increase the output force, while eliminating the backdrivability and the transparency of the device mechanically. Still, the high transparency, which is one of the most important requirements of haptic applications, can be ensured by utilizing $1DoF$ simple strain gauges to measure interaction forces along the actuators and implementing active backdrivability.

## 3.4 OVERALL SYSTEM EXPERIMENTS

This section illustrates the performance of the algorithm proposed in Section 3.2. The robotic setup has been used to assist arm, wrist, and finger movements performing reaching and grasping tasks. The aim of this experiment is to test whether the tracking module can provide accurate position reference to the robotic devices so that the subject can perform the grasping.

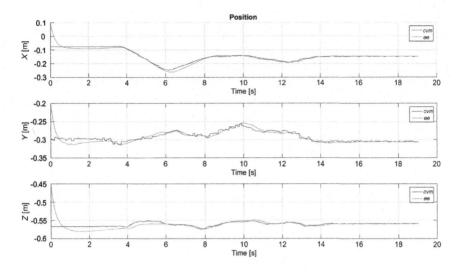

**Figure 3.15** Temporal plot of the position of the object and the robot.

A healthy subject was involved in the test and asked to remain completely passive during the session (male, 29 yrs old with no known neuromuscular disorder affecting his upper limb). During the experiment, a cylindrical object was placed in the scene. In order to emphasize the efficacy of the tracking performance of the algorithm and the robotic devices, another subject was chosen to act as the therapist during a rehabilitation session and to move the object within the scene limited by the robotic workspace. The output of the conducted experiment is presented in the coordinate space in Figs. 3.15 and 3.16 and in the joint space for the robotic devices in Figs. 3.17 and 3.18. The experiment lasted for about 20 s.

Figs. 3.15 and 3.16 show the temporal trend of the object to be tracked and of the end effector of the device in terms of the position and orientation. In particular, the labels *cvm* and *ee* are used to define the trend of the object provided by the computer vision module and the trend of the device end effector, respectively.

A lower refresh rate may be observed in the *cvm* trend, due to the frequency difference between the computer vision module (10 Hz) and the low-level control of the robot (100 Hz).

Figs. 3.17 and 3.18 show the temporal trend of all the joints for the ALEx and WRES devices, respectively. When the wrist is mounted on the arm exoskeleton, ALEx's non-actuated joints are replaced by the three

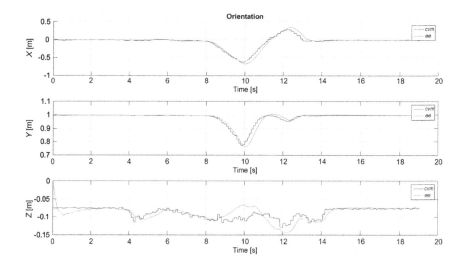

**Figure 3.16** Temporal plot of the orientation of the object and the robot.

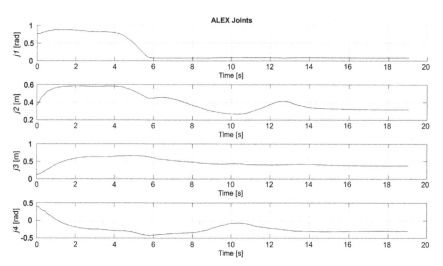

**Figure 3.17** Temporal plot of the ALEx device joints.

actuated DoFs of WRES; therefore, only the first four (actuated) joints of the arm are reported in the plot.

Finally, the grasping activation trend is reported in Fig. 3.19. A value of 0 indicates no grasping, whereas a value of 1 indicates grasping. The hand grasp is triggered only after the exoskeleton reaches the object; in particular, only if the device end effector remains stable in the position

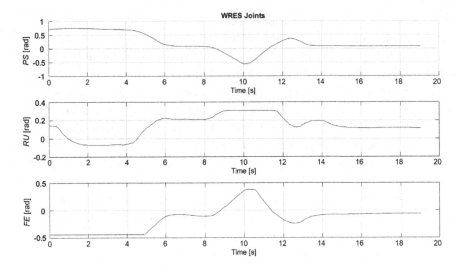

**Figure 3.18** Temporal plot of the WRES device joints.

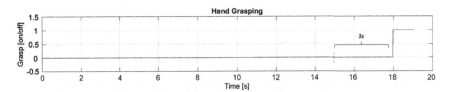

**Figure 3.19** Temporal plot of the hand grasping procedure. (For interpretation of the colors in this figure, the reader is referred to the web version of this chapter.)

of the object for more than 3 s when the object is still. The red dotted line in the plot indicates the moment where the grasping was requested (after 14.96 s of execution), which was indeed triggered when the effector position was considered stable (after 17.96 s of execution).

## 3.5 DISCUSSION AND CONCLUSION

In this work, a novel RGB-D-based algorithm was developed to track generic unknown objects in real time to be used for robot-assisted grasping tasks in neurorehabilitation applications. Even though the proposed algorithm can be used to track any object with no shape or size requirement, its functionality has been limited to cylindrical objects only to simplify the object identification process with high efficacy and to smooth the grasping tasks under the guidance of the robotic hand.

The novelty of the proposed tracking algorithm comes from combining different features to achieve the object recognition and tracking. In this way, the issues regarding the object tracking literature discussed in Section 3.1 have been solved. The most important feature of the given algorithm is getting rid of the requirement of using predefined objects only and to adopt its functionality for any object with unknown physical properties. Furthermore, the algorithm achieves a fast and robust reconstruction of occluded objects, the absence of drift over time, and a high level of transparency.

The efficacy and the robustness of the developed algorithm were tested with a healthy subject using a rehabilitation robot setup supporting arm, wrist, and finger movements. The experiment results prove the performance of the given tracking approach in a real scenario through a reaching and grasping task experiment. The results of the first experiment might lead us to re-perform the experiment with the given setup with additional healthy subjects and stroke patients with disabilities in the future.

## REFERENCES

[1] S. Beckelhimer, A. Dalton, C. Richter, V. Hermann, S. Page, Computer-based rhythm and timing training in severe, stroke-induced arm hemiparesis, The American Journal of Occupational Therapy 65 (1) (2011) 96.
[2] D. Lloyd-Jones, R. Adams, T. Brown, M. Carnethon, S. Dai, G. De Simone, et al., Executive summary: heart disease and stroke statistics—2010 update: a report from the American Heart Association, Circulation 121 (7) (2010) 948.
[3] H. Krebs, B. Volpe, D. Williams, J. Celestino, S. Charles, D. Lynch, et al., Robot-aided neurorehabilitation: a robot for wrist rehabilitation, IEEE Transactions on Neural Systems and Rehabilitation Engineering 15 (3) (2007) 327–335.
[4] A. Frisoli, L. Borelli, A. Montagner, S. Marcheschi, C. Procopio, F. Salsedo, et al., Arm rehabilitation with a robotic exoskeleton in virtual reality, in: IEEE 10th International Conference on Rehabilitation Robotics, ICORR 2007, IEEE, 2007, pp. 631–642.
[5] A. Frisoli, C. Procopio, C. Chisari, I. Creatini, L. Bonfiglio, M. Bergamasco, et al., Positive effects of robotic exoskeleton training of upper limb reaching movements after stroke, Journal of NeuroEngineering and Rehabilitation 9 (1) (2012) 36.
[6] A. Frisoli, C. Loconsole, D. Leonardis, F. Banno, M. Barsotti, C. Chisari, M. Bergamasco, A new gaze-BCI-driven control of an upper limb exoskeleton for rehabilitation in real-world tasks, IEEE Transactions on Systems, Man and Cybernetics. Part C, Applications and Reviews 42 (6) (2012) 1169–1179.
[7] G. Prange, M. Jannink, C. Groothuis-Oudshoorn, H. Hermens, M. Ijzerman, Systematic review of the effect of robot-aided therapy on recovery of the hemiparetic arm after stroke, Journal of Rehabilitation Research and Development 43 (2) (2006) 171.
[8] T. Nef, R. Riener, Armin-design of a novel arm rehabilitation robot, in: 9th International Conference on Rehabilitation Robotics, IEEE, 2005, pp. 57–60.
[9] A. Frisoli, F. Rocchi, S. Marcheschi, A. Dettori, F. Salsedo, M. Bergamasco, A new force-feedback arm exoskeleton for haptic interaction in virtual environments, in: Eu-

rohaptics Conference and Symposium on Haptic Interfaces for Virtual Environment and Teleoperator Systems, World Haptics 2005, First Joint, IEEE, 2005, pp. 195–201.

[10] R. Sanchez, J. Liu, S. Rao, P. Shah, R. Smith, T. Rahman, et al., Automating arm movement training following severe stroke: functional exercises with quantitative feedback in a gravity-reduced environment, IEEE Transactions on Neural Systems and Rehabilitation Engineering 14 (3) (2006) 378–389.

[11] C. Loconsole, F. Banno, A. Frisoli, M. Bergamasco, A new Kinect-based guidance mode for upper limb robot-aided neurorehabilitation, in: IEEE/RSJ International Conference on Intelligent Robots and Systems (IROS), 2012, IEEE, 2012, pp. 1037–1042.

[12] F. Jurie, M. Dhome, Real time tracking of 3D objects with occultations, in: Proceedings, vol. 1, International Conference on Image Processing, 2001, IEEE, 2001, pp. 413–416.

[13] L. Masson, M. Dhome, F. Jurie, Robust real time tracking of 3D objects, in: Proceedings of the 17th International Conference on Pattern Recognition, vol. 4, ICPR 2004, IEEE, 2004, pp. 252–255.

[14] A.D. Sappa, V. Bevilacqua, M. Devy, Improving a genetic algorithm segmentation by means of a fast edge detection technique, in: Proceedings, vol. 1, International Conference on Image Processing, 2001, IEEE, 2001, pp. 754–757.

[15] T. Lindeberg, Scale invariant feature transform, Scholarpedia 7 (5) (2012) 10491.

[16] C. Tomasi, T. Kanade, Detection and Tracking of Point Features. Shape and Motion from Image Streams: A Factorization Method—Part 3, 1991.

[17] P. Alimi, Object Persistence in 3D for Home Robotics, Ph.D. thesis, University of British Columbia, 2012.

[18] A.E. Johnson, Spin-Images: A Representation for 3-D Surface Matching, Ph.D. thesis, Citeseer, 1997.

[19] R.B. Rusu, G. Bradski, R. Thibaux, J. Hsu, Fast 3D recognition and pose using the viewpoint feature histogram, in: IEEE/RSJ International Conference on Intelligent Robots and Systems (IROS), 2010, IEEE, 2010, pp. 2155–2162.

[20] R.B. Rusu, N. Blodow, M. Beetz, Fast Point Feature Histograms (FPFH) for 3D registration, in: IEEE International Conference on Robotics and Automation, 2009, ICRA'09, IEEE, 2009, pp. 3212–3217.

[21] B. Steder, R.B. Rusu, K. Konolige, W. Burgard, NARF: 3D range image features for object recognition, in: Workshop on Defining and Solving Realistic Perception Problems in Personal Robotics at the IEEE/RSJ International Conference on Intelligent Robots and Systems (IROS), vol. 44, 2010.

[22] R.B. Rusu, S. Cousins, 3D is here: Point Cloud Library (PCL), in: IEEE International Conference on Robotics and Automation (ICRA), 2011, IEEE, 2011, pp. 1–4.

[23] C. Loconsole, F. Stroppa, V. Bevilacqua, A. Frisoli, A robust real-time 3D tracking approach for assisted object grasping, in: International Conference on Human Haptic Sensing and Touch Enabled Computer Applications, Springer, 2014, pp. 400–408.

[24] A. Elgammal, C. Muang, D. Hu, Skin detection—a short tutorial, in: Encyclopedia of Biometrics, 2009, pp. 1–10.

[25] K. Jack, Video Demystified: A Handbook for the Digital Engineer, Elsevier, 2011.

[26] M.A. Fischler, R.C. Bolles, Random sample consensus: a paradigm for model fitting with applications to image analysis and automated cartography, Communications of the ACM 24 (6) (1981) 381–395.

[27] K. Lai, L. Bo, X. Ren, D. Fox, A large-scale hierarchical multi-view RGB-D object dataset, in: IEEE International Conference on Robotics and Automation (ICRA), 2011, IEEE, 2011, pp. 1817–1824.

[28] E. Pirondini, M. Coscia, S. Marcheschi, G. Roas, F. Salsedo, A. Frisoli, M. Bergamasco, S. Micera, Evaluation of the effects of the arm light exoskeleton on movement execution and muscle activities: a pilot study on healthy subjects, Journal of Neuro-Engineering and Rehabilitation 13 (1) (2016) 1.

[29] M. Bergamasco, An exoskeleton structure for physical interaction with a human being, pct application n. WO2013186701 (A1), 2013.

[30] M. Sarac, M. Solazzi, E. Sotgiu, M. Bergamasco, A. Frisoli, Design and kinematic optimization of a novel underactuated robotic hand exoskeleton, Meccanica (2016) 1–13.

# CHAPTER 4

# Computer Vision and Machine Learning for Surgical Instrument Tracking

## Focus: Random Forest-Based Microsurgical Tool Tracking

**Nicola Rieke\*, Federico Tombari\*, Nassir Navab\*,†**

\*Computer Aided Medical Procedures, Technical University of Munich, Garching, Germany
†Johns Hopkins University, Baltimore, MD, United States of America

## Contents

## Abstract

In recent years, computer vision has become a remarkable tool for various computer-assisted medical applications, paving the way towards the use of augmented reality and advanced visualization in the medical domain. In this chapter we outline how computer vision can support the surgeon during an intervention via the example of surgical instrument tracking in retinal microsurgery, which incorporates challenges and

**Computer Vision for Assistive Healthcare.**
DOI: https://doi.org/10.1016/B978-0-12-813445-0.00004-6

requirements that are common for the employment of this technique in various medical applications. In particular, we show how to derive algorithms for simultaneous tool tracking and pose estimation based on random forests and how to increase robustness in problems associated with retinal microsurgery images, such as wide variations in illumination and high noise levels. Furthermore, we elaborate on how to evaluate the overall performance of such an algorithm in terms of accuracy and describe the missing steps that are necessary to deploy these techniques in real clinical practice.

## Keywords

Instrument tracking, Pose estimation, Random forest, Machine learning, Medical computer vision, Closed-loop approach, Online adaptation

## 4.1 INTRODUCTION

Assistive Technologies have great potential to overcome limitations in various applications [1]. The medical field is a particularly relevant area of research for assistive computer vision and has already found its way into common clinical practice. Pre-operative planning using image data of multiple imaging devices such as Computed Tomography, Ultrasound, and Magnetic Resonance Imaging would not be possible without prior registration of the different information sources. The tremendous advances in this field have even enabled real-time applications such as advanced intra-operative navigation for prostate biopsy [2]. In the field of assistive diagnosis, computer vision enables for example the precise comparison of mole shape and growth over time and thereby allows identification of any malignant mutations at an early stage [3]. Another example is the vision-based classification of development disorders [4] which may have the potential to objectively characterize fine behavior via the analysis of eye movements. Also, more futuristic approaches such as medical collaborative robotics [5] require a fast and robust tracking of imaging probes. The range of applications is broad and is still growing.

In this chapter we discuss the general ideas and difficulties of medical assistive computer vision in the context of surgical instrument tracking in Retinal Microsurgery (RM) and show how to tackle the problem with the example of an approach based on Random Forests (RF). First, we outline the potential benefit of surgical instrument tracking in this field including possible applications (Section 4.1.1) and describe the resulting common challenges for computer vision (Section 4.1.2). In the second part of the chapter, a brief overview is given of how these problems are addressed in current state-of-the-art approaches (Section 4.2). Then, we focus on a specific method (Section 4.3) and show step-by-step how to build a robust

 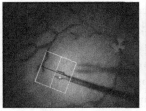 

(A) Testing Setup with Phantom and Pig Eye    (B) Microscope View    (C) iOCT Slices

**Figure 4.1** *Retinal microsurgery setup.* The surgeon observes the scene indirectly through a microscope and illuminates the inner part of the eye with a manually held light pipe, as depicted in (A). The microscope view, as shown in (B) with a phantom eye, is the usual visual feedback for the surgeon. The additional iOCT slices (C) provide cross-sectional views along the indicated lines and are limited to a small area due to the real-time requirement.

pipeline for instrument tracking by breaking down the problem into two separate tasks and building on machine learning. Furthermore, we demonstrate how to evaluate the performance of an algorithm in terms of accuracy, present how the method's components influence the overall outcome, and compare the method to other state-of-the-art methods (Section 4.4). Finally, we elaborate on the missing steps for clinical translation and discuss future work (Section 4.5). Major parts of this chapter are based on previous publications [6–8].

### 4.1.1 Potential Benefit of Surgical Instrument Tracking in Retinal Microsurgery

Retinal Microsurgery is an ophthalmic surgery within the vitreous cavity of the eye. A typical setup is depicted in Fig. 4.1. During the procedure, the surgeon manipulates anatomical structures of the retina with micronscale maneuvers while observing the scene indirectly via a microscope. The surgical tool and the fiber optic light source are inserted into the eye via an artificial opening, called the trocar. As a side effect, the resulting instrument shadows can serve as visual cues for the surgeon to estimate the distance between the tool and the retina surface. This is crucial information; penetration into deeper layers of the retina, for example during membrane peeling, causes irrevocable injury. However, shadows can only provide a rough depth estimation and a complete understanding of correct depth perception requires years of experience. An additional imaging technology, known as intraoperative Optical Coherence Tomography (iOCT) [9], allows real-time cross-sectional imaging of subretinal structures and therefore depth information in high resolution. Nevertheless, the spatial capture range is very limited in order to meet the real-time requirement and it has to be positioned manually to the region of interest which further complicates the workflow of the surgeon and is time consuming. Instrument tracking has great potential to simplify the surgical workflow here by repositioning the iOCT automatically according to the position of the instrument tips [10]. This would be valuable not only during membrane peeling, but also during targeted drug delivery. Furthermore, it paves the way for advanced augmented reality techniques used to estimate and visualize the proximity to the retina [11]. Another application is based on the observation that the surgeon's center of attention is usually close to the surgical instrument tips. The visualization of information close to this area would reduce the surgeon's distraction. Furthermore, instrument tracking is a key component of robotic assistance and visual servoing [12] or can be employed for recognizing the current action of the surgical workflow [13]. Since instrument manipulation can be a crucial indicator of the level of surgical skill [14], instrument tracking can also serve as objective and automatic assessment of surgical performance.

## 4.1.2 Challenges of Computer Vision in Medical Applications

One of the main requirements of computer-assisted medical interventions is that the method supports the surgeon without interfering with the surgical workflow. Physicians develop expertise and automatisms over years of training, and a modification may result in a distraction rather than an aid. Most crucial is the insight that it is rarely possible to ask a surgeon to perform an intervention differently in order to simplify the problem for the computer assistance, such as avoiding a certain angle during an acquisition in order to guarantee a field of view for optical tracking or performing an extensive calibration before the procedure. In this context, the vision-based tracking of the instruments in RM holds great potential because the microscope images are already available and a modification of the instrument in terms of markers is not required. However, the high level of noise and the limited field of view create a challenging computer vision problem. In particular, the hand-held light source and the resulting problems, such as specular reflections on the metallic instrument and strong shadows, complicate the task. It has been shown that machine learning approaches are able to tackle these problems in general computer vision applications, but patient data security and the need for expert annotations implies that datasets in the medical field are usually very limited. And of course, an algorithm also has to be robust and run in real time in order to meet the requirements of the operating room.

## 4.2 OVERVIEW OF THE STATE OF THE ART

In recent years there has been significant progress towards robust vision-based tracking of instruments during in vivo sequences, as recently summarized by Bouget et al. [15]. Prior work addressing the challenges has considered the use of geometric models such as the approach proposed by Baek et al. [16] to track the forceps by generating a database of the projected contours of a 3D CAD model of the robotic forceps. The likelihood to the projected contour of the microscopic image is measured and finally the full state of the forceps is estimated via particle filtering. Reiter et al. [17] presented a tracking method that relies on the appearance of natural landmarks. They trained an efficient multi-class classifier and as the locations of the natural landmarks are known in the tool's CAD model, they are used to compute the final pose. The algorithm was tested on five endoscopic sequences. Color-based approaches [18,19] were presented for the related field of laparoscopic tool tracking. Other relevant work (e.g. [20]) presents results on both phantom and in vivo data, using a two-stage procedure: brute-force search of the tool tip in the surroundings of the instrument coordinates in the previous frame, and weighted mutual information to optimize the initial guess.

Other works build on learning-based approaches like [21] which uses natural features of surgical instruments for tracking and adopts a spiking neural network to recognize the instrument tip in laparoscopic surgeries. Li et al. [22] propose an online learning approach for tool tracking in RM. The system starts with a manual initialization and gradually builds the database for tracking by adding new positive and negative tool samples, which are collected through a filtering process. The algorithm provides an accurate bounding box around the forceps' central point, but does not localize the two tips of the instrument. The proposed method of Sznitman et al. [23] utilizes a parametrization of the surgical tool considering the following three criteria: the location of the insertion point, the angle between image boundary and tool, and the tool length. Afterwards, tracking is considered as a Bayesian filtering estimation problem. To compute the necessary posterior distribution, they use a strategy based on active testing (ATF). In their most recent work [24], they propose a robust and efficient algorithm which uses a multi-class classifier based on boosted regression trees. Each class represents a different part of the instrument (e.g. center, insertion point, shaft) or background. In order to provide both accuracy and a good frame rate, an early stopping method was also implemented using a probabilistic model, which evaluates the reliability of current classification and stops when more computation is necessary. A deep learning approach for instrument detection was introduced by Sarikaya et al. [25]; however, it only infers the bounding box around the instrument tip and does not regress a precise localization of the instrument landmarks, which would be necessary for most of the applications (see Section 4.1.1).

As this is only a brief overview of different approaches, we refer the reader to the survey paper by Bouget et al. [15] for more details.

## 4.3 METHOD

In this section, we derive the surgical instrument tracking method in in vivo microsurgery images that was originally introduced in [6–8]. The algorithm is inspired by recent computer vision approaches [26,27] and handles application specific difficulties (as described in Section 4.1.2) by modeling the problem as a combination of two different tasks: tracking and 2D pose estimation. In this way we focus the algorithm on exploiting different available information that is essential for solving the two distinct problems. For both steps, we will leverage the machine learning RF technique (Section 4.3.1), which is shown to be reliable for noisy data and generalized to unseen scenarios. The tracking RF (Section 4.3.3) regresses the bounding box around the instrument tip and is a fast multi-template tracking algorithm. In order to achieve low computational costs, it solely employs the color information in the current frame extracted at the previous template position. In the second step, the pose estimation algorithm determines the location of the instrument parts inside the bounding box via Histogram of Oriented Gradients (HOG) features for the pose RF (Section 4.3.4). These two steps can be combined to a feed-forward pipeline (Section 4.3.5) to achieve the overall goal of localizing the instrument landmarks for every frame in real time. In this model, the pose estimation is highly dependent on the output of the tracker and both algorithms rely on the assumption that the training dataset captures the broad variety of illumination changes during in vivo sequences. In order to increase robustness, the two algorithms can be combined in a closed-loop approach by employing the output of the pose estimation to correct the predicted template position of the tracker. Depending on the confidence of the RF predictions, the final template is adaptively influenced by the more reliable one. Furthermore, the current predictions are employed to learn new trees for the tracking RF. In this way, we fuse the intensity-based and gradient-based predictions in a synergic way and adapt them online to the present scenario (Section 4.3.6).

## 4.3.1 Random Forests

A crucial part of the method is the random forest, which is a machine learning method used in the tracking stage and in the pose estimation stage of the algorithm. Considering an input $\mathbf{X}$ and output $\mathbf{Y}$, the RF is used to learn the relation between $\mathbf{X}$ and $\mathbf{Y}$ such that, given the input $\mathbf{X}$, the forest can predict the output $\mathbf{Y}$. A forest itself consists of an ensemble of decision trees which can output a class (classification) or, in our case, real numbers (regression). In particular, RFs are used to correct the tendency of trees to overfit the training set. Moreover, while predictions of a single tree are highly sensitive to noise in the training set, the average of many trees is not, under the hypothesis that they are independent. Each classifier at each node of the tree represents a "weak learner", while the ensemble of all the weak classifiers allows the forest to be a strong classifier to predict $\mathbf{Y}$. Further information on RFs can be found in the original work of [28].

A tree is composed of nodes that can either be a branch that has two children (i.e. left and right) or a leaf, which is the terminal node. Training a tree requires a learning dataset $P = \{(\mathbf{X}_d, \mathbf{Y}_d)\}_{d=1}^{n_d} = \{(\mathbf{X}_1, \mathbf{Y}_1), (\mathbf{X}_2, \mathbf{Y}_2), \ldots, (\mathbf{X}_{n_d}, \mathbf{Y}_{n_d})\}$ with the input $\mathbf{X}_d$ and its corresponding ground truth observation $\mathbf{Y}_d$. During *training* at each node, the data is split into two subsets, based on a splitting criterion $\theta$, that are passed down to its children $(P_l, P_r)$. The splitting criterion is a pool of random tests and has the goal, at each step, of finding the best split of the set. The information gain can be used to evaluate the best split and is given by

$$g(\theta) = H(P_n) - \sum_{i \in \{l,r\}} \frac{|P_i(\theta)|}{|P_n|} H\big(P_i(\theta)\big) , \qquad (4.1)$$

where $P_n$ is the set of samples that reach the node $n$, $|P|$ is the number of samples in the set $P$, and $H(\cdot)$ evaluates the randomness of $P$. Since we consider regression forests, $H(\cdot)$ can be estimated by standard deviation of the multivariate Gaussian distribution. Starting from the root node, the dataset is iteratively split into two subsets and is passed down to the node's children until one of the following stopping criteria is true:

1.  the maximum depth of tree is reached;

**2.** the number of samples that reach the node is insufficient to split;

**3.** the information gain of the best split is too small.

Then, the leaf stores the distribution of the parameters of **Y** that typically employ a normal distribution with its mean and standard deviation. As a result of learning, each branch of the tree stores the parameters of the splitting function with respect to the input **X**, while each leaf stores a distribution of the output **Y**. To ensure the independence of the trees in the forest, each random tree selects a random subset of elements in **X** or a random subset of the learning dataset.

During *testing*, a new sample of **X** traverses the tree. At each branch, it moves to the left or right child node depending on the splitting function, eventually ending up at a leaf node that contains the prediction to be associated with the sample. Finally, the results of the leaf nodes at different trees are aggregated in order to robustly obtain the final prediction.

### 4.3.2 Template Definition

The template is the bounding box around the instrument tip, which will be followed by the tracker and used as an input for the pose estimator. Since the features used for the pose estimator are sensitive to scale changes, we can define the template based on the landmark positions of the instrument, as in [8]. Let us denote $(L, R, C)^\top \in \mathbb{R}^{2 \times 3}$ as the central landmark of the instrument and the right and left tips of the tool. The midpoint of the instrument tips is then given by $M = \frac{L+R}{2}$ and the 2D similarity transform from the patch coordinate system to the frame coordinate system can be defined as

$$\mathbf{H} = \begin{bmatrix} s \cdot \cos(\theta) & -s \cdot \sin(\theta) & C_x \\ s \cdot \sin(\theta) & s \cdot \cos(\theta) & C_y \\ 0 & 0 & 1 \end{bmatrix} \begin{bmatrix} 1 & 0 & 0 \\ 0 & 1 & 30 \\ 0 & 0 & 1 \end{bmatrix}$$

with $s = \frac{b}{100} \cdot \max\{\|L - C\|_2, \|R - C\|_2\}$ and $\theta = \cos^{-1}\left(\frac{M_y - C_y}{\|M - C\|_2}\right)$ for a patch **P** of fixed size of $100 \times 150$ pixel and $b \in \mathbb{R}$ defining the relative size. In this way, the template is axis-aligned to the shaft and the instrument tips are on the upper third of the bounding box.

### 4.3.3 Tracking

Template tracking determines the transformation parameters that optimally map a defined region of interest from the previous frame to the current frame. Mathematically it can be modeled as follows: Considering an image at time $t$, $I_t : \Omega \subset \mathbb{R}^2 \to \mathbb{R}^3, p \to I(p)$ and $n_s$ randomly selected sample points $\{\mathbf{x}^s\}_{s=1}^{n_s} \subset \mathbf{H} * \mathbf{P} = \Omega_T \subset \Omega$, the template $I_B : \Omega_T \to \mathbb{R}^3, p \to I_B(p) \subset I_t$ can be represented as the intensity vector $\mathbf{i} = [I(\mathbf{x}^s)]_{s=1}^{n_s}$. Tracking the template from the previous frame $I_{t-1}$ to the current frame $I_t$ corresponds then to finding the 2D translation vector $\delta\boldsymbol{\mu}$ that updates the location of the bounding box around the instrument tip based on the intensities in the current frame $\mathbf{i}_t = [I_t(\mathbf{x}_{t-1}^s)]_{s=1}^{n_s}$ using the location of sample points from the previous frame $\{\mathbf{x}_{t-1}^s\}_{s=1}^{n_s}$. As presented in [27], an RF can be employed to learn the relation of the changes in the intensities $\delta\mathbf{i} = \mathbf{i}_t - \mathbf{i}_0$ and the transformation parameters $\delta\boldsymbol{\mu}$ by setting the input $\mathbf{X} = \delta\mathbf{i}$ and the output $\mathbf{Y} = \delta\boldsymbol{\mu}$. However, in the context of surgical instrument tracking, the definition of a single unique template is not possible because of the opening and closing movement of the tool. And the variation of lighting and the strong reflections on the metallic surface complicate further the task. Therefore, the tracker utilizes multiple correlated templates and thereby learns a generalized model. This is possible because the color information $\mathbf{i}_0$ in the initial frame $I_0$ is constant and the splitting function only considers an index of the vector $\delta\mathbf{i}$. Consequently the forests can be simplified by learning the relation of $\mathbf{i}$ and $\delta\boldsymbol{\mu}$ instead of $\delta\mathbf{i}$ and $\delta\boldsymbol{\mu}$. In this way, we can use multiple templates within the same forests [6].

Since the tree recursively subdivides the dataset into two subsets by lowering the resulting lower standard deviation, a reliable prediction is identifiable by an average standard deviation less than a threshold $\tau_\sigma$ and the template is updated using $\delta\boldsymbol{\mu}$. Otherwise, its previous location is propagated to the next frame in order to avoid possible loss during tracking. Since the tracker solely relies on the directly available color information, it only requires 2 ms per frame using only one CPU core. For more details on the tracker, we refer the reader to [6] and [7].

## 4.3.4 Two-Dimensional Pose Estimation

The aim of 2D pose estimation is to precisely detect the instrument landmarks for every frame. By modeling the instrument as an articulated object, we can employ parametric models similar to those successfully proposed in the field of human pose estimation [26]. In general, the number of landmarks for these approaches is variable. However, in the case of surgical forceps, let us define the set of joints as the left tip of the instrument (L), the right tip of the instrument (R), and the center joint (C) connecting these two parts, as introduced in Section 4.3.2. More specifically, the tip landmarks are set to be the innermost and top visible point of the instrument

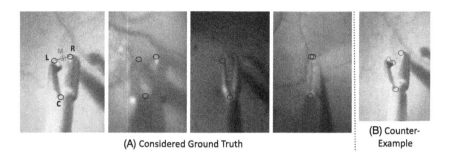

(A) Considered Ground Truth

(B) Counter-Example

**Figure 4.2** *Ground truth and template definition.* The template is a bounding box around the instrument tip and is defined by three landmarks. The examples in (A) show the considered ground truth, where the landmarks at the instrument tip are closest to the microscope and touch during a closing movement. In (B) a counterexample is depicted.

part, which tend to touch each other in the case of a closed forceps (see Fig. 4.2) in order to have a well-defined ground truth. For 2D pose estimation, these landmarks are inferred on the basis of a set of extracted image features and for a precise localization, we suggest employing a computationally more expensive representation than color information. Histograms of oriented gradients features [29] have shown their robustness in fields such as object detection [30], image retrieval [31], and classification [32]. The main idea behind the feature descriptor is that an object can be described by the local distribution of its gradients. Again, we can employ an RF (Section 4.3.1) to learn the mapping between these features and the offsets to the 2D position of the landmarks in an holistic approach. Therefore, HOG feature vectors can be extracted at random pixel positions within the considered image region $\Omega_T \subset \Omega$. The ground truth prediction during training is given by the associated offset to the landmark positions. Consequently, for 2D pose estimation the input $\mathbf{X}$ is given by the HOG feature space and the output $\mathbf{Y}$ is the offset to the joints. The binary split function $\theta$ is a threshold on one dimension of the HoG feature vector and the function $H(\cdot)$ is defined by the Sum of Squared Distances (SSD)

$$H(P) = \sum_{i \in I} \sum_{j} \|o_{i,j} - \mu_j\|_2^2 , \qquad (4.2)$$

where $I$ denotes the image patch, the 2D vector $o_{i,j}$ contains the offset of the joint $j \in J$ from the image patch center, and $\mu_j$ is the mean for each joint offset. The leaves of the RF contain the associated offsets $o_j = \mathbf{Y} \subset \mathbb{R}^2$ of all instrument joints $j \in J = \{L, R, C\} \subset \mathbb{R}^2$. The overall prediction of the separate trees is accumulated by a greedy dense-window algorithm that models the estimation as a density function. Similar to the work of [26], the 2D predictions for every joint $j \in J$ are therefore discretized on a fixed grid. For every grid cell, the number of contained votes is known and an integral matrix can be created. Consequently, all the cells form an integral image and the final estimation corresponds to the region with the maximum number of points, which is found via a sliding window approach. In this way, a direct mapping between the extracted HOG features and the location of the instrument joints can be modeled.

### 4.3.5 Feed-Forward Pipeline

In the two-stage pipeline approach [6], the multi-template tracker (Section 4.3.3) and the 2D pose estimation (Section 4.3.4) can be combined via a feed-forward connection. As an input, we assume RGB-valued image $I : \Omega \subset \mathbb{R}^2 \to \mathbb{R}^3, p \to I(p)$ along with the initial localization of the tool. For every new frame $I_t$, the tracking algorithm estimates the transformation that updates the location of a bounding box $I_B : \Omega_T \subset \mathbb{R}^2 \to \mathbb{R}^3, p \to I(p) \subset I_t$ containing the entire instrument tip. $I_B$ is subsequently employed as an input to the 2D pose estimation, which localizes the instrument landmarks within this region of interest.

Theoretically, it is possible to solve the task without using any kind of template tracker and directly regressing the joint positions via upper 2D pose estimation by considering the image information of the entire frame. However, the combination of the two algorithms has several advantages: the template tracker simplifies the problem for the 2D pose estimation by limiting the region of interest to a bounding box $I_B \subset I_t$ around the instrument tip, and therefore drastically reduces the computational cost. This is a key component for enabling the real-time requirement since in this second step of the pipeline, it is possible to make use of the computationally more expensive gradient information. Furthermore, misleading information from strong gradients of the retinal blood vessels are reduced. Another main advantage is that the tracked template, as defined in Sect 4.3.2, yields a bounding box around the tool tip, which is aligned with the direction of the instrument shaft at the time of the initialization. Since the utilized HOG features are non-rotationally invariant, this template orientation avoids the extensive training data that would be necessary for capturing the in-plane rotational changes in the case of an upright bounding box. An overview of the pipeline is depicted in Fig. 4.3.

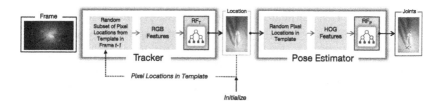

**Figure 4.3** *Feed-forward pipeline.* The tracker (Section 4.3.3) and the pose estimator (Section 4.3.4) can be combined via a feed-forward connection. In this way, the computational cost is reduced and real-time performance is enabled. Furthermore, the tracker significantly reduces the problem complexity for the pose estimator by providing a shaft axis aligned bounding box.

The drawbacks of this approach are that the 2D pose estimation completely relies on the output of the tracking algorithm. If the template tracker fails and provides a template that does not contain the instrument tip, the pose estimation has no chance of regressing the precise locations of the landmarks. Furthermore, there is no temporal–spatial regularization for the pose estimation and every bounding box is considered a still image. Another disadvantage is that both RFs are learned offline and require a training dataset that captures all variations in terms of instrument appearance and lighting, for example.

## 4.3.6 Robust Pipeline via Online Adaptation and Closed Loop

Based on the feed-forward pipeline, let us build a robust closed-loop framework which is able to withstand photometric distortions and motion blur and is also capable of generalizing to unseen instruments and background [8]. Instead of providing a single static image for the pose estimation, we can enforce temporal–spatial constraints for all landmark locations via a Kalman filter [33] by employing a constant velocity model which considers four dynamic parameters (2D location of the joints in the frame coordinate system and their 2D velocity) and two measurement parameters (2D location of joints). In this way, the output of the 2D pose estimation is subject to temporal regularization, but still solely relies on the gradient information. While gradients are usually more reliable, they nearly vanish in the case of motion artifacts or blur. In this scenario the color information is still reliable for predicting the movement. Therefore, a fusion of the intensity-based and the gradient-based predictions in a synergistic way could be advantageous. Both the tracker and the pose estimator employ RFs, but regress different information; however, since the template is directly defined by the instrument landmarks, both algorithms provide a template prediction and their outcomes can be fused. In the ideal scenario both algorithms would be free from error and we could simply average the predictions. In the real world scenario, however, this is rarely the case. Instead, we can compute a weighted average depending on the confidence for the current prediction of the separate RFs. A confident prediction of an RF can be identified by a distinct peak of votes and therefore by a low average standard deviation. Henceforth, we can define the scale $s^*$ and the translation $t^*$ of the joint similarity transform as the weighted average

$$s^* = \frac{s_T \cdot \sigma_P + s_P \cdot \sigma_T}{\sigma_T + \sigma_P} \quad \text{and} \quad t^* = \frac{t_T \cdot \sigma_P + t_P \cdot \sigma_T}{\sigma_T + \sigma_P},$$

where $\sigma_T$ and $\sigma_P$ are the average standard deviation of the tracking prediction and pose prediction, respectively, and the $t^*$ is set to be greater than or equal to the initial translation. The position of the corrected template $I_T^*$, which is defined by $H^*$, is then used for the regression of the bounding box in the next frame $I_{T+1}$. This closed-loop approach allows us to improve the tracking accuracy via pose information.

The evaluation of the prediction quality of the forest enables an additional important refinement of the method: a reliable online adaptation to the specific conditions at hand in order to enhance the robustness the tracker. In particular, new trees can be added incrementally to the tracking RF by using the template location on the current frames that have been predicted reliably by the pose RF. To this end, random synthetic transformations are applied on $I_T^*$ in order to build the online learning dataset composition of feature and translation vector pairs, such that the transformations imitate the movement of the template between two subsequent frames. The resulting learned trees are then added to the existing tracking RF that were learned offline. In this way, the tracker can predict the template location for the succeeding frames based on both the generalized and environment-specific trees.

In order to further improve robustness, a piecewise constant velocity from consecutive frames can be assumed. Therefore, given the image $\mathbf{I}_t$ at time $t$ and the translation vector of the template from $t-2$ to $t-1$ as $\mathbf{v}_{t-1} = (v_x, v_y)^\top$, the input to the tracking forest is a feature vector concatenating the intensity values on the current location of the template $\mathbf{I}_t(\mathbf{x}_p)$ with the velocity vector $\mathbf{v}_{t-1}$, assuming a constant time interval.

As depicted in Fig. 4.4 the result is a robust closed-loop approach, which enables an implicit handshake between tracker and pose estimation, and

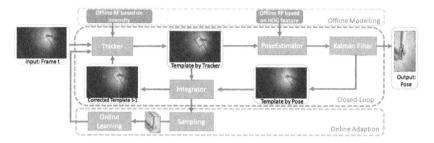

**Figure 4.4** *Robust pipeline.* Robustness to photometric distortions and unseen scenarios can be achieved via a closed-loop approach and an online adaptation. The Kalman Filter enforces spatial–temporal constraints for the regression of the instrument landmarks and the Integrator fuses the template predictions of the two algorithms depending on the confidence of the random forests. The Online Adaption allows us to add newly learned trees to the tracking random forest during testing. Figure adapted from [8].

allows online adaption to the conditions that are specific to a particular situation.

## 4.4 PERFORMANCE EVALUATION

In this section, we analyze and compare the performance of the feed-forward pipeline (Section 4.3.5, originally introduced in [6]) and the robust pipeline (Section 4.3.4, originally introduced in [8]) in terms of localization accuracy of the tool landmarks.

The experiments are performed on two different RM datasets: the dataset introduced by Sznitman et al., referred to as the *public* dataset [34], is a publicly available and fully annotated dataset of in vivo surgeries. It consists of three different sequences with a total number of 1171 images of 640 × 480 pixel resolution. The main difficulty of this dataset consists in the dominant blue and green coloring of the sequences and variable illumination conditions. The present instrument is the same for all sequences. The dataset utilized in [6,8], referred to as the *appearance* dataset, features different types of forceps and comprises four in vivo surgery sequences. Every sequence consists of 200 subsequent frames with 1920 × 1080 pixels each. In addition to the varying instrument appearances, it includes different illumination conditions, microscope zoom factors, and strong motion blur.

In order to compare the performance of the proposed pipelines to each other and to state-of-the-art approaches, we can employ two different metrics, as suggested in [8]:

**Keypoint Threshold (KT).** The metric, as employed by [34], addresses the quality of the keypoint predictions as a pixel-wise measure separately for every landmark $j \in J$. An estimated joint location $j \in \mathbb{R}^2$ is evaluated as correct if the Euclidean distance to the ground truth annotation $\hat{j} \in \mathbb{R}^2$ is lower than a fixed pixel threshold $T \in \mathbb{R}$:

$$\| j - \hat{j} \| < T .$$

**Strict Percentage of Correct Pose (strict PCP).** This measure evaluates the part-based quality of the prediction for an articulated object. Let the part be connected by two joints $j_1, j_2 \in \mathbb{R}^2$. Then, the part prediction is evaluated as correct only if *both* Euclidean distances of the estimated landmark locations $j_1, j_2$ to their respective ground truths $\hat{j}_1, \hat{j}_2$ are lower than a threshold defined as a function of the ratio $\alpha \in \mathbb{R}$ times the ground truth length of the part. In other words, both of the following equations have to be fulfilled:

$$\| j_1 - \hat{j}_1 \| < \alpha \cdot \| \hat{j}_1 - \hat{j}_2 \|$$
$$\| j_2 - \hat{j}_2 \| < \alpha \cdot \| \hat{j}_1 - \hat{j}_2 \| .$$

In the field of human pose estimation, the threshold value is usually set to $\alpha = 0.5$ [35]. For RM sequences, a suitable threshold value was determined in [6–8].

In the following, we present the results of the experiments that are performed for both pipeline approaches (Sections 4.3.5 and 4.3.6). For the initialization of the method, we use the ground truth annotation of the preceded frame.

## 4.4.1 Comparison to the State of the Art

On the *public dataset*, we compare the performance of the suggested methods with the state-of-the-art methods DDVT [34], MI [20], and ITOL [22]. In the first experiment, we consider only one sequence per evaluation by splitting the respective video into two halves. The forests are learned on the first half of the sequence and evaluated on the second half. In this way, we can evaluate whether the algorithms can generalize to unseen situations with very limited background variations. Analogously to the reported results of this dataset [34,22], we evaluated keypoint metrics **KT** for the center landmark **C** for thresholds between 15 and 40 pixels. Throughout the experiments the proposed methods outperformed the

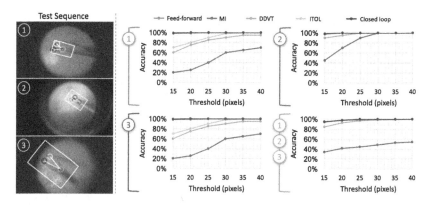

**Figure 4.5** *Comparison to the state of the art.* The feed-forward pipeline and the closed-loop approach are compared to the state-of-the-art methods MI [20], DDVT [34], and ITOL [22] by learning the forests on the first halves of the sequences and testing on the remaining halves. Please note that for an accuracy over 93%, the results are so close that the single graphs are not distinguishable.

state-of-the-art methods, as depicted in Fig. 4.5, reaching an accuracy of over 94% in every sequence. In the second experiment the first half of each of the sequences is included for training, and the learned forests are evaluated on each of the second halves. In this way, it is guaranteed that the forests cannot specialize on a particular sequence. Also, in this experiment setup both pipelines estimate the location of the center landmark of the instrument more precisely than the state-of-the-art methods. However, the difference between the feed-forward pipeline (Section 4.3.5) and the robust pipeline (Section 4.3.6) is marginal in both experiments.

## 4.4.2 Comparison of the Suggested Pipelines

On the *appearance* dataset, a cross-validation instead of a sequence-wise evaluation is performed in order to increase the problem complexity for the presented methods. In this setting, the training dataset for the forest is given by three of the sequences and the resulting forests are evaluated on the entire remaining one. In this way, we can investigate the generalization capacity of the proposed algorithm regarding different illumination conditions, noise levels, and even different instrument shapes. As depicted in Fig. 4.6, the advantages of the robust pipeline with online adaption (Section 4.3.6) are more evident now. In all test settings, this results in higher accuracy for all landmarks. Also in terms of the strict PCP score (Table 4.1), the results indicate that the closed-loop approach provides significant ben-

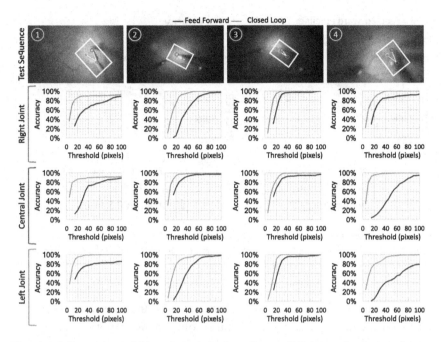

**Figure 4.6** *Comparison of the suggested pipelines.* Cross-validation evaluation on the *appearance* dataset; the offline forests are learned on three sequences and tested on the unseen one.

**Table 4.1**  Strict PCP for cross-validation of *Instrument* Dataset for **L**eft and **R**ight Fork

| Methods | Set I (L/R) | Set II (L/R) | Set III (L/R) | Set IV (L/R) |
|---|---|---|---|---|
| Closed loop | **89.0/88.5** | **98.5/99.5** | **99.5/99.5** | **94.5/95.0** |
| Feed forward | 69.7/58.5 | 93.94/93.43 | 94.47/94.47 | 46.46/57.71 |

efits as this method is twice as accurate as the feed–forward pipeline in one case.

### 4.4.3 Component Analysis for Robustness

In this section, we investigate the contribution of the components of the closed-loop approach to an increased robustness. To this end, we again carry out the experiments on the *appearance* dataset, where sequences I to III are used for training and sequence IV is used for testing. As depicted in Fig. 4.7, each component leads to an improved accuracy and contributes to a robust performance when combined. The results suggest that the implicit handshake between the pose estimator and the tracker via the weighted template averaging provides the strongest boost. However, it

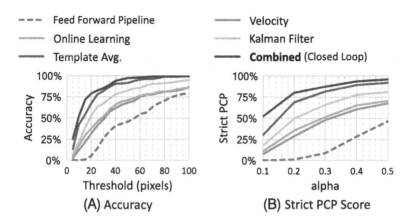

**Figure 4.7** *Component evaluation.* All added components of the robust pipeline contribute to an improved performance in terms of keypoint accuracy (A) and part-based accuracy (B).

has to be considered that the utilized sequences of 200 frames are rather limited and consequently the online learning component cannot realize its full potential in terms of contribution.

## 4.5 CONCLUSION AND FUTURE WORK

In this chapter we elaborated the potential benefit and the specific requirements of assistive computer vision in medicine in the context of surgical instrument tracking. We showed examples of how to tackle the problem by breaking down the complexity into the two separate tasks of template tracking and 2D pose estimation. Based on these building blocks, we first presented a feed-forward pipeline and then showed how to increase the robustness. The result is an algorithm that runs in real time and is able to withstand the challenging conditions of RM, such as a high level of noise and various illumination changes.

Although the algorithms show promising performance in terms of accuracy, the translation to clinical practice is still challenging. One of the major drawbacks of the presented method is the manual initialization and the lack of a recovery procedure if there are tracking failures. By construction, the pose estimation relies on the bounding box provided by the tracker. Due to the template correction via template averaging, this dependency is reduced in the robust pipeline approach and the performance of the tracker is improved by the pose information. However, if the instrument tip is not captured at all by the bounding box, the pose estimation cannot infer the location of the instrument landmarks. A similar problem may occur if the instrument is not present for several subsequent frames. During the surgery, these scenarios imply that a manual user input would be necessary, which is not feasible in a real clinical setting. Somebody would have to select the bounding box around the instrument to start the tracking, and also would have to take care of re-initialization in the case of tracking failure. Consequently, it would be preferable to extend the approach to a fully automated system, as presented by Page et al. [36].

Another interesting future direction is to leverage more advanced machine learning techniques such as Deep Learning. In comparison to classical computer vision fields, the available datasets in medical applications are rather limited mainly due to patient data security and the need of an expert for annotations. Consequently a complete training dataset that captures the entire variety range, for example of possible backgrounds, illumination changes, and instrument appearances, is unlikely and poses challenges for a Deep Learning approach. A notable effort in this direction has been made in the sub-challenge on instrument segmentation and tracking of laparoscopic and robotic MIS instruments, which was part of the endoscopic vision challenge at MICCAI 2015. And in fact, successful Deep Learning approaches have been introduced [13,25,37,38], but have not been tested in real clinical settings yet. Another difficulty arises from the fact that the current clinically established microscope systems do not provide a powerful GPU, which may be necessary for the use of offline learned networks.

However, these problems can be solved, and assistive computer vision already has improved—and most likely will even increase its potential to improve—the clinical outcome by simplifying the surgical workflow and enabling advanced technologies to provide additional, valuable information during surgery.

## ACKNOWLEDGMENTS

The authors would like to thank Dr. Abouzar Eslami, Carl Zeiss Meditec AG, Munich, for the many topical discussions and suggestions.

## REFERENCES

[1] M. Leo, G. Medioni, M. Trivedi, T. Kanade, G. Farinella, Computer vision for assistive technologies, Computer Vision and Image Understanding 154 (2017) 1–15.

[2] O. Zettinig, A. Shah, C. Hennersperger, M. Eiber, C. Kroll, H. Kübler, et al., Multi-modal image-guided prostate fusion biopsy based on automatic deformable registration, International Journal of Computer Assisted Radiology and Surgery 10 (12) (2015) 1997–2007.

[3] F. Bogo, J. Romero, E. Peserico, M.J. Black, Automated detection of new or evolving melanocytic lesions using a 3D body model, in: International Conference on Medical Image Computing and Computer-Assisted Intervention, Springer, 2014, pp. 593–600.

[4] G. Pusiol, A. Esteva, S.S. Hall, M. Frank, A. Milstein, L. Fei-Fei, Vision-based classification of developmental disorders using eye-movements, in: International Conference on Medical Image Computing and Computer-Assisted Intervention, Springer, 2016, pp. 317–325.

[5] M. Esposito, B. Busam, C. Hennersperger, J. Rackerseder, N. Navab, B. Frisch, Multimodal US-gamma imaging using collaborative robotics for cancer staging biopsies, International Journal of Computer Assisted Radiology and Surgery (2016) 1–11.

[6] N. Rieke, D.J. Tan, M. Alsheakhali, F. Tombari, C. Amat di San Filippo, V. Belagiannis, et al., Surgical tool tracking and pose estimation in retinal microsurgery, in: Medical Image Computing and Computer-Assisted Intervention, MICCAI 2015, Springer, 2015, pp. 266–273.

[7] N. Rieke, D.J. Tan, C. Amat di San Filippo, F. Tombari, M. Alsheakhali, V. Belagiannis, et al., Real-time localization of articulated surgical instruments in retinal microsurgery, Medical Image Analysis (2016) 34.

[8] N. Rieke, D.J. Tan, F. Tombari, J. Page Vizcaino, C. Amat-di San Filippo, A. Eslami, N. Navab, Real-time online adaption for robust instrument tracking and pose estimation, in: MICCAI, 2016.

[9] J. Ehlers, P.K. Kaiser, S.K. Srivastava, Intraoperative optical coherence tomography using the RESCAN 700: preliminary results from the DISCOVER study, British Journal of Ophthalmology (2014) 1329–1332.

[10] N. Rieke, S. Duca, N. Navab, A. Eslami, Automatic iOCT positioning during membrane peeling via real-time high resolution surgical forceps tracking, in: International Society of Imaging in the Eye Conference, ARVO 2016, Association for Research in Vision and Ophthalmology, Seattle, USA, 2016.

[11] H. Roodaki, K. Filippatos, A. Eslami, N. Navab, Introducing augmented reality to optical coherence tomography in ophthalmic microsurgery, in: IEEE International Symposium on Mixed and Augmented Reality, ISMAR 2015, Fukuoka, Japan, September 29–October 3, 2015, 2015, pp. 1–6.

[12] C. Bergeles, B.E. Kratochvil, B.J. Nelson, Visually servoing magnetic intraocular microdevices, IEEE Transactions on Robotics 28 (4) (2012) 798–809.

[13] C. Rupprecht, C. Lea, F. Tombari, N. Navab, G.D. Hager, Sensor substitution for video-based action recognition, in: Intelligent Robots and Systems, IEEE, 2016.

[14] N. Stylopoulos, S. Cotin, S. Dawson, M. Ottensmeyer, P. Neumann, R. Bardsley, M. Russell, P. Jackson, D. Rattner, CELTS: a clinically-based Computer Enhanced Laparoscopic Training System, in: Medicine Meets Virtual Reality 11: NextMed: Health Horizon, vol. 94, 2003, p. 336.

[15] D. Bouget, M. Allan, D. Stoyanov, P. Jannin, Vision-based and marker-less surgical tool detection and tracking: a review of the literature, Medical Image Analysis (2017) 35.

[16] Y.M. Baek, S. Tanaka, H. Kanako, N. Sugita, A. Morita, S. Sora, et al., Full state visual forceps tracking under a microscope using projective contour models, in: Proceedings of IEEE ICRA, 2012, pp. 2919–2925.

[17] A. Reiter, P.K. Allen, T. Zhao, Feature classification for tracking articulated surgical tools, in: N. Ayache, H. Delingette, P. Golland, K. Mori (Eds.), MICCAI 2012, Part II, in: Lecture Notes in Computer Science, vol. 7511, Springer, Heidelberg, 2012, pp. 592–600.

[18] M. Allan, S. Ourselin, S. Thompson, D.J. Hawkes, J. Kelly, D. Stoyanov, Toward detection and localization of instruments in minimally invasive surgery, IEEE Transactions on Biomedical Engineering 60 (2013) 1050–1058.

[19] M. Allan, P.L. Chang, S. Ourselin, D.J. Hawkes, A. Sridhar, J. Kelly, D. Stoyanov, Image based surgical instrument pose estimation with multi-class labelling and optical flow, in: Medical Image Computing and Computer-Assisted Intervention, MICCAI 2015, Springer, 2015, pp. 331–338.

[20] R. Richa, M. Balicki, E. Meisner, R. Sznitman, R. Taylor, G. Hager, Visual tracking of surgical tools for proximity detection in retinal surgery, in: IPCAI, 2011, pp. 55–66.

[21] C.J. Chen, W.W. Huang, K.T. Song, Image tracking of laparoscopic instrument using spiking neural networks, in: ICCAS 2013, 2013, pp. 951–955.

[22] Y. Li, C. Chen, X. Huang, J. Huang, Instrument tracking via online learning in retinal microsurgery, in: P. Golland, et al. (Eds.), MICCAI 2014, in: Lecture Notes in Computer Science, vol. 8673, Springer, Heidelberg, 2014, pp. 464–471.

[23] R. Sznitman, A. Basu, R. Richa, J. Handa, P. Gehlbach, R.H. Taylor, et al., Unified detection and tracking in retinal microsurgery, in: G. Fichtinger, A. Martel, T. Peters (Eds.), MICCAI 2011, Part I, in: Lecture Notes in Computer Science, vol. 6891, Springer, Heidelberg, 2011, pp. 1–8.

[24] R. Sznitman, C. Becker, P. Fua, Fast part-based classification for instrument detection in minimally invasive surgery, in: P. Golland, et al. (Eds.), MICCAI 2014, in: Lecture Notes in Computer Science, vol. 8673, Springer, Heidelberg, 2014, pp. 692–699.

[25] D. Sarikaya, J.J. Corso, K.A. Guru, Detection and localization of robotic tools in robot-assisted surgery videos using deep neural networks for region proposal and detection, IEEE Transactions on Medical Imaging (2017).

[26] V. Belagiannis, C. Amann, N. Navab, S. Ilic, Holistic human pose estimation with regression forests, in: F.J. Perales, J. Santos-Victor (Eds.), AMDO 2014, in: Lecture Notes in Computer Science, vol. 8563, Springer, Heidelberg, 2014, pp. 20–30.

[27] D.J. Tan, S. Ilic, Multi-forest tracker: a chameleon in tracking, in: CVPR 2014, 2014, pp. 1202–1209.

[28] L. Breiman, Random forests, Machine Learning 45 (1) (2001) 5–32.

[29] N. Dalal, B. Triggs, Histograms of oriented gradients for human detection, in: IEEE Computer Society Conference on Computer Vision and Pattern Recognition, vol. 1, CVPR 2005, IEEE, 2005, pp. 886–893.

[30] P.F. Felzenszwalb, R.B. Girshick, D. McAllester, D. Ramanan, Object detection with discriminatively trained part-based models, IEEE Transactions on Pattern Analysis and Machine Intelligence 32 (9) (2010) 1627–1645.

[31] M. Eitz, K. Hildebrand, T. Boubekeur, M. Alexa, Sketch-based image retrieval: benchmark and bag-of-features descriptors, IEEE Transactions on Visualization and Computer Graphics 17 (11) (2011) 1624–1636.

[32] M.E. Nilsback, A. Zisserman, Automated flower classification over a large number of classes, in: Sixth Indian Conference on Computer Vision, Graphics & Image Processing, 2008, ICVGIP'08, IEEE, 2008, pp. 722–729.

[33] S.S. Haykin, Kalman Filtering and Neural Networks, J. Wiley & Sons, Inc., 2001.

[34] R. Sznitman, K. Ali, R. Richa, R.H. Taylor, G.D. Hager, P. Fua, Data-driven visual tracking in retinal microsurgery, in: N. Ayache, H. Delingette, P. Golland, K. Mori (Eds.), MICCAI 2012, Part II, in: Lecture Notes in Computer Science, vol. 7511, Springer, Heidelberg, 2012, pp. 568–575.

[35] M.R. Pickering, A.A. Muhit, J.M. Scarvell, P.N. Smith, A new multi-modal similarity measure for fast gradient-based 2D–3D image registration, in: EMBC 2009, 2009, pp. 5821–5824.

[36] J. Page Vizcaíno, N. Rieke, D.J. Tan, F. Tombari, A. Eslami, N. Navab, Automatic initialization and failure detection for surgical tool tracking in retinal microsurgery, in: Bildverarbeitung für die Medizin 2017: Algorithmen-Systeme-Anwendungen, Proceedings des Workshops vom 12. bis 14. März 2017, Springer, Heidelberg, 2017, p. 346.

[37] I. Laina, N. Rieke, C. Rupprecht, J. Page Vizcaíno, A. Eslami, F. Tombari, N. Navab, Concurrent segmentation and localization for tracking of surgical instruments, in: MICCAI, 2017, pp. 664–672.

[38] T. Kurmann, P. Marquez, X. Du, P. Fua, D. Stoyanov, S. Wolf, R. Sznitman, Simultaneous recognition and pose estimation of instruments in minimally invasive surgery, in: MICCAI, 2017, pp. 505–513.

# CHAPTER 5

# Computer Vision for Human–Machine Interaction

**Qiuhong Ke\*, Jun Liu†, Mohammed Bennamoun\*, Senjian An\*, Ferdous Sohel‡,\*, Farid Boussaid\***

\*The University of Western Australia, Perth, WA, Australia
†Nanyang Technological University, Singapore
‡Murdoch University, Murdoch, WA, Australia

## Contents

## Abstract

Human–machine interaction (HMI) refers to the communication and interaction between a human and a machine via a user interface. Nowadays, natural user interfaces such as gestures have gained increasing attention as they allow humans to control machines through natural and intuitive behaviors. In gesture-based HMI, a sensor such as Microsoft Kinect is used to capture the human postures and motions, which are processed to control a machine. The key task of gesture-based HMI is to recognize the

**Computer Vision for Assistive Healthcare.**
DOI: https://doi.org/10.1016/B978-0-12-813445-0.00005-8

meaningful expressions of human motions using the data provided by Kinect, including RGB (red, green, blue), depth, and skeleton information. In this chapter, we focus on the gesture recognition task for HMI and introduce current deep learning methods that have been used for human motion analysis and RGB-D-based gesture recognition. More specifically, we briefly introduce the convolutional neural networks (CNNs), and then present several deep learning frameworks based on CNNs that have been used for gesture recognition by using RGB, depth and skeleton sequences.

### Keywords

Human–machine interaction (HMI), Motion analysis, Gesture recognition, Deep learning

## 5.1 BACKGROUND OF HUMAN–MACHINE INTERACTION

Human–machine interaction (HMI) refers to the interaction and communication between humans and machines [1]. HMI is a multidisciplinary field, which includes human–robot interaction, human–computer interaction, artificial intelligence, and robotics. HMI is traditionally applied in industrial plants for efficiency, quality, and process safety. Nowadays, HMI is widely used for medical, transportation, and entertainment systems.

### 5.1.1 Human–Machine Interfaces

Humans interact with machines via interfaces. Traditional user interfaces are physical parts of machines that can be seen and touched. Users have to undergo training before being able to operate the machines. Take the computer as an example. In early digital computers, the interfaces were non-interactive batch interfaces consisting of punch cards. Later on, with the advent of command-line interfaces, users could interactively operate the computers using request–response transactions. In the early 1970s, the first graphical user interface was created to make the interaction between humans and computers more efficient and easy. Compared to the command-line interface, the graphical user interface does not require users to memorize commands [2].

Another type of user interface is invisible natural user interfaces (NUIs). An NUI is a user interface that allows users to rely on their natural and intuitive everyday behavior to perform interactions. Instead of forcing users to learn rules of operation, NUIs focus on understanding the expressions of the users' behaviors [2].

## 5.1.2 Gesture-Based Human–Machine Interaction

One example of an NUI is gestures. Gesture-based HMI has been used since the 1980s. At that time, humans needed to wear gloves such as the DataGlove and Z-Glove [3] to perform gesture-based interactions. These gloves are used to measure the motion, orientation, and position of the fingers using the sensors inside the gloves.

Gesture-based HMI has been used in military applications. For example, the robot is controlled to mimic the motions of the operator and perform dangerous tasks such as bomb disposal, thus minimizing injuries in these tasks [4]. Gesture-based HMI is also very important in industrial applications. Machines with gesture-based HMI enable users to perform open/close, on/off, and other operations using gestures, which reduces contamination risks and helps cleaning efforts [5]. Gesture-based HMI is now very popular in video game consoles. For example, the Nintendo Wii [6] has a remote controller that includes an infrared camera and accelerometers to track human motion. Bluetooth is used to wirelessly transmit data, thus allowing the users to manipulate objects on the screen using gestures. Similar to the Wii Remote, the Sony PlayStation Move [7] contains a motion controller that tracks the users in three dimensions using an accelerometer, a gyroscope, and a magnetic sensor. The system also contains a video camera to handle the cursor movement, aiming, and other motions. Compared to the Wii and PlayStation Move, the Microsoft Kinect for Xbox 360 [8] is camera-only and allows users to interact with machines without a physical controller.

In this chapter, we focus on the task of gesture-based HMI, which aims to recognize the whole-body human gestures captured by the Kinect. The system of gesture-based HMI consists of the acquisition of human gestures and gesture recognition algorithms. We first briefly explain the data acquisition process using Kinect, and then introduce current deep learning frameworks to solve the gesture recognition task for HMI.

## 5.2 DATA ACQUISITION FOR GESTURE RECOGNITION

There are two main methods for acquiring data for gesture recognition. One is to use accelerometer-based devices such as data gloves to obtain data for gesture recognition [9–12]. Accelerometer-based devices need complex calibration approaches. Another method is to use vision-based devices to capture human gestures. Vision-based devices are less invasive, more human-centered, and more flexible [13]. The most popular vision-based device is Kinect due to its ability to acquire different modalities, including RGB, depth, and skeleton sequences [14]. The first version of Kinect uses a structured light method to estimate the depth of the scene. More specifically, the Kinect device contains a near-infrared (NIR) laser projector and an NIR camera. The NIR projector sends a fixed and known dot pattern to the scene. The NIR camera captures the deformed patterns on the surface of the object in the scene, which is then used to estimate the depth information using triangulation techniques. The second generation of Kinect computes the depth information based on the Time-of-Flight (ToF) method, i.e. the depth information is estimated by measuring the time the laser light travels from the projector to the scene. The ToF provides depth information in real time and it is easy to capture human motions for gesture recognition in HMI [15]. Once the depth map is obtained, it can then be used to generate the human skeleton. The skeleton is estimated in two steps [16]: (1) the body part image is estimated from the depth image, and (2) the skeleton is estimated from the body part image. To obtain the body part image from the depth image, a motion capture system is first used to capture 10,000 depth images with known skeletons. The data samples are extended to one million by rendering each image to more images using computer graphics techniques. The samples are used to learn a randomized decision forest that maps the depth images to body parts. The body part images are transformed to a skeleton using the mean shift algorithm. For each frame, 20 skeleton joints are estimated. Each joint is provided with the $x$, $y$, and $z$ coordinates.

## 5.3 COMPUTER VISION-BASED GESTURE RECOGNITION

The RGB, depth, and skeleton sequences obtained by Kinect provide the gesture information of the whole human body. In this section, we introduce RGB-D gesture recognition using computer vision algorithms. The aim of computer vision is to provide computers with the same high-level understanding of videos and images as humans have [17]. Computer vision tasks such as gesture recognition are very challenging due to a number of intraclass variations (e.g. viewpoint differences and temporal variations). Traditional methods focus on designing features such as SIFT [18] and HOG [19] to represent the raw data (i.e. pixels), and then use classifiers to solve the problems. These traditional hand-crafted features rely on expert knowledge and extensive human labor as it is difficult to know what is the best feature. Nowadays, deep learning has achieved great success in computer vision. The performance of many tasks has been dramatically improved [20]. Compared to traditional methods, deep learning acquires representations directly from the data in a hierarchical way and exploits more useful information. One of the most widely used deep learning networks in the computer vision community is Convolutional Neural Networks (CNNs) [21]. Compared to traditional neural networks with fully connected layers, CNNs are easier to train and are capable of generalizing well [20]. In this section, we introduce deep learning frameworks based on CNNs to learn features of sequences for gesture recognition. Since the methods are based on CNNs, we first briefly introduce some concepts of CNNs, and then present RGB-based, depth-based, and skeleton-based gesture recognition methods using CNNs.

## 5.3.1 Convolutional Neural Networks

A CNN is made up of multiple layers of neurons, each of which is a non-linear operation on a linear transformation of the preceding layer's outputs. The layers mainly include convolutional layers and pooling layers. The convolutional layers have weights that need to be trained, while the pooling layers transform the activation using a fixed function.

### 5.3.1.1 Convolutional Layers

A convolutional layer contains a set of filters whose parameters need to be learned. The height and weight of the filters are smaller than those of the input volume. Each filter is convolved with the input volume to compute an activation map made of neurons. In other words, the filter is slid across the width and height of the input and the dot products between the input and filter are computed at every spatial position. The output volume of the convolutional layer is obtained by stacking the activation maps of all filters along the depth dimension. Since the width and height of each filter is designed to be smaller than the input, each neuron in the activation map is only connected to a small local region of the input volume. In other words, the receptive field size of each neuron is small, and is equal to the filter size. The local connectivity is motivated by the architecture of the animal visual cortex [22] where the receptive fields of the cells are small. The local connectivity of the convolutional layer allows the network to learn filters which maximally respond to a local region of the input, thus exploiting the spatial local correlation of the input (for an input image, a pixel is more correlated to the nearby pixels than to the distant pixels). In addition, as the activation map is obtained by performing convolution between the filter and the input, the filter parameters are shared for all local positions. The weight sharing reduces the number of parameters for efficiency of expression, efficiency of learning, and good generalization.

### 5.3.1.2 Pooling Layers

A pooling layer is usually incorporated between two successive convolutional layers. The pooling layer reduces the number of parameters and computation by down-sampling the representation. The pooling function can be max or average. Max pooling is commonly used as it works better [23].

## 5.3.2 RGB-Based Gesture Recognition

Complex human gestures might last for a long time and contain multiple temporal stages with dynamic human postures and motions. Given a gesture video with multiple frames, the temporal information needs to be exploited for a good understanding of the human gesture. Regular CNNs learn information from a single image and are incapable of learning long-range temporal structures. Their inability to model the long-range information in the videos leads to unexpected results of gesture recognition. Recently, CNNs have been successfully designed to model the temporal information in the videos. In this section, we introduce the two methods for gesture recognition. Both methods consist of two streams of CNNs, i.e. a spatial stream, which is trained using the original RGB frames, and a temporal stream, which is trained using the optical flow images. The optical flow images are computed from the consecutive frames. We first briefly introduce the method of extracting optical flow, and then introduce the temporal segmentation networks [24] and temporal evolution networks [25] for gesture recognition.

### 5.3.2.1 Extracting Optical Flow

Optical flow represents the motion of the scene context relative to an observer [26]. For a human gesture, a single frame with only a static scene can introduce ambiguity which makes the inference of the gesture class difficult. The motion of human bodies needs to be exploited for a good understanding of the gesture class. There are many methods of estimating the optical flow between two frames, including differential-based, region-based, energy-based, and phase-based methods [27]. The most widely used method is the differential method [28], which is based on the assumption of image brightness constancy. Given a video sequence, let the intensity of a voxel at position $(x, y)$ of the $t$th frame be $I(x, y, t)$. According to the brightness constancy assumption, the intensity of the voxel remains the same despite small changes of position and time period. More specifically,

$$I(x, y, t) = I(x + \delta x, y + \delta y, t + \delta t), \tag{5.1}$$

where $(\delta x, \delta y, \delta t)$ is the small change of the movement. $I(x + \delta x, y + \delta y, t + \delta t)$ can be expressed with a Taylor series expansion:

$$I(x + \delta x, y + \delta y, t + \delta t) \approx I(x, y, t) + \frac{\partial I}{\partial x}\delta x + \frac{\partial I}{\partial y}\delta y + \frac{\partial I}{\partial t}\delta t. \tag{5.2}$$

Therefore,

$$\frac{\partial I}{\partial x}\delta x + \frac{\partial I}{\partial y}\delta y + \frac{\partial I}{\partial t}\delta t \approx 0, \tag{5.3}$$

$$\frac{\partial I}{\partial x}U_x + \frac{\partial I}{\partial y}U_y + \frac{\partial I}{\partial t} \approx 0, \tag{5.4}$$

where $U_x = \frac{\delta x}{\delta t}$, $U_y = \frac{\delta y}{\delta t}$ are two components of the optical flow of the voxel $(x, y, t)$.

Once the optical flow is computed, it can be used to learn video-level representation for gesture recognition. Traditional methods used histograms of optical flow to represent the temporal information [29]. Using the histogram can result in the lose the structure of the data [30]. We represent optical flow as images that are then fed to a CNN to learn high-level temporal features for gesture recognition.

### 5.3.2.2 Temporal Segmentation Networks

An overall architecture of temporal segmentation networks is shown in Fig. 5.1. The goal of temporal segmentation networks is to model the temporal dynamics of an entire video, and to perform a video-level prediction. The networks are comprised of two streams of CNNs: a spatial stream CNN and a temporal stream CNN. The input of the two streams are snippets consisting of several video frames and optical flow images. The snippets are sampled from multiple temporal segments of an entire video, i.e. a video is divided into several segments and a snippet is randomly sampled from each segment. For each stream, the network outputs a frame-level preliminary prediction for each snippet. The preliminary prediction of all

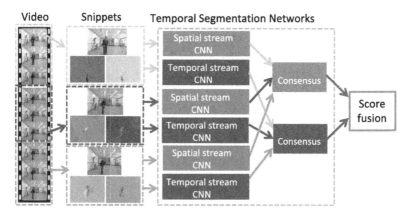

**Figure 5.1** Overall architecture of temporal segmentation networks. An entire video is divided into several segments. A snippet is randomly sampled from each segment. The snippets include the original RGB frames and the optical flow images. The frames and the optical flow images are separately fed to the spatial stream and the temporal stream CNNs with weight sharing. The class scores of each snippet are combined using a consensus function to generate a video-level prediction. During testing, the predictions of different modalities are weight averaged for the final determination of the class. (Figure adapted from [24].)

snippets are combined to generate a video-level prediction using a consensus function. During training, the video-level predictions rather than the preliminary frame-level predictions are used to compute the loss value to update the network parameters.

More specifically, given a video $D$, this method first segments the video into $n$ segments. From each segment, a snippet is randomly selected. Let them be $X_1, X_2, \cdots, X_n$. The snippets are fed to $n$ parallel CNNs with weight sharing. Let the output of the $t$th CNN be $\mathbf{z}_t = f(W, X_i)$. $W$ represents the parameters of each CNN. The outputs of all segments are preliminary predictions, which are further combined using a consensus function to generate a video-level prediction. In the paper the maximum, weighted averaging, and evenly averaging functions are used as the consensus functions. Of the three, the evenly averaging function achieves the best performance. The video-level prediction is fed to a softmax layer to predict the probability of each class. The network loss is computed as

$$\mathcal{L}(\mathbf{s}, \mathbf{y}) = \sum_{j=1}^{m} y_j \left( \log \sum_{i=1}^{m} \exp s_i - s_j \right), \qquad (5.5)$$

where $\mathbf{s} = \frac{1}{n} \sum_{t=1}^{n} \mathbf{z}_t$, which represents the video-level prediction; $m$ is the number of classes; and $y_j$ is the ground truth label for class $j$.

The network parameters are optimized using a standard back-propagation algorithms. Since the predictions of all snippets are used to jointly update the parameters, the networks learn information from the entire video, thus exploiting the long-term video structure. During training, each video is divided into three segments. Thus, only three frames are used for learning which can reduce the computation cost. The architecture of CNNs are adapted from the Inception with Batch Normalization (BN-Inception) [31] because it has a good balance between efficiency and accuracy [24]. The input of the spatial stream is a single RGB frame image, which is a 3D volume with three channels. For the temporal stream, the input is the combination of optical flow images. More specifically, the optical flow fields $x$ and $y$ computed from every two consecutive frames are transformed into two gray images with values ranging from 0 to 255 using a linear transformation. The two gray images are considered a set of optical flow images. Then 10 sets of optical flow images are stacked as an input. Thus, the input of the temporal stream is a 3D volume of 20 channels.

To avoid overfitting in the training of the deep networks, the ImageNet [32] is used to pre-train the spatial stream CNN. For the temporal stream CNN, cross-modality pre-training is adopted. More specifically, the spatial network is used to initialize the temporal network. Since the input channel of the temporal network is 20, the weights of the first convolutional layer of the spatial network is averaged and repeated for 20 times to handle the input of the temporal stream. In addition to network pre-training, data augmentation is also performed to train the spatial and temporal networks. The augmentation process includes corner cropping, horizontal flipping, and scale jittering [33]. First, each frame image and optical flow image are resized to $256 \times 340$. Then five patches are cropped from the four corners and the center with size randomly selected from $[256, 224, 192, 168]$, followed by random horizontal flipping. The cropped patches are then resized to $224 \times 224$ for network training.

During testing, 25 RGB frames and 25 stacking optical flow images are first sampled from each video. Each of the RGB frames and optical flow images are then separately fed to the spatial and temporal network to produce frame-level results. For the spatial network and temporal network, the 25 results are averaged before Softmax normalization to output the video-level prediction. Finally, the prediction scores of the spatial stream

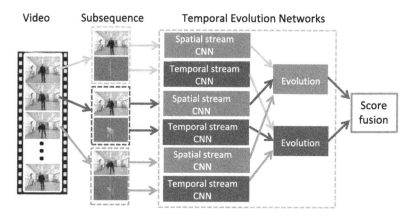

**Figure 5.2** Overall architecture of temporal evolution networks. The inputs are sub-sequences of multiple consecutive frames and optical flow images. Each frame and optical flow image is fed to deep networks with weight sharing. The outputs are high-level feature representations, which are then fused in an evolution block to output a sequence-level prediction. During testing, the prediction scores of different modalities are combined to determine the final class. (Figure adapted from [25].)

and temporal stream are averaged as the final prediction to determine the gesture class.

### 5.3.2.3 Temporal Evolution Networks

Temporal segmentation networks aim to learn the global temporal information of the entire video, while temporal evolution networks [25] focus on leaning the local temporal information. Temporal evolution networks are motivated by the observation that learning the local evolution of human postures provides more useful information than a single frame that contains only static human postures. The inputs of temporal evolution networks are local partial videos. Thus, temporal evolution networks are particularly useful for online recognition and prediction when the entire video is not available.

An overall architecture of the temporal evolution networks is shown in Fig. 5.2. The original temporal evolution is only learned from optical flow images for human interaction prediction [25]. We describe temporal evolution networks which consist of spatial stream and temporal stream CNNs in order to learn both the spatial and temporal information for a better understanding of the video class. The inputs of the two stream CNNs are sub-sequences consisting of multiple consecutive frames and optical flow images. The local sub-sequences contain fine details of human postures and motions which are very important for gesture recognition. The goal of temporal evolution networks is to use sub-sequences to learn the local evolution of human postures for gesture recognition. Each frame and optical flow image is separately fed to spatial stream and temporal stream networks to learn a frame-level feature representation. The frame-level representations of all the frames are then fused in an evolution block. It includes a temporal convolutional layer to generate a sequence-level representation, followed by a fully connected layer and a softmax layer to output class scores.

More specifically, let an input sub-sequence be $X_1, X_2, \cdots, X_t$. The frame-level feature representation of each frame is $\mathbf{u}_1, \mathbf{u}_2, \cdots, \mathbf{u}_t$, where $\mathbf{u}_i = f(W, X_i)$. $W$ denote the parameters of the CNN. The multiple frame-level feature vectors are then fused with another convolutional layer to output a sequence-level representation $\mathbf{v}$ which is given by

$$
\mathbf{v} = \begin{bmatrix} \mathbf{u}_1 & \mathbf{u}_2 & \cdots & \mathbf{u}_t \end{bmatrix} \begin{bmatrix} \gamma_1 \\ \gamma_2 \\ \vdots \\ \gamma_t \end{bmatrix}, \tag{5.6}
$$

where $\boldsymbol{\gamma} = [\gamma_1, \gamma_2, \cdots, \gamma_k]^T$ denotes the parameters of the convolutional filter which need to be learned; $\mathbf{v}$ is then fed to a fully connected layer and Softmax layer to generate class scores.

During training, the length of sub-sequence is set to 3 and 7. Among them, 7 achieves the best performance. Longer sub-sequences perform better, but also increase the computational cost. The CNN network is adapted from the CNN-M-2048 network [34]. The input of the spatial stream is the consecutive RGB frame images. For the temporal stream, the optical flow components $x$ and $y$ are combined to a flow coding image by computing the magnitude and the angle of each optical flow vector. Another simple alternative is to set the two components of the optical flow as two channels of an image and to set the third channel to 0. Thus, both the spatial and temporal stream have the same network architecture, which are pretrained using the ImageNet [32] to avoid overfitting. During testing, the scores of all the sub-sequences are averaged as the video-level prediction. The importance of spatial and temporal information might be different in gesture inference. The weights between the two streams can be learned using ranking SVM to effectively fuse the two models [35].

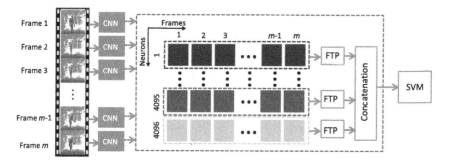

**Figure 5.3** Overall architecture of depth-based gesture recognition. Each frame of the depth sequence is fed to a view-invariant CNN which is learned using samples of different viewpoints. The output of the CNN for each frame is a high-level feature vector which is robust to viewpoints. The feature vector of each frame contains 4096 neurons. The time series of each neuron is used to learn Fourier Temporal Pyramid (FTP) features. The FTP features of all neurons are concatenated as the final representation of the video, which is then used to perform classification using SVM. (Figure adapted from [36].)

### 5.3.3 Depth-Based Gesture Recognition

Given a depth sequence, the RGB-based sequence learning methods (i.e. temporal segmentation network [24] and temporal evolution networks [25]) introduced in Section 5.3.2 can also be used for gesture recognition. In this section, we describe another method which focuses on multiview gesture recognition [36]. An overall architecture of the method is shown in Fig. 5.3. The main idea is to learn a feature representation of each frame using a view-invariant CNN, and then model the temporal structure of the entire video using a Fourier Temporal Pyramid (FTP) [37].

#### 5.3.3.1 View-Invariant Pose Representation

Human gestures might be captured from different viewpoints, which results in large intra-class variations and makes it challenging for gesture recognition. In this section, we describe a view-invariant CNN, which is capable of extracting a high-level view-invariant representation of human postures. The training of the view-invariant CNN requires that the training data contain a large number of viewpoints. Due to the lack of this dataset, a skeleton dataset is used to generate synthetic training data to train the CNN model. More specifically, the skeleton data is rendered from 180 different viewing directions. Each skeleton is fitted with a smooth surface and is then converted to depth images by normalizing the value into the interval between 0 and 255. The generated depth images are then used to train a deep CNN network. The architecture is adapted from a network used in [38]. The CNN classifies a posture of different viewpoints as the same class in order to learn a high-level view-invariant representation.

### 5.3.3.2 Temporal Modeling

As shown in Fig. 5.3, given a depth sequence, the first step is to feed each frame to the trained view-invariant CNN to extract a high-level view-invariant representation. The output of the fully connected layer fc7 is used as the feature representation of the input depth image. The feature vector of each frame contains 4096 neurons. The features of all frames produce 4096 time series of neurons. The next step is to feed the time series of each neuron to FTP to learn the temporal information because the representation of each frame contains only the information of the static human posture. The temporal information of the entire video needs to be exploited for gesture recognition.

More specifically, given the CNN features of a depth video, let it be $[\mathbf{v}_1, \mathbf{v}_2, \cdots, \mathbf{v}_t]$; $t$ is the number of frames of the video; $\mathbf{v}_i = [q_{i,1}, q_{i,1}, \cdots, q_{i,m}]^T$ is the CNN feature of the $i$th frame, which consists of $m$ neurons. The output of the fc7 layer is a 4096D feature vector, thus $m = 4096$. Combing all the frames, the time series of the $j$th neuron is given by $[q_{1,j}, q_{2,j}, \cdots, q_{t,j}]$. The sequence of each neuron is used for the Fourier transform [39] in hierarchical levels. Take the sequence of the $j$th neuron as an example. In the first level, the whole sequence $\mathbf{v}_0 = [q_{1,j}, q_{2,j}, \cdots, q_{t,j}]$ is used to perform a Fourier transform and the first $p$ coefficient is used as the descriptor. Let the descriptor of the first level be $\mathbf{d}_{j,1}$. In the second level, $\mathbf{v}_0$ is divided into two segments $\mathbf{v}_{11} = [q_{1,j}, q_{2,j}, \cdots, q_{s,j}]$ and $\mathbf{v}_{12} = [q_{s+1,j}, q_{s+2,j}, \cdots, q_{t,j}]$; $s = [\frac{t}{2}]$; $\mathbf{v}_{11}$ and $\mathbf{v}_{12}$ are separately used to perform the Fourier transform. For each segment, the first $p$ coefficient is used as the descriptor. The descriptors of the two segments are then concatenated as the representation of the second level. Let it be $\mathbf{d}_{j,2}$. Each of the two segments $\mathbf{v}_{11}$ and $\mathbf{v}_{12}$ are then divided into two sub-segments in the next level. Thus, in the third level, there are four segments. As in the previous levels, the four segments are also used to perform a Fourier transform and the first $p$ Fourier coefficient of each segment is extracted. The representation of the third level $\mathbf{d}_{j,3}$ is the concatenation of the descriptors of the four segments. The segment can be divided into sub-segments to extract features. The final representation of the sequence of the $j$th neuron $\mathbf{d}_j$ is given by the concatenation of the descriptors of all levels $\mathbf{d}_j = [\mathbf{d}_{j,1}, \mathbf{d}_{j,2}, \cdots, \mathbf{d}_{j,l}]^T$; $l$ is the level number; $\mathbf{d}_{j,l} \in R^{c \times 1}$. $c = p \times 2^{l-1}$. All

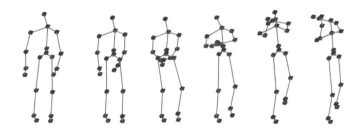

**Figure 5.4** Example of a human skeleton sequence (from G3D dataset [45]).

the $m$ neurons are concatenated as the final spatial–temporal representation of the video sequence $\mathbf{d} = [\mathbf{d}_1, \mathbf{d}_j, \cdots, \mathbf{d}_m]^T$. The representations of all the videos can be used for classification using SVM.

## 5.3.4 Skeleton-Based Gesture Recognition

Compared to the RGB and depth data, the skeleton data only provides the trajectory of discrete human joints. However, skeleton data can also be used for human gesture recognition [40–43]. As shown in Fig. 5.4, the human skeleton joints in each frame depicts the information of five body parts including the trunk, the left hand, the right hand, the left leg, and the right leg. Each part has its own specific structure to form a human posture. In this section, we introduce a SkeletonNet [44] for skeleton-based gesture recognition. The main idea is to extract fine-grained local information of each body part, and then combine the local information of all the parts using CNN to generate a high-level representation for gesture recognition.

### 5.3.4.1 Robust Features of Body Parts

The structural layout of the five body parts (i.e. the trunk, the left hand, the right hand, the left leg, and the right leg), and the spatial relationship between two body parts is useful information for recognizing a human posture (e.g. the spatial structure of the hand in waving and the relationships of two hands in clapping). Given a human skeleton, any two joints can be connected as a vector (referred to as joint-joint vector). The structural feature of each body part can be described using the geometric relationships between the joint-joint vectors in the same body part, while the spatial relationship between two parts can be represented using the geometric relationships between the joint-joint vectors in the two body parts. As shown in Fig. 5.5, the geometric relationships, including the angle and the magnitude ratios between two joint-joint vectors, are different for two postures.

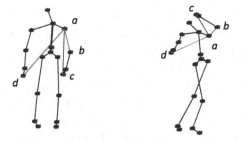

**Figure 5.5** Two different human postures (standing and bowling from the G3D dataset [45]) for the two with-part vectors $v_{ab}$ and $v_{ac}$ and the two between-part vectors $v_{ab}$ and $v_{ad}$. The magnitude ratio and angle between the two vectors are different in the two postures.

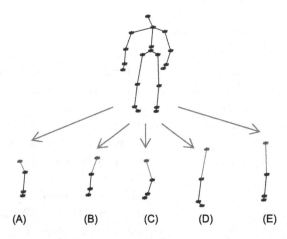

**Figure 5.6** A human skeleton is divided into five body parts: (A) the trunk, (B) the left hand, (C) the right hand, (D) the left leg, and (E) the right leg. A starting joint (shown in red) and a reference vector (shown in magenta) are selected for each body part. (Figure adapted from [44].) (For interpretation of the references to color in this figure, the reader is referred to the web version of this chapter.)

The geometric relationships of the joint–joint vectors in the same body part and different body parts depict a human posture, which provides an important clue for gesture recognition. In this section, we introduce robust body features based on the relationships between the joint–joint vectors in the same body part and different body parts.

As shown in Fig. 5.6, a starting joint (shown in red) is selected for each body part, including the head, the left shoulder, the right shoulder, the left hip, and the right hip. The starting joint of each part is connected with other joints to generate two types of joint-joint vectors: (1) within-part vectors, where the starting joint of each part is connected to the other joints of the same part, and (2) between-part vectors, where the starting joint of each part is connected to the other joints of a different part. A reference vector is also selected in each part, which is shown in magenta in Fig. 5.6. For the $k$th body part, let the set of within-part vectors be $\mathcal{V}_w^{(k)} \in \mathbb{R}^{m-1}$ and between-part vectors be $\mathcal{V}_b^{(k)} \in \mathbb{R}^{n-1}$; $m$ and $n$ are the numbers of joints in that body part and the whole human body. The cosine distances between all vectors of $\mathcal{V}^{(k)}$ and all vectors of $\mathcal{V}_w^{(k)} \cup \mathcal{V}_b^{(k)}$ are concatenated as the CD feature for the $k$th body part. The magnitude ratios of all vectors of $\mathcal{V}_w^{(k)} \cup \mathcal{V}_b^{(k)}$ to the reference vector of the $k$th body part are concatenated as the NM feature for the $k$th body part.

### 5.3.4.2 High-Level Feature Learning

Given a skeleton sequence, the CD and NM features of all frames are separately combined as a 2D array with size $p \times t$ for each body part; $p$ denotes the dimension of CD or NM and $t$ is the frame number of the sequence. The CD array and NM array of each body part represents a low-level spatial–temporal information of the sequence. A deep CNN is used to combine the CD and NM arrays of the five body parts. The CNN network is adapted from the CNN-M-2048 network [34], which is pre-trained using the ImageNet [32].

More specifically, the last fully connected layer and output layer of the CNN-M-2048 network [34] are discarded. Each CD and NM array is fed to the CNN model to generate a compact representation. The output feature vectors of the five CD arrays and the five NM arrays are separately concatenated as two feature vectors. The two features are separately fed to two fully connected layers. The two outputs are concatenated to generate a compact representation, which is fed to another fully connected layer and a softmax layer to output the class scores.

## 5.4 CONCLUSION

In this chapter, we focused on the task of gesture-based HMI and introduced several deep learning frameworks based on CNNs for gesture recognition using the RGB, depth, and skeleton sequences provided by Kinect. One challenge of gesture recognition from sequences is the temporal variance. The long-term temporal structure of the entire sequence is very important to gesture recognition. For RGB sequences, we introduced temporal segmentation networks [24] and temporal evolution networks [25] to learn the temporal structure of the sequence for gesture recognition. For depth sequences, we introduced a multiview CNN [36] to learn the posture from each frame, followed by a Fourier Temporal Pyramid (FTP) [37] for temporal modeling of the whole sequence for gesture recognition. For skeleton sequences, we introduced robust part features to exploit human postures from each frame and a deep CNN to generate a compact spatial–temporal representation for gesture recognition [44]. The prediction scores of RGB, depth, and skeleton can be fused using simple averaging or a linear SVM for a better performance. The application of gesture-based HMI requires both accuracy and efficiency of gesture recognition. Future works are needed for faster recognition models.

## ACKNOWLEDGMENTS

This work was partially supported by Australian Research Council grants DP150100294, DP150104251, and DE120102960.

## REFERENCES

[1] G. Johannsen, Human–machine interaction, in: Control Systems, Robotics and Automation, Volume XXI: Elements of Automation, 2009, p. 132.

[2] B. Blazica, The inherent context awareness of natural user interfaces: a case study on multitouch displays, Informatica 38 (4) (2014) 385.

[3] T.G. Zimmerman, J. Lanier, C. Blanchard, S. Bryson, Y. Harvill, A hand gesture interface device, in: ACM SIGCHI Bulletin, vol. 18, ACM, 1987, pp. 189–192.

[4] Robo Sally – bomb disposal robot, https://www.xsens.com/customer-cases/robo-sally-bomb-disposal-robot/ (Accessed 2 March 2017).

[5] https://www.renesas.com/en-sg/about/web-magazine/edge/solution/22-hmi-gesture-recognition.html (Accessed 2 March 2017).

[6] http://gizmodo.com/294642/unearthed-nintendos-pre-wiimote-prototype (Accessed 2 March 2017).

[7] https://en.wikipedia.org/wiki/PlayStation_Move#cite_note-E3_2010_PR-1 (Accessed 2 March 2017).

[8] https://en.wikipedia.org/wiki/Kinect (Accessed 2 March 2017).

[9] E.S. Choi, W.C. Bang, S.J. Cho, J. Yang, D.Y. Kim, S.R. Kim, Beatbox music phone: gesture-based interactive mobile phone using a tri-axis accelerometer, in: IEEE International Conference on Industrial Technology, ICIT 2005, IEEE, 2005, pp. 97–102.

[10] J. Wu, G. Qiao, J. Zhang, Y. Zhang, G. Song, Hand motion-based remote control interface with vibrotactile feedback for home robots, International Journal of Advanced Robotic Systems 10 (6) (2013) 270.

[11] J.K. Min, B. Choe, S.B. Cho, A selective template matching algorithm for short and intuitive gesture UI of accelerometer-builtin mobile phones, in: Second World Congress on Nature and Biologically Inspired Computing (NaBIC), 2010, IEEE, 2010, pp. 660–665.

[12] J. Rao, T. Gao, Z. Gong, Z. Jiang, Low cost hand gesture learning and recognition system based on hidden Markov model, in: Second International Symposium on Information Science and Engineering (ISISE), 2009, IEEE, 2009, pp. 433–438.

[13] Z. Ren, J. Meng, J. Yuan, Depth camera based hand gesture recognition and its applications in human–computer-interaction, in: 8th International Conference on Information, Communications and Signal Processing (ICICS), 2011, IEEE, 2011, pp. 1–5.

[14] I. Ben Abdallah, Y. Bouteraa, C. Rekik, Kinect-based sliding mode control for Lynxmotion robotic arm, Advances in Human-Computer Interaction 2016 (2016) 1.

[15] H. Sarbolandi, D. Lefloch, A. Kolb, Kinect range sensing: structured-light versus time-of-flight Kinect, Computer Vision and Image Understanding 139 (2015) 1–20.

[16] J. Shotton, T. Sharp, A. Kipman, A. Fitzgibbon, M. Finocchio, A. Blake, M. Cook, R. Moorse, Real-time human pose recognition in parts from single depth images, Communications of the ACM 56 (1) (2013) 116–124.

[17] T. Huang, Computer Vision: Evolution and Promise, Report, European Organization for Nuclear Research, CERN, 1996, pp. 21–26.

[18] D.G. Lowe, Distinctive image features from scale-invariant keypoints, International Journal of Computer Vision 60 (2) (2004) 91–110.

[19] N. Dalal, B. Triggs, Histograms of oriented gradients for human detection, in: IEEE Computer Society Conference on Computer Vision and Pattern Recognition, vol. 1, CVPR 2005, IEEE, 2005, pp. 886–893.

[20] Y. LeCun, Y. Bengio, G. Hinton, Deep learning, Nature 521 (7553) (2015) 436–444.

[21] D. Graupe, Deep Learning Neural Networks: Design and Case Studies, World Scientific Publishing Co. Inc., 2016.

[22] D.H. Hubel, T.N. Wiesel, Receptive fields and functional architecture of monkey striate cortex, Journal of Physiology 195 (1) (1968) 215–243.

[23] https://en.wikipedia.org/wiki/Convolutional_neural_network (Accessed 2 March 2017).

[24] L. Wang, Y. Xiong, Z. Wang, Y. Qiao, D. Lin, X. Tang, L. Van Gool, Temporal segment networks: towards good practices for deep action recognition, in: European Conference on Computer Vision, Springer, 2016, pp. 20–36.

[25] Q. Ke, M. Bennamoun, S. An, F. Boussaid, F. Sohel, Human interaction prediction using deep temporal features, in: European Conference on Computer Vision Workshops, Springer, 2016, pp. 403–414.

[26] S. Akpinar, F.N. Alpaslan, Video action recognition using an optical flow based representation, in: Proceedings of the International Conference on Image Processing, Computer Vision, and Pattern Recognition (IPCV), The Steering Committee of The World Congress in Computer Science, Computer Engineering and Applied Computing (WorldComp), 2014, p. 1.

[27] J.L. Barron, D.J. Fleet, S.S. Beauchemin, Performance of optical flow techniques, International Journal of Computer Vision 12 (1) (1994) 43–77.

[28] A. Bruhn, J. Weickert, C. Schnörr, Lucas/Kanade meets Horn/Schunck: combining local and global optic flow methods, International Journal of Computer Vision 61 (3) (2005) 211–231.

[29] R. Chaudhry, A. Ravichandran, G. Hager, R. Vidal, Histograms of oriented optical flow and Binet–Cauchy kernels on nonlinear dynamical systems for the recognition of human actions, in: IEEE Conference on Computer Vision and Pattern Recognition, CVPR 2009, IEEE, 2009, pp. 1932–1939.

[30] Q. Ke, Y. Li, Is rotation a nuisance in shape recognition? in: Proceedings of the IEEE Conference on Computer Vision and Pattern Recognition, 2014, pp. 4146–4153.

[31] S. Ioffe, C. Szegedy, Batch normalization: accelerating deep network training by reducing internal covariate shift, arXiv preprint, arXiv:1502.03167, 2015.

[32] J. Deng, W. Dong, R. Socher, L.J. Li, K. Li, L. Fei-Fei, ImageNet: a large-scale hierarchical image database, in: IEEE Conference on Computer Vision and Pattern Recognition, CVPR 2009, IEEE, 2009, pp. 248–255.

[33] K. Simonyan, A. Zisserman, Very deep convolutional networks for large-scale image recognition, arXiv preprint, arXiv:1409.1556, 2014.

[34] K. Chatfield, K. Simonyan, A. Vedaldi, A. Zisserman, Return of the devil in the details: delving deep into convolutional nets, in: British Machine Vision Conference, 2014, arXiv:1405.3531.

[35] Q. Ke, M. Bennamoun, S. An, F. Bossaid, F. Sohel, Leveraging structural context models and ranking score fusion for human interaction prediction, arXiv preprint, arXiv:1608.05267, 2016.

[36] H. Rahmani, A. Mian, 3D action recognition from novel viewpoints, in: Proceedings of the IEEE Conference on Computer Vision and Pattern Recognition, 2016, pp. 1506–1515.

[37] J. Wang, Z. Liu, Y. Wu, Learning actionlet ensemble for 3D human action recognition, in: Human Action Recognition with Depth Cameras, Springer, 2014, pp. 11–40.

[38] S. Gupta, R. Girshick, P. Arbeláez, J. Malik, Learning rich features from RGB-D images for object detection and segmentation, in: European Conference on Computer Vision, Springer, 2014, pp. 345–360.

[39] V.O. Alan, W.S. Ronald, R. John, Discrete-Time Signal Processing, Prentice Hall Inc., New Jersey, 1989.

[40] J. Liu, G. Wang, P. Hu, L.Y. Duan, A.C. Kot, Global context-aware attention LSTM networks for 3D action recognition, in: Proceedings of the IEEE Conference on Computer Vision and Pattern Recognition, 2017.

[41] J. Liu, A. Shahroudy, D. Xu, G. Wang, Spatio-temporal LSTM with trust gates for 3D human action recognition, in: European Conference on Computer Vision, Springer, 2016, pp. 816–833.

[42] A. Shahroudy, J. Liu, T.T. Ng, G. Wang, NTU RGB+D: a large scale dataset for 3D human activity analysis, in: Proceedings of the IEEE Conference on Computer Vision and Pattern Recognition, 2016, pp. 1010–1019.

[43] Q. Ke, M. Bennamoun, S. An, F. Sohel, F. Boussaid, A new representation of skeleton sequences for 3D action recognition, in: Proceedings of the IEEE Conference on Computer Vision and Pattern Recognition, 2017.

[44] Q. Ke, M. Bennamoun, S. An, F. Sohel, F. Boussaid, SkeletonNet: mining deep part features for 3D action recognition, IEEE Signal Processing Letters (2017).

[45] V. Bloom, V. Argyriou, D. Makris, Hierarchical transfer learning for online recognition of compound actions, Computer Vision and Image Understanding 144 (2016) 62–72.

# CHAPTER 6

# Computer Vision for Ambient Assisted Living

Monitoring Systems for Personalized Healthcare and Wellness That Are Robust in the Real World and Accepted by Users, Carers, and Society

**Sara Colantonio\*, Giuseppe Coppini†, Daniela Giorgi\*, Maria-Aurora Morales†, Maria A. Pascali\***

\*National Research Council, Institute of Information Science and Technologies, Pisa, Italy
†National Research Council, Institute of Clinical Physiology, Pisa, Italy

## Contents

## Abstract

The Ambient Assisted Living (AAL) paradigm proposes advanced technologies and services to improve the quality of life, health, and wellbeing of citizens by making their daily-life activities easier and more secure, by monitoring patients under specific treatment, and by addressing at-risk subjects with proper counseling. The challenges brought by AAL range from robust, accurate, and nonintrusive data acquisition in daily-life settings to the development of services that are easy to use and appealing to the

Computer Vision for Assistive Healthcare.
DOI: https://doi.org/10.1016/B978-0-12-813445-0.00006-X

users and that support long-term engagement. This chapter offers a brief survey of existing vision-based monitoring solutions for personalized healthcare and wellness, and introduces the Wize Mirror, a multisensory platform featuring advanced algorithms for cardiometabolic risk prevention and quality-of-life improvement.

## Keywords

Ambient assisted living (AAL), Self-monitoring, Unobtrusiveness, User engagement, Physiological monitoring, Wellness index

## 6.1 INTRODUCTION

The term Ambient Assisted Living (AAL) concerns ICT[1] solutions embedded in the living environment to monitor the setting and the behavior of its occupants in real time. AAL solutions can go beyond observing; they can move towards interacting by communicating with the user via prompts, triggering assistance from emergency services, and providing up-to-date information to families and caregivers.

AAL technologies trace their roots back to home automation and assistive domotics; they are now applied in many different scenarios, from telecare to pervasive wellness [1,2]. The main aim is improving the quality of life, health, and wellbeing of citizens; making their daily-life activities easier and more secure; monitoring and curing ill or at-risk subjects; and even performing primary prevention in healthy subjects.

The three main targets of AAL solutions are

- monitoring (and acting on) environmental conditions (e.g. lights, temperature, opening of doors, gas detection) for the occupants' safety and comfort;
- monitoring human activity and behavior, from basic motion tracking and human activity recognition up to long-term behavioral analysis;
- monitoring human physiological signs (physical, clinical, vital, emotional parameters) for personalized healthcare and wellness, including telerehabilitation, cure management, and prevention.

The general scheme of AAL solutions is to enrich the environment with a distributed sensor system. The sensors include *ambient sensors* (e.g. magnetic switches, photosensors) as well as *body* or *wearable sensors* that are fixed to the human body or clothing (e.g. gyroscopes, pulse oximeters). A large network of sensors are often needed to gather enough data to accomplish the complex tasks required for AAL applications; therefore, systems may be

---

[1] Information and Communication Technology.

costly to maintain, highly sensitive to a sensor's performance, and obtrusive. Obtrusiveness is an issue especially for wearable sensors, due to their weight, effects on the skin, and burden on the subject during daily-life activities.

Recently, video cameras (including standard and depth cameras) and computer vision techniques came into the spotlight as alternative solutions for AAL applications [3,4]. Video-based AAL solutions benefit from the maturity of the research in computer vision: decades of studies, boosted by needs in domains such as security and surveillance, have brought robust techniques for interpreting the rich information provided by cameras. Also, video-based solutions can favor the acceptability of AAL systems: video cameras are expected to reduce the possible anxiety towards new technologies, even for the elderly, as they are familiar objects. Moreover, cameras may be perceived as less invasive than other sensors as they do not require any physical contact with the subject, they do not have to be worn (with the exception of wearable cameras), and they can be fitted in the home setting by taking aesthetics into account.

Another main advantage of video and depth cameras is that they can register a great deal of information compared to other sensors, and allow the analysis of complex scenarios. A single camera can capture most of the activities performed in a room; therefore, in principle, it could replace many sensors. This would cut costs, also thanks to the fact that high-quality cameras are now available at affordable prices. Moreover, the same hardware setup can serve for multiple applications by just changing the software analyzing the visual data. Therefore, computer vision supports flexible and adaptive solutions, which can be easily extended and updated on demand.

## 6.1.1 Chapter Scope

There is a huge corpus of work on the use of computer vision for ambient monitoring and navigation and for activity recognition [5]. These topics are also covered in other chapters of this book. Therefore, after a brief introduction to the field of computer vision for AAL (Section 6.2), this chapter focuses on the use of computer vision techniques for monitoring human physiological signs in AAL solutions for personalized healthcare and wellness. In Section 6.3, we discuss the monitoring through videos, images, and 3D data, of

- vital signs (heart rate and respiratory rate);

- posture and movement, especially for rehabilitation;
- morphological parameters (e.g. body mass index);
- emotional state (e.g. stress and anxiety).

Subsequently, in Section 6.4, we identify barriers and challenges for vision-based AAL solutions to reach their full maturity and applicability, especially within the healthcare domain.

Finally, as a possible solution, we introduce the Wize Mirror, a multisensory platform for health and wellbeing monitoring (Section 6.5). The Wize Mirror features advanced computer vision algorithms for cardiometabolic risk assessment through facial analysis.

Our target audience includes researchers and healthcare professionals, who will find information about the aspects that may affect their practice.

## 6.2 COMPUTER VISION FOR AAL

The traditional three-step pipeline for video-based AAL solutions includes image/video acquisition, preprocessing (segmentation of the region of interest, filtering for noise reduction, or contrast enhancement), and feature extraction and interpretation. The acquisition devices include traditional RGB cameras (including omnidirectional cameras and wearable cameras for egocentric vision); thermal cameras that acquire the infrared radiation of the scene; MultiSpectral Imaging (MSI) systems [6]; and depth sensors, based either on time of flight or on structured light [7]. Depth sensors in particular have became very popular lately because of their decreasing cost and the advantages they provide, including robustness to changing lighting conditions and protection of privacy since the appearance of a person is not recognized in depth images.

The architecture of a video-based AAL solution traditionally consists of a set of cameras for data capture connected to a server; processing modules for data processing and analysis; and an alert/decision module, which may or may not include a human operator. This solution can ease the integration of data from different sources, but usually requires high bandwidth for data transmission, unless compressed video is used for transmission. An alternative band-saving solution is given by distributed smart camera networks that analyze the data locally and only transmit the alert/decision [8].

One of the main challenges of video-based AAL systems is the robustness with respect to the operating conditions in real-life settings. A first issue for robust data acquisition is what cameras are able to see, which depends on the viewing angle and user positioning with respect to the camera, but also on occlusions and clutter. The problem of the viewing angle of the scene can be solved via hardware or software solutions. A solution often adopted in ambient surveillance systems is based on omnidirectional cameras which can capture virtually a 360 degree viewing angle of data. Alternatively, view-invariant computer vision algorithms are required, based either on single cameras or on image fusion strategies from multiple cameras. Multiple cameras and image fusion strategies may also help to solve the problem of occlusions that are likely to occur in a room. Finally, face recognition and the ability to track (multiple) people are needed in case of multiple occupants of the monitored space.

Another major concern for video-based assistive technologies is privacy. AAL systems imply the collection of information in private spaces about individual subjects and their lives. With image and video recordings, the privacy protection issue becomes even more prominent since the occupants' identities are shown. Privacy protection is considered urgent especially if image data are stored and transmitted to a server, and if they are expected to be analyzed by designated carers. The consequences can be low acceptance by the end users, and even techniques for avoiding the camera to sabotage the monitoring [9]. Striking a happy medium between the benefits of image and video recording and the potential privacy loss requires focusing on the perspective of the end user. Some studies [10,11] have suggested that people would be willing to accept video-based activity monitoring systems, in principle, if they felt that the technology would make a real difference to their lives, and if they were really in need of assistance. Also, improving the reliability of systems so that they do not require images to be analyzed by humans could improve confidentiality. Again, a possible solution is given by smart cameras, which can perform a part of the image and video processing locally, and either only transmit decision information to carers or filter the data to obscure individual identities [8].

## 6.3  MONITORING IN PERSONALIZED HEALTHCARE AND WELLNESS: THE STATE OF THE ART

The home can be a focal point for ensuring healthy living if it is equipped with the right infrastructure: AAL systems based on computer vision solutions can support health status monitoring for the management of chronic diseases and rehabilitation therapies, but also for the prevention of disease and improvement of individual wellbeing through the achievement and maintenance of a healthy lifestyle. These systems may become part of the medical toolbox for healthcare professionals, who could benefit from innovative communication and monitoring facilities.

The parameters to be monitored include vital signs such as heart rate and respiratory rate (Section 6.3.1); measures related to posture and movement, especially during rehabilitation (Section 6.3.2); morphological parameters of the body and face related to risk factors such as overweight and obesity (Section 6.3.3); and descriptors of the affective state, such as cues about the onset and progression of different diseases (Section 6.3.4).

## 6.3.1 Vital Signs

The monitoring of vital signs includes the estimation of

- heart rate (HR);
- heart rate variability (HRV);
- respiratory rate (RR).

### Heart Rate

Heart rate monitoring is important by virtue of the significance of this vital sign in both health and disease. Physiological changes in HR can be assessed during exercise or during sleep when a reduction in HR may occur due to prevalence of vagal drive. In disease, there is an association between HR and outcome in heart failure patients, and a baseline HR is considered a cardiovascular risk factor. A special application of HR monitoring can be reported in the aging population as HR can also be a sign of improper movements. Since an increasing number of older subjects wish to live in their home environments rather than moving into care facilities, the opportunity to unobtrusively monitor HR is challenging. Therefore, several studies have been performed to evaluate whether the HR can be assessed from video streams [12], thus avoiding the use of wearable sensors.

A commonly adopted approach to HR assessment is based on the processing of RGB video images acquired by RGB cameras or webcams. Blood Volume Pulses (BVPs) are expected to produce changes of the intensity of spectral components of the video signal [13]. In analogy with standard PhotoPlethysmoGraphy (PPG), the term video PhotoPlethysmoGraphy (vPPG) has been introduced. The basic steps of vPPG are

- acquiring the video signal, possibly in a specific wavelength band, and averaging it in a region of interest to produce a time sequence reflecting BVP changes;

- enhancing BVPs by specific processing, and locating them, e.g. by peak detection.

Concerning the signal acquisition, webcams [14–18] and standard video cameras [19,13,20,21] are both used. Webcams provide low-cost and easily available setups, while standard cameras are expected to produce better quality signals, due to higher spatial and temporal resolution, along with multispectral imaging capabilities.

For the processing methods, a common framework is based on Blind Source Separation (BSS) techniques modeling the video signal as a mixture of contributions including BVP propagation, the various motion sources, and external illumination changes. Poh et al. [14] process the RGB component by Independent Components Analysis (ICA) to separate BVP from other contributions. An alternative approach based on BSS by Principal Components Analysis (PCA) was suggested by Lewandowska et al. [18]. The limitations of BSS (ICA- and PCA-based) have led several researchers to investigate more robust processing methods able to cope with motion artifacts. Wang et al. [19] exploit image redundancy to counteract the effects of facial movements. Feng et al. [17] adopted a simplified model of skin optical properties to compensate for head motion. Tarassenko et al. [20] proposed a vPPG system exploiting autoregressive modeling of video time series to compute HR together with RR and oxygen saturation.

In general, HR estimation from vPPG is in good agreement with reference techniques (usually standard PPG or ECG in a few studies), high correlation coefficients being reported by most authors. It is worth noting that, to date, experimental results have been from small populations of volunteers.

Though video signal intensity is the most utilized source of information, an alternative method based on head motion related to BVP propagation and recorded by video is reported in [22].

### Heart Rate Variability

In addition to heart rate, the temporal variation of Heart Rate Variability (HRV) is an indicator of health status in the general population, of adaptation to stress in athletes [23], and of fatigue in drivers [24].

The assessment of HRV from video is usually more demanding than simply assessing HR. In fact, while HR estimation only requires BVP detection to compute the average number of pulses per minute, a precise temporal localization of pulses is needed for HRV assessment. Most methods for HR assessment from video can be adapted to estimate HRV. For example, in [14] vPPG is used to compute the related tachogram (i.e. the time series of inter-beat interval duration) and to compute standard HRV indices both in time and frequency domain. A high correlation with the parameters derived by standard PPG on 15 subjects is reported. An alternative solution based on zero-phase component analysis has been reported by Iozzia et al. [25] to evaluate the suitability of vPPG for assessing autonomic response.

### Respiratory Rate

Computer vision may support the assessment of RR and related breathing disorders. A recent application that may have important clinical impact is the assessment sleep apnea. Obstructive Sleep Apnea (OSA) is a sleep breathing disorder characterized by partial or complete obstruction of the upper airway during sleep. Left untreated, OSA has been linked to an increase in motor vehicle and occupational accidents, hypertension, cardiovascular disease, and diabetes. Since present diagnostic methods are complicated and require in-hospital stay, computer vision technologies offer a novel and interesting solution for diagnosing OSA [26]. Monitoring RRs has been approached by computer vision systems also related to positions in bed. This aspect is particularly important for the aging population since posture changes in bed may reflect sleeping disorders; disturbances of sleep rhythm could be related to nocturnal falls, one of the most important causes of morbidity in elderly subjects.

A direct approach to video-based RR estimation is based on the analysis of chest movements. Benetazzo et al. [27] used a RGB camera equipped with a depth sensor to evaluate RR from depth data. As already mentioned, Tarassenko et al. [20] proposed a direct RR estimation from video streams by autoregressive models of video time series. Indirect approaches to RR estimation are based on the power spectrum of HRV signals: indeed, respiration introduces a peak in the high-frequency region of the spectrum [14].

## 6.3.2 Posture and Movement

In recent decades, studies of quantitative analysis of human posture and movements have been stimulated by several fields, including rehabilitation, sports, children's motor skills, and aging. We focus in particular on
- gait analysis and fall detection;
- movement for rehabilitation.

## Gait Analysis and Fall Detection

A relation exists between gait, cognitive decline, and risk of falls [28]. In older people, higher stages of cognitive impairment were associated with reduced ability to increase speed and walk quickly [29]. In particular, preventing falls in the elderly is a challenging task currently being investigated by many researchers [30]. Wearable sensors are mostly used in gait studies [31]; however, in recent years vision-based solutions have been introduced to improve unobtrusive monitoring. In this perspective, several researchers have proposed vision based systems to assist the elderly at home and reduce the risk of falls. The vast majority of reported approaches exploit standard single-view cameras with a wide field of view able to image the entire person [32–35], multiview camera-based systems [36], or omnidirectional surveillance cameras [37]; the use of depth sensors is also reported [38].

In general, the idea underlying most of these systems is predicting/recognizing falls by dynamic analysis of image sequences of the monitored subject. This usually relies on a model of the observed motion coupled with the classification of the inferred activities. To obtain a compact motion description, a common choice is based on the silhouette of the monitored subject. It is usually extracted following some sort of background suppression so as to reduce interference from the surrounding scene. Even though this operation can dramatically simplify the motion description, it can be prone to noise and can lead to false alarms, a matter that needs to be dealt with to achieve a reasonable system acceptability. Subject motion can be described by features extracted from the bounding box of the silhouette tracked over time.

Various solutions have been proposed which include: Hidden Markov Models [39] or Layered Hidden Markov Models [36], Finite State Models [35], and the use of Historical Motion Image [32,33] computed by time integration of image features possibly using eigenspace methodology to reduce dimensionality [33]. Falls can be recognized by standard machine learning methods such as multilayer perceptrons [33], algorithms based on distance between feature histograms of the silhouette bounding-box [40], and specific motion quantification coefficients [32]. As some processing steps can be prone to structured and nonstructured noise (e.g. background suppression), attempts to reduce possible inaccuracies have been faced [40]. The integration of visual and nonvisual cues (e.g. acoustic signals) has been proposed to improve fall detection accuracy [39].

## Movement for Rehabilitation

Rehabilitation is a dynamic process which allows persons to restore or at least improve their functional capability to normal. Physiotherapy exercise is a medical treatment aimed at returning patients to a normal life. During rehabilitation plans, patients may encounter difficulties such as time and cost of traveling, waiting for the availability of specialists, and ineffective personal exercise. These difficulties could be greater for those patients living in the areas that are distant from medical centers or that lack medical staff and experts, and for the frail elderly. Therefore, the concept of tele-physiotherapy has been developed to allow patients and medical experts to carry on their sessions through telecommunication networks as if they were in the same place. In general, the implementation of home-based rehabilitation interventions is an emerging area of scientific interest and where the use of sensors plays a fundamental role.

Computer vision-based approaches in particular are expected to ease the optimization of rehabilitation strategies at home. In fact, they enable objective and quantitative monitoring of physiotherapy exercises: the user's motion can be accurately tracked in 3D and compared to the planned therapy, thus enabling the fine-tuning of the therapeutic feedback on an individual basis. Also, the combination of sensing technology and interactive gaming or virtual reality can facilitate the implementation of rehabilitation exercise programs [41].

One of the main fields of interest is represented by stroke survivors. Approximately 800,000 new cases of stroke are reported each year in the United States. About 80% of acute stroke survivors lose arm and hand movement skills [42]. Movement impairment after a stroke typically requires intensive treatment. Due to resource reduction in healthcare all over the world, in-hospital rehabilitation tends to be limited. Thus, stroke rehabilitation in home environments is an appealing solution. Computer vision systems that allow individuals with stroke to practice arm movement exercises at home with periodic interactions with a therapist have been developed by different research groups [43,44].

It has been demonstrated that combining telemedicine with in-home robot-assisted therapy (telerehabilitation) for people with residual impairment following stroke is able to reduce barriers. Moreover, it is cost-effective, providing high-quality treatment to patients with limited access to rehabilitation centers. Interestingly, patients and caregivers report overall satisfaction and acceptance of telerehabilitation intervention, with a marked reduction in drop-outs [45].

## 6.3.3  Anthropometric Parameters

Anthropometry is the branch of anthropology dealing with the quantitative measurement of the human body, including size, proportions, and composition [46]. The most often used anthropometric body parameters are weight, height, skin-fold thickness, mass, triceps skin-fold, neck circumference, waist circumference, hip circumference, and mid-arm circumference. Derived indices such as Body Mass Index (BMI) [47], waist-to-hip circumference ratio, and waist-to-height ratio are also commonly used. Anthropometric measurements are used by the World Health Organization to describe physical trends in large-scale population studies [48,49]. Cut-off values and ranges, inferred on the basis of these statistics, are used to classify the individual status. Recently, many works have focused on correlating anthropometric parameters or derived indices with risk factors of cardiovascular and metabolic risk, such as overweight, obesity, and body fat distribution [50–58].

Manually collected anthropometric measurements are intrinsically affected by inter- and intra-observer variability; instead, computer vision technologies may support the automation and standardization of the acquisition, analysis, and recording of the physical parameters. Therefore, vision-based AAL solutions for the automatic computation and monitoring over time of anthropometric parameters may offer great support to longitudinal health studies.

Nowadays, there are many devices that can acquire people's 3D shape and appearance, including low-cost depth sensors (e.g. Microsoft Kinect, Asus Xtion), whole-body scanners (e.g. Cyberware, $TC^2$), mixed solutions (e.g. the turnable station designed by Styku for fitness and health), portable scanners (e.g. Artec3D). These 3D scanners can capture highly accurate 3D body maps, including size, shape, and skin-surface area, in 1 to 10 seconds [59–63]. Standard photogrammetry has also been used to obtain a 3D reconstruction of a human body from a sequence of RGB images [64]. The above technologies provide the automatic extraction of hundreds of measurements from a body or facial reconstruction, avoiding manual measurement and transcription errors, and ensuring the repeatability of the measurements. Also, the use of acquired 3D data overcomes the 1D nature of tape measurements, by supporting volume and surface measurements, potentially meaningful with respect to the subject's overall health status.

Additional resources for 3D digital anthropometry include publicly or commercially available datasets. For example, the Civilian American and European Surface Anthropometry Resource project (CAESAR) [65] dataset contains thousands of full-body textured 3D scans, labeled with anthropometric landmarks, and supporting the study of shape spaces in [66]. Also, several databases of human faces are available, the face being a rich source of information about an individual's psycho-physical status. The repositories of 3D facial data include CASIA HFB [60], FRGC v2.0 [67], Magna Database [68], USF Human ID 3-D [69], EURECOM Kinect Face dataset [70], and the Basel Face Morphable model [71].

In the following, we revised main methods to assess

- body parameters,
- face parameters.

## Body Parameters

The analysis of reconstructed 3D bodies includes the surface skeleton extraction; the location of feature points; the analysis of local properties (such as curvature); and the measurement of lengths, areas, and volumes of specific curves or regions. The registration and alignment of 3D reconstructions with reference models is either based on the location of a very few feature points characterized by local shape descriptors (e.g. curvature maps [72], auto diffusion function [73], integral invariants [74], salient geometric features [75]) or on dense correspondence (e.g. multidimensional scaling methods based on geodesic distances [76]).

The correlation between 3D shape measurements of the human body and health issues has been studied in [77–81] in comparison with classical anthropometric measures and indices. The use of body landmarks to perform shape measurements requires the precise location of the body landmarks themselves [82], which can be difficult, especially in the case of points that are poorly geometrically characterized. Therefore, many digital anthropometry methods have focused on overcoming the precise location of body landmarks. Recent works have focused on finding relevant correlations between geometric parameters automatically computed and body fat estimates, or cardiovascular and metabolic risk factors. Among these, Giachetti et al. presented in [83] a pipeline for processing heterogeneous 3D body scans (skeletonization and body segmentation) and extracting geometric parameters which are independent of pose and robust against noise. Another representative example of an automatic system for the extraction of a health-related index is described in the 2009 US patent [84], where the *Barix* index is introduced and is defined as

$$Barix = \frac{\text{Torso Height} * \text{Torso Surface Area}}{\text{Torso Volume}}.$$

The clinical value of the index was assessed through a study of hundreds of human scans, showing significant correlation with body fat composition.

## Face Parameters

The face has always been a very rich source of information, for example in studies on Human Computer Interaction, automatic detection of user's feelings, and early detection of numerous diseases, such as obstructive sleep apnea.

A number of studies have focused on using 2D images to detect morphological facial correlates of body fat. In [85], Ferrario et al. observed an increase in some facial dimensions in a study on facial morphology of obese adolescents. Djordjevic et al. [86] reported an analysis of facial morphology of a large population of adolescents under the influence of confounding variables: though the statistical univariate analysis showed that four principal components of the face (face height, asymmetry of the nasal tip and columella basis, asymmetry of the nasal bridge, depth of the upper eyelids) correlated with insulin levels, the regression coefficients were weak and no significance persisted in the multivariate analysis. In [87], Lee et al. proposed a prediction method, based on BMI, of normal and overweight for adult females using geometrical facial features extracted from 2D images. The features include Euclidean distances, angles, and facial areas defined by selected soft-tissue landmarks. The study was extended in [88] focusing on the association of visceral obesity with several facial characteristics. The authors determined statistically the best predictor of normal waist and visceral obesity among the considered facial characteristics. Cross-sectional data were obtained from a population of over 11 thousand adult Korean men and women, aged between 18 and 80 years.

Recently, 3D data have started to be used. One of the first results concerning the relation between 3D facial shape and a syndrome was presented by Banabilh et al. in [89]: the authors showed that craniofacial obesity, assessed via 3D stereo-photogrammetry, is correlated with obstructive sleep apnea syndrome. In [90], Giorgi et al. defined a shape descriptor based on the theory of persistent homology, and tested it on a synthetic dataset of 3D faces to assess the relation of facial morphology with obesity and overweight.

## 6.3.4 Emotions, Expressions, and Individual Wellness

The intertwining of emotional states and the onset and progression of diseases has been under investigation for a long time [91]. Studies in modern neuroscience have shed new light on the topic, leading to an extended reconsideration of the classical view of the interaction of soma and psyche [92]. On the one hand, chronic adverse emotional states, such as stress, anxiety, and tiredness, are among the main risk factors of serious illnesses, including metabolic syndrome and cardiovascular diseases [93–95]; conversely, the effect of positive states may reduce the incidence of cardiovascular disease [96]. On the other hand, chronic illness is coupled with an emotional dimension that can affect therapy outcome; for example, distress, depression, and anxiety may reduce one's motivation to access medical care and follow treatment plans [97]. Also, many studies have investigated the relation between adverse affective states (stress and depression) and the response of the immune system [98,99]. It is therefore not surprising that there is a great deal of interest in the medical community for novel tools that can assist the individual in self-managing stressful and anxiogenic situations and reducing disease risks.

Computer vision may support the implementation of unobtrusive, continuous monitoring systems of the emotional state of an individual, not only by supporting the estimation of emotion-related physical parameters (e.g. HR and RR; see Section 6.3.1), but also by analyzing facial expressions and body language evoked by emotions [100]. Indeed, assessing individual psychological wellness implies the recognition of complex mental states such as fatigue, frustration, pain, depression, and mood, which require the integration of multiple cues: physiological, acoustic, and visual cues. The combined use of physiological parameters computed by sensors and expressive features is documented in [101].

The visual clues to the affective state include general facial expression, micro-expressions, and other features such as eye gaze and head orientation. In general, the interpretation of facial codes in terms of underlying emotional states remains an open problem. Coding schemes such as the Facial Action Coding Scheme (FACS) [102] are generally related to primary affective states (e.g. anger, fear, happiness) rather than to complex states. Additional information about complex states may derive from micro-expressions, which are brief (lasting between 1/25 and 1/3 of a second) low-intensity facial expressions believed to reflect repressed feelings [103]. Due to their subtlety, they are hard to detect; for this reason the use of local dynamic appearance representations extracted from high-frequency video is usually reported [104]. The video-based analysis of multiple visual clues, including head, eyebrows, eye, and mouth movements, led in [105] to a technique for the identification of stress and anxiety.

For a detailed analysis of computer vision techniques for affective state recognition, we point the reader to Chapter 10 of this book. Here we confine ourselves to observing how moving emotional analysis in AAL settings is an exciting though extremely challenging task, which can pursue the objective of assessing emotional wellness. In particular, at-home monitoring can be implemented to evaluate stress, fatigue, and anxiety in different conditions, including resting states and a wide variety of everyday tasks (e.g. watching TV, listening to music, using a PC, or doing homework), which should ensure a detailed and possibly complete description of the individual responses to many external stimuli. In addition, AAL favors the correlation of monitoring data with life-style habits. This can be the key towards a true holistic representation of individual wellbeing in naturalistic contexts.

## 6.4 METHODOLOGICAL, CLINICAL, AND SOCIETAL CHALLENGES

We have seen how computer vision holds the potential to boost a new generation of noninvasive, effective systems for continuous health monitoring. The possibility of close follow-up by means of technological solutions at home would be an additional time-saving support to busy health professionals. These tools could help physicians not only in managing chronic diseases, but also in counseling on risk factors in the general population, especially for primary prevention. To make this scenario come true, some methodological, technological, clinical, and societal challenges still need to be faced.

### Computational Demand

In physiological monitoring, high computational demand could result from the need to acquire and process data in real time, including real-time tracking, data alignment, and data analysis. Also, the spatial and temporal resolution of the acquisition may be high, depending on the parameters to be computed; the resolution requirements should be balanced with complexity and cost requirements, depending on the application at hand.

### Acquisition Robustness in Real-Life Scenarios

Acquiring physiological parameters in uncontrolled settings requires methods that are robust to occlusions, clutter, or pose, for example the presence of a beard or make-up while analyzing the facial skin, or slight movements of the user while recording vital signs. Also, robustness to lighting conditions can be an issue in several applications, and computer-controlled external light sources may be required.

To operate in real-life scenarios, it is also mandatory to capture data without requiring too much effort from the subject, i.e. without asking the subject to stay still in front of a camera for several minutes. The challenge here is how to automate the acquisition process so that data are captured when the environmental conditions are most favorable.

### Reliability of Measurements for Clinical Purposes

It is mandatory that physiological monitoring systems provide a description of the individual status coherent with the clinical view. In particular, a crucial requirement is the reliability of measurements and their usability by doctors and health professionals. Even though it is not expected that monitoring systems produce results in the form of validated clinical tests, the data provided must be adequate to properly drive the physician's reasoning and action planning. This requires intelligent image processing algorithms, ensuring high repeatability, adequate sensitivity, and specificity in measuring data in comparison to standard reference methods used in clinical practice.

### Clinical Evaluation

Evaluating and validating in-home monitoring systems is a nontrivial task that requires an interdisciplinary team and a close involvement of the medical experts. The validation issue also calls for shared benchmark datasets.

The literature reports few and small case studies, which means that new studies involving a greater number of subjects become mandatory. In the particular case of physiological monitoring, satisfactory clinical evidence should be provided regarding the reliability of measurements, and also regarding the improvement in the quality of life, wellness, and health conditions brought by technological solutions.

### Acceptability and Long-Term Engagement

To be applicable on a large scale, video-based assistive technologies must be accepted by the end users, including monitored subjects, their families, and carers.

Stimulating initial adoption of the technology and long-term engagement is the key to making a true impact in real-life scenarios. To this end, the monitoring system design should adapt to different scenarios that range from primary prevention for healthy youngsters to the management of chronic diseases in the elderly. While the physiological parameters computed could be the same in these scenarios, the best way to acquire data, interact, and communicate should be tailored to the specific needs of the users and their carers.

### Privacy and Security

The monitoring of sensitive physiological data bears with it the issue of privacy. On the one hand, people should be made aware of system functioning in order to augment their sense of trust and confidence and the feeling that they are retaining control of their data. On the other hand, cross-age and cross-gender studies are needed to assess individual levels of user acceptance since the concept of privacy is evolving as technology evolves [106] and the perception of privacy shows cultural, gender, and age differences. The goal is to support the development of privacy-by-design video-based systems, which take into account user privacy requirements from the very beginning [107].

Another issue is that sensitive data can be intercepted by third parties and used for malicious purposes. The recording of images and video data makes preventing sensitive data from attacks especially urgent. The problem of cyber-security calls for solutions from different fields, including cryptography, data management, data mining, and related areas.

## 6.5  A POSSIBLE SOLUTION: THE WIZE MIRROR

The European Community's Seventh Framework Programme Project SE-MEOTICONS developed an innovative AAL solution that extensively leverages computer vision methods into a noninvasive device for wellbeing monitoring. The multisensory device, which looks like a common mirror and is called the Wize Mirror, was conceived as an effective technology-assisted intervention to prevent cardiometabolic diseases.

Cardiometabolic diseases (i.e. cardiovascular diseases and type 2 diabetes) are the leading causes of mortality worldwide and their spread on a global scale is putting a strain on social resources and health systems. An estimated 17.7 million people died from cardiovascular diseases in 2015, corresponding to 31% of all global deaths [108]. This figure is expected to increase in the coming years due to population aging and the increasing incidence of obesity and diabetes [109]. Clinical evidence has demonstrated that the majority of cardiometabolic diseases can be prevented by limiting exposure to the main risk factors; by exercising regularly, eating well, controlling stressful conditions, avoiding smoking, and moderating alcohol intake, about 90% of type 2 diabetes, 80% of coronary heart disease, and 70% of strokes could be prevented [110]. This awareness has fostered the promotion of educational and guidance programs for lifestyle improvement that call for assistive and personal technologies as a powerful ally to address large shares of the population.

The Wize Mirror responds to this call coming as a self-monitoring device able to blend seamlessly into life's daily routines. The guiding principle was to have a device able to minimize invasiveness, obtrusiveness, and attention theft, while maximizing usability, trustfulness, and user acceptance. This appears, in fact, to be the most promising way to address and engage various shares of the at-risk population, who have assorted needs and expectations and have diverse digital skills. The Wize Mirror implements an AAL solution based on a virtuous cycle underpinning three main elements:

• self-measurement;
• education and coaching;
• user experience, with particular emphasis on contact with healthcare professionals.

Combining together these features into a loop has demonstrated to be the key to ensuring the acceptance, effectiveness, and long-term impact of self-monitoring AAL interventions [111]. The Wize Mirror blends together these functionalities by taking into account the diverse challenges reported in Section 6.4.

## 6.5.1 Self-Measurement

The Wize Mirror seamlessly integrates a contactless sensing framework, including different types of cameras, and a data processing platform able to scan the person in front of it and assess physiological markers of cardiometabolic risk. The main cardiometabolic risk factors include hypertension; dyslipidemia; glucose dysmetabolism; obesity and overweight; noxious habits such as smoking and alcohol abuse; and adverse psychological states such as chronic conditions of stress, anxiety, and fatigue [108]. Stemming from the principles of medical semeiotics [112], the Wize Mirror analyzes physical and expressive traits of the face and the composition of the breath to detect both perceptible and subtle signs correlated to the factors listed above. The sensing framework relies on

• an inexpensive 3D scanner based on depth cameras;

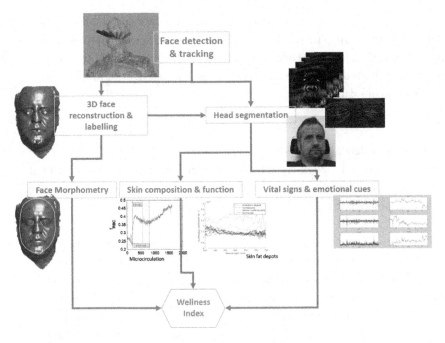

**Figure 6.1** Scheme of the Wize Mirror workflow. Data from sensors are preprocessed, analyzed, and finally integrated in the virtual individual model.

- a MSI system, made up of five compact monochrome cameras with band-pass filters at selected wavelengths, and two computer-controlled LED strips (white and UV light sources);
- high-resolution RGB cameras;
- and a portable gas-sensor device connected to the Wize Mirror.

The sensors acquire depth and multispectral images, videos, and breath signals that are processed with cutting-edge computer vision and data-processing methods to evaluate anthropometric and morphometric parameters, facial skin compositional and functional markers, vital signs and emotional cues, and breath composition. The data workflow in the Wize Mirror is summarized in Fig. 6.1.

Details on the methods developed have been reported in other publications [113,7]. Here we report and discuss their main features with respect to the challenges introduced in Section 6.4. It is worth noting that the technological development of the Mirror required a careful organization of the sensory framework to meet data quality and robustness requirements. Position, displacement, and tilt of the cameras resulted from several tests and prototyping trials that explored several options with respect to the user position in front of the Mirror. This was done to achieve the best view and overcome disturbances. Similarly, the lighting system was devised to counteract ambient light disturbances and ensure the homogeneous illumination of the face during data acquisitions.

### 3D Data

The 3D scanner data are processed to enable

- face detection and recognition, used to detect and recognize the user in front of the Mirror, as reported in [114];
- 3D head pose tracking, facial segmentation, and labeling;
- 3D reconstruction.

These are core functionalities that serve as a common asset for the other data processing facilities of the Mirror. The 3D data are in fact used to enable face detection and labeling on the 2D data obtained from the other camera systems. This choice responds to the need for robustness and efficiency of data processing methods. Face detection and tracking is performed on the data from the depth sensors only once rather than several times on the various data streams: this ensures robustness to varying illumination conditions and optimizes the processing time. The user is first detected in the 3D space; then, by fitting a face mask on the depth sensor data, the position and the orientation of the user's face are detected. After that, selected facial landmarks are localized and their 3D coordinates are projected into the 2D frames of the other camera streams. This is done by using the intrinsic and extrinsic parameters of the cameras within a camera calibration and registration procedure done at system setup [113]. The camera's synchronization procedure allows the system to meet the requirement for real-time processing.

The 3D model of the face serves the anthropometric and morphometric analyses. It results from a reconstruction algorithm adapted from the Kinect fusion method to meet the needs posed by the fact that in the Mirror the depth sensors remain in a fixed position while the face is moving [7]. A re-meshing algorithm runs on the 3D point cloud to ensure producing a manifold, without holes and degenerate elements [115].

### Facial Morphometry

Facial morphometry analyzes the 3D face model to detect signs related to overweight and obesity. Four shape parameters are computed on the 3D

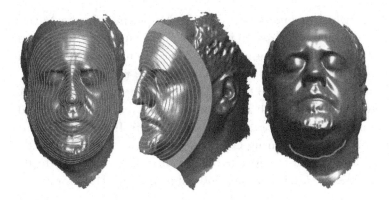

**Figure 6.2** Facial morphology description in the Wize Mirror.

manifold as strictly correlated to weight, BMI, waist circumference, hip circumference, and neck circumference [116]. They correspond to

- the length of the maximal curve among those resulting from the intersection of a family of concentric Euclidean spheres centered on the nose tip and the face manifold (see the left panel in Fig. 6.2);
- the geodesic analogue of the previous length;
- the area of an annulus computed at the border of the face manifold (see the mid panel in Fig. 6.2);
- the length of the geodesic path in the neck area that connects two points under the ears (see the right panel in Fig. 6.2).

In addition to their relevance with respect to risk factors, these parameters represent a good trade-off between processing time and accuracy. Indeed, their estimation does not require the detection of a large number of landmarks, which can be cumbersome and time-consuming, and their value is invariant to rotation, translation, and scale, and robust to noise and pose estimation errors.

### Skin Compositional and Functional Markers

Multispectral imaging data enable the analysis of skin composition and function to detect signs correlated to dyslipidemia, glucose dysmetabolism, and endothelial dysfunction. These are completely innovative techniques, developed for the first time in SEMEOTICONS, which rely on noninvasive, contactless data acquisition through a camera system.

The analysis of skin composition focuses on the detection of fat depots and the accumulation of Advanced Glycation End-products (AGEs). Skin fat depots turned out to be anti-correlated to low levels of HDL cholesterol in the blood, which is one of the signs of dyslipidemia. In the Mirror, they are detected and measured on images acquired on a single wavelength (560 nm or 580 nm) by estimating the droplets with higher skin reflection in an area underneath the eye (where fat accumulates most) [113].

AGE-products are linked to the metabolism of glucose. Their accumulation in the skin increases with aging, but when above a certain threshold, it signals an increased risk of cardiometabolic disorders. In the Mirror, AGE accumulation is estimated during UV exposure from a 365 nm LED as the ratio of the fluorescence intensity (evaluated on a 475 nm image) to the illumination intensity (evaluated on a 360 nm image). See [117] for further details.

The analysis of skin function aims to assess endothelium function. The endothelium regulates vasodilation in response to different blood-flow needs. A dysfunction of this tissue may be a consequence or a cause of several pathologies, such as hypertension, hypercholesterolemia, and diabetes. In the Mirror, endothelial function is measured on MSI data by assessing the response of facial skin microcirculation after heating the cheek via a computer-controlled remote skin heater. An index of the function is defined based on two parameters calculated on 475–650 nm images: hemoglobin oxygenation and the fraction of red blood cells in the skin before and after heating [113].

## Vital Signs and Emotional Cues

By processing 2D videos, the Mirror estimates vital signs and emotional cues connected to adverse psychological statuses. Vital signs include HR, RR, and HRV, which are indeed the most informative parameters associated with the psycho-physical status of an individual. In the Mirror these parameters are analyzed through video-processing methods based on blind source separation through Independent Component Analysis, as described in [118].

Cues of emotional status derive from the analysis of short video recordings processed to spot micro-expressions and facial gestures typical of stress, anxiety, and fatigue. For stress and anxiety, these cues generally correspond to facial-muscle hyperactivity, which may be discretized in head movements, eye movements (in terms of eye gazing and frequent focus, pupil dilation, and blinking rates), and mouth movements (in terms of jaw clenching, lip trembling, and biting). Fatigue is here intended as a sense of tiredness, lack of energy, and a feeling of exhaustion. Signs of fatigue are mainly yawing and saccadic movements. In the Mirror, all these signs are extracted from the videos though different feature extraction methods and are fed into an Artificial Neural Network model trained to classify emotional traits [113,105]. Considering how challenging and complex a proper recognition is, the neural model also takes in vital signs (i.e. HR and RR) and uses as reference a video shot at baseline during the user's registration when the user is shown a video to induce a relaxed condition.

## Breath Analysis

The gas sensing device connected to the Mirror analyzes the composition of the breath and supplies feedback about the effect of noxious habits. The device, called the Wize Sniffer, detects molecules such as carbon monoxide, ethanol, hydrogen, oxygen, and carbon dioxide. The presence and variation of these substances can be correlated to smoking, alcohol consumption, and metabolic disorders. The Sniffer acquires a breath sample through a corrugated tube connected to the Mirror, analyzes its composition and provides the Mirror with a grading of the risk the users are exposed to due to their habits (no risk, moderate risk, or high risk) [119].

## Acquisition Timing

The Mirror requires just a few seconds without requiring too much effort to scan the faces of the users while they are standing or sitting in front of it as part of their daily routine. Overall, facial morphometry, skin composition, vital signs, and emotional analyses all together take one minute of data acquisition. Breath analysis takes the time of a deep breath and it entails the simple interaction of breathing into a mouthpiece. Thanks to a parallel implementation on a multiprocessor board, results are provided in real time, in only a few seconds. Only HRV and endothelial function require a five- and six-minute acquisition, respectively. This time is unavoidable and is necessary to ascertain the dynamics of the underlying physiological phenomenon.

It is worth underlining that apart from the breath analysis, all the other analyses rely on image and video processing. This means that they can be integrated, with proper customization, into any device or equipment that already hosts or is able to host some of the sensors behind the Mirror such as a TV set, a mobile phone or tablet, a personal computer, etc. This flexibility increases the potential of the Wize Mirror to address users with different needs and preferences.

## 6.5.2  Education and Coaching

The Wize Mirror is an interactive device with a touchscreen Graphical User Interface (GUI) that conveys information to and from the users. Through the GUI, the Mirror trains on and guides the users through the self-measurement procedure and informs them about the significance of the parameters measured. It also provides the users with high-quality educational materials on cardiometabolic risks and the importance of primary prevention. In the current implementation, during data acquisition, the Mirror displays some short and sharp messages promoting behavioral changes, complementing them with captivating and instructive images or short videos.

Although properly explained, the set of measurements that the Mirror is able to assess covers a wide range of psycho-physiological parameters that if displayed separately may overwhelm the users and mislead their understanding of their own status. To avoid this, the Mirror integrates the measurements into a Wellness Index, which measures the users' wellness with respect to the risk of a cardiovascular disease on a scale from 1 to 100. The Index results from the application of a Structural Equation Model [120] to the measurements taken by the device and other data inputted by the users through validated questionnaires on their habits and attitudes [113]. The Index is displayed on the Mirror (see Fig. 6.3), organized in three main components that cover the main facets of individual wellness: (i) the physical component summarizes the physical conditions of the user as the outcome of physiological measurements (i.e. facial morphometry and skin composition and function); (ii) the emotional component measures psychological conditions as a combination of the emotional cues and the vital signs; and (iii) the lifestyle component scores the users' habits in terms of noxious habits, diet, and physical activity by leveraging outcomes from the Sniffer and the data provided by the questionnaires. The approach behind the Wellness Index is highly innovative, since it merges objective measures (obtained by the device) with subjective information provided by the users on their perceived status. Other solutions currently in use, such as the WHO-5 index [121], use only the subjective evaluation and this can obviously hide potential biases.

As shown in Fig. 6.4, the Wellness Index is traced over time and stored in the Mirror in a diary to be consulted by the users. Its evolution over time is the basis of the personalized guidance that counsels the users on lifestyle improvements that increase their WI and ameliorate their physical and emotional health. The guidance relies on the definition of the user profile in

**Figure 6.3** The Wellness Index as presented in the Wize Mirror GUI.

**Figure 6.4** The Wellness Index can be traced over time and stored in a personal diary.

terms of attitudes, habits, and preferences. These pieces of information are gathered through the questionnaires the users answer when registering the first time or whenever they want via a dedicated mobile app connected to the Mirror. The guidance addresses the major lifestyle targets, including diet, physical activity, smoking, alcohol consumption, sleep, and stress and anxiety management. Recommendations are tailored to the users' traits, in terms of frequency, intensity, and linguistic style. Tailoring relies on a set of modulators that estimate initial health conditions, reported self-efficacy, and emotional strength via the set of standardized questionnaires [113].

Education and guidance are essential ingredients of AAL solutions to drive a long-term effect. In the era of doctor Google, the provision of high-quality information based on scientific evidence is becoming more and more urgent and mandatory. Overall, educational and counseling messages contribute to improving users' knowledge and risk perception, ameliorating their health literacy, and increasing their involvement and degree of comfort in making healthy choices [122].

The approach adopted by the Wize Mirror stems from these considerations and is meant to captivate the users by offering a holistic approach to wellness. As a matter of fact, more and more people are seeking ways not only to lose weight or look better, but also to improve their quality of life and overall sense of wellness.

### 6.5.3 User Experience

A pleasant user experience is crucial to stimulating the initial adoption and periodic utilization of self-monitoring devices. The Wize Mirror features a touchscreen and an intuitive interface that makes it usable and appealing to people with different digital skills. It also offers a range of user applications and services that span from sharing data with healthcare professionals and general practitioners to playing music, social network connections, email consultation, and web surfing. The Wize Mirror can indeed be seen as a big tablet integrated into the bathroom mirror. Future extensions include the connection to wearable devices or mobile apps to automatically upload data on physical activity, sleep, and diet.

Among the different services, the link with healthcare professionals is of paramount importance: on the one hand, it reinforces the impact of the device by making the users feel more secure and cared about; on the other hand, it enables care providers to gather data and insight never available before.

To meet privacy and security requirements, the Mirror features an authentication facility that is based on the automatic user recognition, but it always requires a confirmation code. A privacy-by-design approach was adopted to design the storage system along with data encryption. The data shared with care professionals mainly consist of the Wellness Index and measurement values. Images and videos are never transferred from the Mirror over the Internet.

### 6.5.4 Wize Mirror Validation

The Wize Mirror underwent a validation campaign to verify the accuracy, repeatability, reproducibility, and effectiveness of its measurements and to check acceptability by the users. Three prototypes were deployed in three clinical sites in Italy and France between July and October 2016. A human study involved 72 volunteers who underwent Mirror scans every 15 days for three months. Reference data were acquired contextually with diagnostic devices used in clinical practice to measure body composition and metabolic, homeostatic, and vital parameters. A comparison showed a statistically significant correlation of Mirror measurements and standard clinical measures. Moreover, it was observed that both originally motivated and unmotivated volunteers were able to significantly modify their physiological conditions, and that there was an evident decrease in their BMI. Details on these outcomes are being reported in a dedicated paper to be published within the medical literature.

Overall, the validation demonstrated the reliability of the Mirror's measurements and interventions. This is key to nurturing the device acceptance by both the end users and the clinical professionals. In particular, the clinicians' trust in the device plays a key role in the promotion of the Mirror among the at-risk population.

## 6.6 CONCLUSION

In recent decades healthcare systems have experienced an exponential growth in costs that is related to different social, cultural, and economic factors. The need for sustainable healthcare systems translates into challenges in ICT for the implementation of autonomous and proactive healthcare services. We believe the synergy between AAL technologies and computer vision may support moving medical and healthcare services from hospitals to home environments, thus cutting down healthcare costs.

As observed in [2], AAL systems could support the third generation of telecare systems. The first generation was the panic-alarm gadgets used to summon help in case of emergency, and the second was sensor-based monitoring systems used to support medical decisions; the third generation of systems shifts from a reactive approach to a proactive strategy for anticipating emergency situations. Computer vision can help embark on this revolutionary path. Science and technology and research are mature, though further research is needed to solve a number of open issues, including robustness, accuracy, and nonintrusiveness of data acquisition; clinical validity of the output delivered by existing techniques; and attention to the needs and demands of real end users in terms of acceptability and long-term engagement.

With the Wize Mirror, we have seen how we can begin to think of a fourth generation of telecare systems, which are explicitly designed to influence human behavior and persuade people to act upon their lifestyles and their health. We believe the convergence of different disciplines, from information technology (including computer vision) to cognitive science, is the way forward.

## ACKNOWLEDGMENTS

This work has been partially supported by the European Community's Seventh Framework Programme (FP7/2013–2016) under grant agreement number 611516 (SEMEOTICONS). The authors would like to acknowledge the SEMEOTICONS project partners (Centre de Recherche en Nutrition Humaine Rhone–Alpes, COSMED SRL, DRACO SYSTEMS SL, FORTH – Foundation for Research and Technology, Hellenic Telecommunications & Telematics Applications Company, INTECS SPA, Linköping University, Norwegian University of Science and Technology, University of Central Lancashire) for their valuable contributions to the development of the Wize Mirror.

## REFERENCES

[1] R. Al-Shaqi, M. Mourshed, Y. Rezgui, Progress in ambient assisted systems for independent living by the elderly, SpringerPlus 5 (624) (2016).

[2] G. Acampora, D.J. Cook, P. Rashidi, A.V. Vasilakos, A survey on ambient intelligence in health care, Proceedings of the IEEE 101 (12) (2013) 2470–2494.

[3] F. Cardinaux, D. Bhowmik, C. Abhayaratne, M.S. Hawley, Video based technology for ambient assisted living: a review of the literature, Journal of Ambient Intelligence and Smart Environments 3 (3) (2011) 253–269.

[4] R. Planinc, A.A. Chaaraoui, M. Kampel, F. Florez-Revuelta, Computer vision for active and assisted living, in: F. Florez-Revuelta, A.A. Chaaraoui (Eds.), Active and Assisted Living: Technologies and Applications, IET, 2016, pp. 57–79.

[5] T. Subetha, S. Chitrakala, A survey on human activity recognition from videos, in: Proceedings of IEEE International Conference on Information Communication and Embedded Systems (ICICES), 2016.

[6] A. Danielis, D. Giorgi, M. Larsson, T. Strömberg, S. Colantonio, O. Salvetti, Lip segmentation based on Lambertian shadings and morphological operators for hyperspectral images, Pattern Recognition 63 (2017) 355–370.

[7] Y. Andreu, F. Chiarugi, S. Colantonio, G. Giannakakis, D. Giorgi, P. Henriquez, et al., Wize Mirror: a smart, multisensory cardio-metabolic risk monitoring system, Computer Vision and Image Understanding 148 (2016) 3–22.

[8] S. Fleck, W. Strasser, Smart camera based monitoring system and its application to assisted living, Proceedings of the IEEE 96 (10) (2008) 1698–1714.

[9] K. Caine, S. Šabanović, M. Carter, The effect of monitoring by cameras and robots on the privacy enhancing behaviors of older adults, in: Proceedings of the 7th ACM/IEEE International Conference on Human–Robot Interaction (HRI), 2012, pp. 343–350.

[10] S.T. Londei, J. Rousseau, F. Ducharme, A. St-Arnaud, J. Meunier, J. Saint-Arnaud, An intelligent videomonitoring system for fall detection at home: perceptions of elderly people, Journal of Telemedicine and Telecare 15 (8) (2009) 383–390.

[11] M. Mataric, J. Eriksson, D. Feil-Seifer, C. Winstein, Socially assistive robotics for post-stroke rehabilitation, Journal of NeuroEngineering and Rehabilitation 4 (5) (2007).

[12] C. Brüser, C.H. Antink, T. Wartzek, M. Walter, S. Leonhardt, Ambient and unobtrusive cardiorespiratory monitoring techniques, IEEE Reviews in Biomedical Engineering 8 (2015) 30–43.

[13] W. Verkruysse, L.O. Svaasand, J.S. Nelson, Remote plethysmographic imaging using ambient light, Optics Express 16 (26) (2008) 21434–21445.

[14] M.Z.Z. Poh, D.J. McDuff, R.W. Picard, Advancements in noncontact, multiparameter physiological measurements using a webcam, IEEE Transactions on Biomedical Engineering 58 (1) (2011) 7–11.

[15] H. Monkaresi, R. Calvo, H. Yan, A machine learning approach to improve contactless heart rate monitoring using a webcam, IEEE Journal of Biomedical and Health Informatics 18 (4) (2013) 2168–2194.

[16] L. Wei, Y. Tian, Y. Wang, T. Ebrahimi, T. Huang, Automatic webcam-based human heart rate measurements using Laplacian Eigenmap, in: K.M. Lee, Y. Matsushita, J.M. Rehg, Z. Hu (Eds.), Computer Vision – ACCV 2012, ACCV 2012, in: Lecture Notes in Computer Science, vol. 7725, Springer, Berlin, Heidelberg, 2013, pp. 281–292.

[17] L. Feng, L.M. Po, X. Xu, Y. Li, R. Ma, Motion-resistant remote imaging photoplethysmography based on the optical properties of skin, IEEE Transactions on Circuits and Systems for Video Technology 25 (5) (2015) 879–891.

[18] M. Lewandowska, J. Ruminski, T. Kocejko, J. Nowak, Measuring pulse rate with a webcam; A non-contact method for evaluating cardiac activity, in: 2011 Federated Conference on Computer Science and Information Systems (FedCSIS), 2011, pp. 405–410.

[19] W. Wang, S. Stuijk, G. De Haan, Exploiting spatial redundancy of image sensor for motion robust rPPG, IEEE Transactions on Biomedical Engineering 62 (2) (2015) 415–425.

[20] L. Tarassenko, M. Villarroel, A. Guazzi, J. Jorge, D.A. Clifton, C. Pugh, Non-contact video-based vital sign monitoring using ambient light and auto-regressive models, Physiological Measurement 35 (5) (2014) 807–831.

[21] G. De Haan, V. Jeanne, Robust pulse rate from chrominance-based rPPG, IEEE Transactions on Biomedical Engineering 60 (10) (2013) 2878–2886.

[22] G. Balakrishnan, F. Durand, J. Guttag, Detecting pulse from head motions in video, in: Proceedings of the IEEE Computer Society Conference on Computer Vision and Pattern Recognition, 2013, pp. 3430–3437.

[23] L. Capdevila, J. Moreno, HRV based health&sport markers using video from the face, in: 34th Annual International Conference of the IEEE EMBS, 2012, pp. 5646–5649.

[24] S.J. Jung, W.Y. Chung, B.G. Lee, Real-time physiological and vision monitoring of vehicle driver for non-intrusive drowsiness detection, IET Communications 5 (17) (2011) 2461–2469.

[25] L. Iozzia, L. Cerina, L. Mainardi, Relationships between heart-rate variability and pulse-rate variability obtained from video-PPG signal using ZCA, Physiological Measurement 37 (11) (2016) 1934–1944, http://stacks.iop.org/0967-3334/37/i=11/a=1934?key=crossref.c843764ea2967b599a70a9124fef61c3.

[26] M.H. Li, A. Yadollahi, B. Taati, Non-contact vision-based cardiopulmonary monitoring in different sleeping positions, IEEE Journal of Biomedical and Health Informatics (2016) 2194, http://ieeexplore.ieee.org/lpdocs/epic03/wrapper.htm?arnumber=7468522.

[27] F. Benetazzo, A. Freddi, A. Monteriù, L. Sauro, Respiratory rate detection algorithm based on RGB-D camera: theoretical background and experimental results, Healthcare Technology Letters 1 (3) (2014) 81–86.

[28] G. Allali, C.P. Launay, H.M. Blumen, M.L. Callisaya, A.M.D. Cock, R.W. Kressig, et al., Falls, cognitive impairment, and gait performance: results from the GOOD initiative, Journal of the American Medical Directors Association 18 (4) (2017) 335–340.

[29] M.L. Callisaya, C.P. Launay, V.K. Srikanth, J. Verghese, G. Allali, O. Beauchet, Cognitive status, fast walking speed and walking speed reserve – the Gait and Alzheimer Interactions Tracking (GAIT) study, GeroScience (2017), http://link.springer.com/10.1007/s11357-017-9973-y.

[30] R. Kenny, C.N. Scanaill, M. McGrath, Falls prevention in the home, in: Intelligent Technologies for Bridging the Grey Digital Divide, IGI Global, 2010, pp. 46–64, http://services.igi-global.com/resolvedoi/resolve.aspx?doi=10.4018/978-1-61520-825-8.ch004.

[31] A. Muro-de-la Herran, B. García-Zapirain, A. Mendez-Zorrilla, Gait analysis methods: an overview of wearable and non-wearable systems, highlighting clinical applications, Sensors 14 (2) (2014) 3362–3394.

[32] C. Rougier, J. Meunier, A. St-Arnaud, J. Rousseau, Fall detection from human shape and motion history using video surveillance, in: Advanced Information Networking and Applications Workshops, 2007.

[33] H. Foroughi, A. Naseri, A. Saberi, H. Sadoghi Yazdi, An eigenspace-based approach for human fall detection using Integrated Time Motion Image and Neural Network, in: 9th International Conference on Signal Processing, 2008, pp. 1499–1503.

[34] T. Lee, A. Mihailidis, An intelligent emergency response system: preliminary development and testing of automated fall detection, Journal of Telemedicine and Telecare 11 (4) (2005) 194–198.

[35] J. Tao, M. Turjo, M.-F. Wong, M. Wang, Y.-P. Tan, Fall incidents detection for intelligent video surveillance, in: 5th International Conference on Information Communications & Signal Processing, IEEE, 2005.

[36] N. Thome, S. Miguet, S. Ambellouis, A real-time, multiview fall detection system: a LHMM-based approach, IEEE Transactions on Circuits and Systems for Video Technology 18 (11) (2008) 1522–1532.

[37] Y. Xiang, Y.P. Tang, B.Q. Ma, H.C. Yan, J. Jiang, X.Y. Tian, Remote safety monitoring for elderly persons based on omni-vision analysis, PLoS ONE 10 (5) (2015) 1–16.

[38] C. Rougier, J. Meunier, A. St-Arnaud, J. Rousseau, Monocular 3D head tracking to detect falls of elderly people, in: Proceedings, Annual International Conference of the IEEE Engineering in Medicine and Biology, 2006, pp. 6384–6387.

[39] B.U. Töreyin, Y. Dedeoğlu, A.E. Çetin, HMM based falling person detection using both audio and video, in: N. Sebe, M. Lew, T.S. Huang (Eds.), Computer Vision in Human–Computer Interaction, HCI 2005, in: Lecture Notes in Computer Science, vol. 3766, Springer, Berlin, Heidelberg, 2005, pp. 1–4.

[40] C.W. Lin, Z.H. Ling, Automatic fall incident detection in compressed video for intelligent homecare, in: Proceedings, International Conference on Computer Communications and Networks, ICCCN, 2007, pp. 1172–1177.

[41] C.Y. Chang, B. Lange, M. Zhang, S. Koenig, P. Requejo, N. Somboon, et al., Towards pervasive physical rehabilitation using Microsoft Kinect, in: 6th International Conference on Pervasive Computing Technologies for Healthcare (PervasiveHealth) and Workshops, 2012, pp. 159–162.

[42] S.S. Rathore, A.R. Hinn, L.S. Cooper, H.A. Tyroler, W.D. Rosamond, Characterization of incident stroke signs and symptoms, Stroke 33 (2002) 2718–2721.

[43] L.E. Sucar, F. Orihuela-Espina, R.L. Velazquez, D.J. Reinkensmeyer, R. Leder, J. Hernández-Franco, Gesture therapy: an upper limb virtual reality-based motor rehabilitation platform, IEEE Transactions on Neural Systems and Rehabilitation Engineering 22 (2014) 634–643.

[44] K.M. Vamsikrishna, D.P. Dogra, M.S. Desarkar, Computer-vision-assisted palm rehabilitation with supervised learning, IEEE Transactions on Biomedical Engineering 63 (2016) 991–1001.

[45] T. Johansson, C. Wild, Telerehabilitation in stroke care – a systematic review, Journal of Telemedicine and Telecare 17 (1) (2011) 1–6.

[46] K. Norton, T. Olds, Anthropometrica: A Textbook of Body Measurements for Sports and Health Courses, UNSW Press, 1996.

[47] A. Keys, F. Fidanza, M. Karvonen, N. Kimura, H. Taylor, Indices of relative weight and obesity, Journal of Chronic Diseases 25 (1972) 329–343.

[48] World Health Organization, Obesity and overweight – fact sheet n. 311, http://www.who.int/mediacentre/factsheets/fs311/en/, 2016.

[49] World Health Organization, Physical activity – fact sheet n. 385, http://www.who.int/mediacentre/factsheets/fs385/en/, 2017.

[50] M.K. Tulloch-Reid, D.E. Williams, H.C. Looker, R.L. Hanson, W.C. Knowler, Do measures of body fat distribution provide information on the risk of type 2 diabetes in addition to measures of general obesity?, Diabetes Care 26 (9) (2003) 2556–2561.

[51] A. Onat, G. Hergenc, H. Yuksel, G. Can, E. Ayhan, Z. Kaya, D. Dursunoglu, Neck circumference as a measure of central obesity: associations with metabolic syndrome and obstructive sleep apnea syndrome beyond waist circumference, Clinical Nutrition 28 (1) (2009) 46–51.

[52] C. Stabe, A. Vasques, M. Lima, M. Tambascia, J. Pareja, A. Yamanaka, B. Geloneze, Neck circumference as a simple tool for identifying the metabolic syndrome and insulin resistance: results from the Brazilian Metabolic Syndrome Study, Clinical Endocrinology 78 (6) (2013) 874–881.

[53] L. Ben-Noun, A. Laor, Relationship between changes in neck circumference and cardiovascular risk factors, Experimental and Clinical Cardiology 11 (1) (2006) 14–20.

[54] D. Brenner, K. Tepylo, K. Eny, L. Cahill, A. El-Sohemy, Comparison of body mass index and waist circumference as predictors of cardiometabolic health in a population of young Canadian adults, Diabetology & Metabolic Syndrome 2 (1) (2010) 1–8, https://doi.org/10.1186/1758-5996-2-28.

[55] I. Janssen, P. Katzmarzyk, R. Ross, Body mass index, waist circumference, and health risk: evidence in support of current national institutes of health guidelines, Archives of Internal Medicine 162 (18) (2002) 2074–2079.

[56] A. Liese, et al., Development of the multiple metabolic syndrome in the ARIC cohort: joint contribution of insulin, BMI, and WHR. Atherosclerosis risk in communities, Annals of Epidemiology 7 (6) (1997) 407–416.

[57] B. Heitmann, et al., Hip circumference and cardiovascular morbidity and mortality in men and women, Obesity Research 12 (2004) 482–487.

[58] R. Huxley, S. Mendis, E. Zheleznyakov, S. Reddy, J. Chan, Body mass index, waist circumference and waist: hip ratio as predictors of cardiovascular risk—a review of the literature, European Journal of Clinical Nutrition 64 (2009) 16–22, https://doi.org/ 10.1038/ejcn.2009.68.

[59] R. Newcombe, S. Izadi, O. Hilliges, D. Molyneaux, D. Kim, A. Davison, et al., Kinectfusion: real-time dense surface mapping and tracking, in: Proceedings of the 10th IEEE International Symposium on Mixed and Augmented Reality (ISMAR), Basel, Switzerland, 2011, pp. 127–136.

[60] S. Li, Z. Lei, M. Ao, The HFB face database for heterogeneous face biometrics research, in: 6th IEEE Workshop on Object Tracking and Classification Beyond and in the Visible Spectrum (OTCBVS, in conjunction with CVPR 2009), Miami, Florida, 2009.

[61] H. Li, E. Vouga, A. Gudym, L. Luo, J. Barron, G. Gusev, 3D self-portraits, ACM Transactions on Graphics 32 (6) (2013) 1–9, https://doi.org/10.1145/2508363. 2508407.

[62] J. Tong, J. Zhou, L. Liu, Z. Pan, H. Yan, Scanning 3D full human bodies using Kinects, IEEE Transactions on Visualization and Computer Graphics 18 (2012) 643–650, https://doi.org/10.1109/TVCG.2012.56.

[63] A. Weiss, D. Hirshberg, M. Black, Home 3D body scans from noisy image and range data, in: 2011 IEEE International Conference on Computer Vision (ICCV), IEEE, pp. 1951–1958, 2011.

[64] F. Remondino, 3-D reconstruction of static human body shape from image sequence, Computer Vision and Image Understanding 93 (1) (2004) 65–85, https://doi.org/10. 1016/j.cviu.2003.08.006.

[65] K. Robinette, H. Daanen, E. Paquet, The CAESAR project: a 3-D surface anthropometry survey, in: IEEE Second International Conference on 3-D Digital Imaging and Modeling, IEEE, 1999, pp. 380–386.

[66] L. Pishchulin, S. Wuhrer, T. Helten, C. Theobalt, B. Schiele, Building statistical shape spaces for 3D human modeling, Pattern Recognition (2017).

[67] P. Phillips, P. Flynn, T. Scruggs, K. Bowyer, K. Hoffman, J. Marques, Jaesik Min, W. Worek, Overview of the face recognition grand challenge, in: IEEE Conference on Computer Vision and Pattern Recognition (CVPR), IEEE, 2005, pp. 947–954.

[68] M.P. Evison, R.W.V. Bruegge, The Magna database: a database of three-dimensional facial images for research in human identification and recognition, Forensic Science Communications 10 (2) (2008).

[69] V. Blanz, T. Vetter, A morphable model for the synthesis of 3D faces, in: SIGGRAPH 1999, ACM, 1999, pp. 187–194.

[70] T. Huynh, R. Min, J. Dugelay, An efficient LBP-based descriptor for facial depth images applied to gender recognition using RGB-D face data, in: ACCV Workshop on Computer Vision with Local Binary Pattern Variants, Springer, 2012, pp. 133–135.

[71] P. Paysan, R. Knothe, B. Amberg, S. Romdhani, T. Vetter, A 3D face model for pose and illumination invariant face recognition, in: 6th IEEE International Conference on Advanced Video and Signal Based Surveillance for Security, Safety and Monitoring in Smart Environments, IEEE, 2009, pp. 296–301.

[72] T. Gatzke, C. Grimm, M. Garland, S. Zelinka, Curvature maps for local shape comparison, in: Proceedings of the International Conference on Shape Modeling and Applications 2005, SMI '05, IEEE Computer Society, Washington, DC, USA, 2005, pp. 246–255.

[73] K. Gębal, J.A. Bærentzen, H. Aanæs, R. Larsen, Shape analysis using the auto diffusion function, Computer Graphics Forum 28 (5) (2009) 1405–1413, https://doi.org/10.1111/j.1467-8659.2009.01517.x.

[74] N. Gelfand, N.J. Mitra, L.J. Guibas, H. Pottmann, Robust global registration, in: Proceedings of the Third Eurographics Symposium on Geometry Processing, SGP '05, Eurographics Association, ISBN 3-905673-24-X, 2005.

[75] R. Gal, D. Cohen-Or, Salient geometric features for partial shape matching and similarity, ACM Transactions on Graphics 25 (1) (2006) 130–150, https://doi.org/10.1145/1122501.1122507.

[76] A. Elad, R. Kimmel, On bending invariant signatures for surfaces, IEEE Transactions on Pattern Analysis and Machine Intelligence 25 (10) (2003) 1285–1295, https://doi.org/10.1109/TPAMI.2003.1233902.

[77] B.K. Ng, B.J. Hinton, B. Fan, A.M. Kanaya, J.A. Shepherd, Clinical anthropometrics and body composition from 3D whole-body surface scans, European Journal of Clinical Nutrition 70 (11) (2016) 1265–1270, https://doi.org/10.1038/ejcn.2016.109.

[78] A. Tahrani, K. Boelaert, R. Barnes, et al., Body volume index: time to replace body mass index?, Endocrine Abstracts 15 (2008) 104.

[79] A. Romero-Corral, V. Somers, F. Lopez-Jimenez, Y. Korenfeld, S. Palin, K. Boclaert, et al., 3-D body scanner, body volume index: a novel, reproducible and automated anthropometric tool associated with cardiometabolic biomarkers, Obesity Research Journal 16 (1) (2008) 266-P.

[80] N.Y. Krakauer, J.C. Krakauer, A new body shape index predicts mortality hazard independently of body mass index, PLoS ONE 7 (7) (2012) 1–10, https://doi.org/10.1371/journal.pone.0039504.

[81] C.Y. Yu, C.H. Lin, Y.H. Yang, Human body surface area database and estimation formula, Burns 36 (5) (2010) 616–629, https://doi.org/10.1016/j.burns.2009.05.013.

[82] M. Kouchi, M. Mochimaru, Errors in landmarking and the evaluation of the accuracy of traditional and 3D anthropometry, Applied Ergonomics 42 (3) (2011) 518–527, https://doi.org/10.1016/j.eswa.2010.08.048.

[83] A. Giachetti, C. Lovato, F. Piscitelli, C. Milanese, C. Zancanaro, Robust automatic measurement of 3D scanned models for the human body fat estimation, IEEE Journal of Biomedical and Health Informatics 19 (2) (2015) 660–667, https://doi.org/10.1109/JBHI.2014.2314360.

[84] D.B. Stefan, D.A. Gilbert, Method for determining a subject's degree of adiposity, 2009, US Patent 7,599,537.

[85] V.F. Ferrario, C. Dellavia, G.M. Tartaglia, M. Turci, C. Sforza, Soft-tissue facial morphology in obese adolescents: a three-dimensional noninvasive assessment, Angle Orthodontist 74 (1) (2004) 37–42.

[86] J. Djordjevic, D.A. Lawlor, A.L. Zhurov, et al., A population-based cross-sectional study of the association between facial morphology and cardiometabolic risk factors in adolescence, BMJ Open 3 (2013) 1–10, e002910.

[87] B.J. Lee, J.H. Do, J.K. Kim, A classification method of normal and overweight females based on facial features for automated medical applications, Journal of Biomedicine and Biotechnology 2012 (2012) 1–9, 834578, https://doi.org/10.1155/2012/834578.

[88] B.J. Lee, J.K. Kim, Predicting visceral obesity based on facial characteristics, BMC Complementary and Alternative Medicine 14 (248) (2014) 1–9, https://doi.org/10.1186/1472-6882-14-248.

[89] S.M. Banabilh, A.H. Suzina, S. Dinsuhaimi, A.R. Samsudin, G.D. Singh, Craniofacial obesity in patients with obstructive sleep apnea, Sleep and Breathing 13 (2009) 19–24, https://doi.org/10.1007/s11325-008-0211-9.

[90] D. Giorgi, M.A. Pascali, P. Henriquez, B.J. Matuszewski, S. Colantonio, O. Salvetti, Persistent homology to analyse 3D faces and assess body weight gain, The Visual Computer 33 (5) (2017) 549–563.

[91] S. Cohen, D. Janicki-Deverts, G.E. Miller, Psychological stress and disease, JAMA: the Journal of the American Medical Association 298 (2007) 1685–1687.

[92] A. Damasio, G.B. Carvalho, The nature of feelings: evolutionary and neurobiological origins, Nature Reviews. Neuroscience 14 (2013) 143–152.

[93] M.F. Piepoli, A.W. Hoes, S. Agewall, et al., European Guidelines on cardiovascular disease prevention in clinical practice, European Heart Journal 2016 (37) (2016) 2315–2381.

[94] J.E. Dimsdale, Psychological stress and cardiovascular disease, Journal of the American College of Cardiology 51 (2008) 1237–1246.

[95] M. Bagherniya, S.S. Khayyatzadeh, A. Avan, M. Safarian, M. Nematy, G.A. Ferns, et al., Metabolic syndrome and its components are related to psychological disorders: a population based study, Diabetes & Metabolic Syndrome: Clinical Research & Reviews (2017), http://linkinghub.elsevier.com/retrieve/pii/S1871402117300784.

[96] K.W. Davidson, E. Mostofsky, W. Whang, Don't worry, be happy: positive affect and reduced 10-year incident coronary heart disease: The Canadian Nova Scotia Health Survey, European Heart Journal 31 (2010) 1065–1070.

[97] J. Turner, Emotional dimensions of chronic disease, Western Journal of Medicine 172 (2000) 124–128.

[98] M.E. Kemeny, M. Schedlowski, Understanding the interaction between psychosocial stress and immune-related diseases: a stepwise progression, Brain, Behavior, and Immunity 21 (2007) 1009–1018.

[99] E.M.V. Reiche, S.O.V. Nunes, H.K. Morimoto, Stress, depression, the immune system, and cancer, Lancet Oncology 5 (2004) 617–625.

[100] H. Gunes, M. Piccardi, Bimodal face and body gesture database for automatic analysis of human nonverbal affective behavior, in: Proceedings, vol. 1, International Conference on Pattern Recognition, 2006, pp. 1148–1153.

[101] M. Soleymani, M. Pantic, T. Pun, Multimodal emotion recognition in response to videos, IEEE Transactions on Affective Computing 3 (2012) 211–223.

[102] P. Ekman, W.V. Friesen, Facial Action Coding System: A Technique for the Measurement of Facial Movement, ISBN 0931835011, 1978.

[103] M. Shreve, S. Godavarthy, V. Manohar, D. Goldgof, S. Sarkar, Towards macro- and micro-expression spotting in video using strain patterns, in: 2009 Workshop on Applications of Computer Vision (WACV), IEEE, 2009, pp. 1–6.

[104] X. Li, X. Hong, A. Moilanen, X. Huang, T. Pfister, G. Zhao, M. Pietikainen, Towards reading hidden emotions: a comparative study of spontaneous micro-expression spotting and recognition methods, IEEE Transactions on Affective Computing 1 (2017), arXiv:1511.00423.

[105] G. Giannakakis, M. Pediaditis, D. Manousos, E. Kazantzaki, F. Chiarugi, P.G. Simos, et al., Stress and anxiety detection using facial cues from videos, Biomedical Signal Processing and Control 31 (2017) 89–101.

[106] C. Hayes, E. Poole, G. Iachello, S. Patel, A. Grimes, G. Abowd, et al., Physical, social, and experiential knowledge in pervasive computing environments, IEEE Pervasive Computing 6 (4) (2007) 56–63.

[107] M. Langheinrich, Privacy by design principles of privacy-aware ubiquitous systems, in: Proceedings of Ubiquitous Computing (Ubicomp), in: Lecture Notes in Computer Science, vol. 2201, 2001, pp. 273–291.

[108] World Health Organization, Cardiovascular diseases, Fact sheet, updated May 2017, http://www.who.int/mediacentre/factsheets/fs317/en/.

[109] A. Pandya, T.A. Gaziano, M.C. Weinstein, D. Cutler, More Americans living longer with cardiovascular disease will increase costs while lowering quality of life, Health Affairs 32 (10) (2013) 1706–1714, NIHMS150003.

[110] F. Sassi, J. Hurst, The Prevention of Lifestyle Related Chronic Disease: An Economic Framework, OECD Health Working Papers 32, 2008.

[111] A.M. Ward, C. Heneghan, R. Perera, D. Lasserson, D. Nunan, D. Mant, P. Glasziou, What are the basic self-monitoring components for cardiovascular risk management?, BMC Medical Research Methodology 10 (2010) 105.

[112] G. Coppini, R. Favilla, A. Gastaldelli, S. Colantonio, P. Marraccini, Moving medical semeiotics to the digital realm: SEMEOTICONS Approach to face signs of cardiometabolic risk, in: Proceedings, HEALTHINF 2014 – 7th International Conference on Health Informatics, Part of 7th International Joint Conference on Biomedical Engineering Systems and Technologies, BIOSTEC 2014, 2014, pp. 606–613.

[113] P. Henriquez, B.J. Matuszewski, Y. Andreu-Cabedo, L. Bastiani, S. Colantonio, G. Coppini, et al., Mirror mirror on the wall... an unobtrusive intelligent multisensory mirror for well-being status self-assessment and visualization, IEEE Transactions on Multimedia 19 (7) (2017) 1467–1481.

[114] S. Klemm, Y. Andreu, P. Henriquez, B.J. Matuszewski, Robust face recognition using key-point descriptors, in: Proceedings of the 10th International Conference on Computer Vision Theory and Applications, SCITEPRESS – Science and Technology Publications, 2015, pp. 447–454.

[115] M. Kazhdan, M. Bolitho, H. Hoppe, Poisson surface reconstruction, in: Proceedings of the Fourth Eurographics Symposium on Geometry Processing, SGP '06, 2006, pp. 61–70.

[116] M.A. Pascali, D. Giorgi, L. Bastiani, E. Buzzigoli, P. Henríquez, B.J. Matuszewski, et al., Face morphology: can it tell us something about body weight and fat?, Computers in Biology and Medicine 76 (2016) 238–249, https://doi.org/10.1016/j.compbiomed.2016.06.006.

[117] M. Larsson, R. Favilla, T. Strömberg, Assessment of advanced glycated end product accumulation in skin using auto fluorescence multispectral imaging, Computers in Biology and Medicine 85 (2015) 106–111.

[118] E. Christinaki, G. Giannakakis, F. Chiarugi, M. Pediaditis, G. Iatraki, D. Manousos, et al., Comparison of blind source separation algorithms for optical heart rate monitoring, in: Proceedings of the 4th International Conference on Wireless Mobile Communication and Healthcare – Transforming Healthcare Through Innovations in Mobile and Wireless Technologies, MOBIHEALTH 2014, 2015, pp. 339–342.

[119] D. Germanese, M. Righi, M. Guidi, M. Magrini, M. D'Acunto, O. Salvetti, A device for self-monitoring breath analysis, in: The Seventh International Conference on Sensor Device Technologies and Applications, SENSORDEVICES 2016, 2016, pp. 77–82.

[120] D. Kaplan, Structural Equation Modeling: Foundations and Extensions, Sage, Newbury Park, CA, USA, 2000.

[121] WHO European Regional Office, Who-Five Well-being Index (WHO-5), https://www.psykiatri-regionh.dk/who-5/Pages/default.aspx.

[122] R. Adams, Improving health outcomes with better patient understanding and education, Risk Management and Healthcare Policy 3 (2010) 61.

# CHAPTER 7

# Computer Vision for Egocentric (First-Person) Vision

**Mariella Dimiccoli**
University of Barcelona and Computer Vision Center, Barcelona, Spain

## Contents

## Abstract

Egocentric (first-person) vision is a relatively new research topic in the field of computer vision that is increasingly attracting the interest of the community. It entails analyzing images and videos captured by a wearable camera, which is typically mounted on the chest or head, and provides a first-person view of the world. This human-centric perspective is naturally suited to gathering visual information about everyday observations and interactions, which in turn can reveal the attention, behavior, and goals of its wearer. By taking advantage of the first-person point of view paradigm, the visual understanding of egocentric images and videos is rapidly advancing. This chapter focuses on those aspects of egocentric vision that can be directly exploited to develop platforms for ubiquitous context-aware personal assistance and health-monitoring, also highlighting potential applications and further research opportunities in the context of assistive technologies.

## Keywords

Egocentric (first-person) vision, Personal assistive technologies, Context-aware assistance, Daily living activity recognition, Daily living activity forecasting, Social interaction analysis

Computer Vision for Assistive Healthcare.
DOI: https://doi.org/10.1016/B978-0-12-813445-0.00007-1

## 7.1 INTRODUCTION

The enormous advances in the field of wearable technology over the past decade have led to hardware miniaturization, increased data storage, and reduced power consumption of wearable sensors [1,2]. In particular, today's wearable cameras are compact and robust digital recording devices that can acquire images and videos at different resolutions and frame rates, in a passive fashion, and from a first-person point of view. Wearable sensors have been used increasingly in health related research for the past several years since they offer several benefits associated with long-term monitoring of individuals in home and community settings, such as objectivity of measurements, real time monitoring, and nonintrusiveness [3,4]. With respect to other wearable sensors, wearable cameras provide richer contextual information [5], which can be used to characterize the social and environmental context of activities. In addition, the first-person perspective is naturally suited to gathering visual information about everyday observations and interactions that can reveal the focus of attention, behavior, and goals of its wearer [6].

The first wearable camera deployed in health research was the Sense-Cam, a still camera today commercialized as the Vicon Revue, usually worn as a necklace, that automatically captures a picture every 20 seconds. Given its low frame rate, it is particularly suitable for capturing image data over long periods of time, commonly called *visual lifelogs*. Several studies have shown the benefit of the SenseCam in the cognitive rehabilitation of those with early dementia [7,8], and as a tool to objectively track lifestyle behaviors and the contexts in which they occur [9]. More recently video cameras such as GoPro and Google Glass have gained attention in the field of health research [10,11]. Because they are worn as glasses or mounted on the head, they show exactly what a person is looking at and hence can predict what tasks that person is trying to perform and can, potentially, take action. Additionally, a head-mounted video camera is an ideal sensor for capturing eye contact; it can be used to characterize behavior in face-to-face interactions [12] and for the early detection of autism spectrum disorder [13].

These past studies suggest that wearable cameras can have application breakthroughs in the field of assistive technologies to support independent and healthy living of older and/or impaired people, and to improve their quality of life [14,15]. However, the deployment of wearable cameras strongly relies on advances in egocentric (first-person) vision, the subfield within computer vision that entails analyzing images and videos captured by a wearable camera. Although egocentric vision is a relatively new field of research, it has matured rapidly over the past few years [16,17] due to the interest raised in the community for both its technical challenges and the opportunities it opens. Taking advantage of the first-person point-of-view paradigm, several advances have recently been possible in areas such as video segmentation and summarization [18–20], activity recognition [21], and social interaction analysis [22–24].

This chapter aims to describe those aspects of egocentric visual understanding that can in turn apply to platforms for ubiquitous context-aware personal assistance and health monitoring. The chapter is organized into four key topics that reflect this goal: Section 7.2 focuses on contextual understanding, Sections 7.3 and 7.4 concentrate, respectively, on activity recognition and activity forecasting, while Section 7.5 is devoted to social interaction analysis. For each topic the latest advances in computer vision are introduced and potential applications discussed. Section 7.6 concludes this chapter with a critical assessment of what first-person vision understanding has achieved so far and with a discussion of the challenges that remain to be addressed to enable cutting-edge assistive technologies.

## 7.2 CONTEXTUAL UNDERSTANDING

In their everyday lives, people act and move in environments with specific functional attributes, and perform daily living activities in prototypical scenes. For example, people typically walk around in the office at work, cook in the kitchen, and brush their teeth in the bathroom at home. This suggests that physical locations encode in their functional attributes the ability to perform specific activities. Therefore, understanding the functional attributes of the environment and recognizing personal locations are valuable properties for context-aware wearable computing. The knowledge of the context—where the user is located in the home (e.g. kitchen vs bedroom), what objects are being employed (fridge vs book)—can be leveraged to help with tasks such as navigating through the inside of a house or picking up a pot and food to cook something.

### Recognizing Personal Locations

In line with this view, Furnari et al. [25] proposed a system that is able to discriminate, from a video captured with a wearable camera, a set of personal locations specified by the user and to reject all frames not related to any of them. They considered personal locations such as car, garage, office, living room, coffee vending machine, etc., that are associated with the

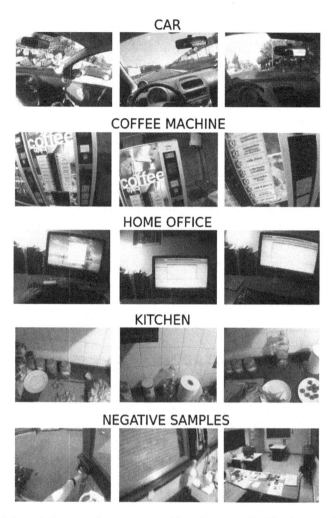

**Figure 7.1** Sample images of some personal locations specified by the user (first four rows) and negative samples not representing any of the specified personal locations (last row). These images were acquired with the Looxcie LX2W wearable camera and downloaded from the author's project page.[1]

daily activities of the user (see Fig. 7.1). Therefore, the proposed system has useful applications not only for helping the visually impaired to recognize their personal objects/locations, but also for monitoring their specific daily activities such as having coffee or driving, which are important in order to characterize healthy habits. The system consists of two modules, one

---

[1] http://iplab.dmi.unict.it/PersonalLocations/.

responsible for multiclass discrimination and the other responsible for rejecting all images not belonging to any of the previously specified locations, hereafter called *negative samples*. The rejection module is formulated as an entropy-based negative rejection mechanism and is based on the assumption that temporally adjacent frames share the same personal location class. The uncertainty of the conditional probability of a generic class given $N$ consecutive observations is used to infer the presence of outliers. Indeed, when the sequence of $N$ observations presents negative samples, the posterior distribution exhibits a large degree of uncertainty since there is no agreement on the identity of the considered samples. The multiclass discrimination module is based on standard supervised techniques, mainly a SVM classifier on the features extracted by a fine-tuned convolutional neural network pretrained on ImageNet. Experimental results showed the importance and the difficulty of a reliable negative-rejection module and achieved a robust performance (accuracy above 85%) in recognizing eight personal locations classes. By testing the proposed approach on data acquired by four different wearable cameras, the authors showed that head-mounted devices with large fields of view such as the wide-angle Looxcie LX2 are most suited to recognizing the user's personal locations.

## Recognizing Generic Scenes and Objects

The ability to recognize generic objects and scenes, other than personal ones, can be employed to assist the user in unfamiliar environments. In an effort to make scene and object identification reliable in egocentric videos, Vaca-Castano el al. [26] leveraged the temporal and the scene contexts. Since people in their daily lives typically spend from minutes to hours in the same scene, the context tends to be constant during several consecutive frames. This observation enables a Conditional Random Field formulation of the scene identification problem that assumes a graph representation of the video, where each node is a frame of the video and is connected temporally through edges to its $r$ neighbor frames. The energy function to minimize has an unary term that favors scene labels with high scores and a pairwise term that penalizes short-term changes of the scene identity over time for frames linked by edge potentials in the graph. Since objects such as the fridge and the kettle are typically found in the kitchen, the identified scene is then used as the scene context to improve object recognition. In practice, the scores of the object detector are reranked so that detections in scenes that typically do not contain the object, for example a kettle in the bathroom, are penalized. To this end a Support Vector Regressor (SVR), trained separately for every object detector, is used to map the object detection score of an object detector to a new score value, considering the scene identity. During test, the output of the regressor associated with the type of object is used as the new object score. In addition to scene context, and independently from its availability, temporal information is employed to improve the results of the existing object detectors by using Long Short-Term Memory (LSTM) recurrent neural networks. During training, the visual feature vector of each frame is mapped to a vector of object probabilities. During the test, the output of the LSTM indicates the likelihood of finding a type of object given the visual information of the frame and its history. The reranking is done by adding the value of this likelihood to the score of the object detector, weighted by a constant indicating the importance of the scene information in the final score. Experiments performed on the Activities of Daily Living (ADL) dataset [27] reported an accuracy above 70% for scene classification. The proposed methods for improving object detection are effective when the base object detector has a mean average precision (mAP) over 20%. Exploiting the temporal context with LSTM is more effective than exploiting the scene context when the latter needs to be estimated from scene classification.

## Discovery of Objects and Their Modes of Interaction

To avoid the need to manually segment and annotate videos, which is the major limitation of assistive video guidance devices in task performance, Damen et al. [28] proposed an unsupervised approach to discover task-relevant objects (TROs). A TRO is an object or part of an object with which a person interacts during task performance. For example, a person operating in the kitchen may interact with different parts of it while performing a task. For each discovered object, three models are built: a location model that guides the user to where an object can be found; an appearance model that recognize the object when visible in the field of view of the camera; and a video-based guidance model consisting of a collection of usage snippets on how different users interacted with the same discovered object. Given a sequence of egocentric images collected from multiple people around a common environment registered jointly with eye tracking information, TROs are defined as a set of $K$ image regions from the sequence, where $K$ is not fixed a priori but estimated. At most, one task-relevant image part is assumed to be present within each image. Image parts are described by combining position features that allow the discovery of static objects, and appearance features (HOG) that allow the discovery of movable objects by relying on appearance feature similarity over time. The proposed online object discovery method works by clustering image parts as they are collected, and by updating clusters incrementally. In online discovery, a TRO is defined as a collection of at least $n$ consecutive and similar image parts, independently of their location in the image. TROs are found by applying the Bhattacharyya distance to merge clusters. Once the model has been trained, discovered TROs can be used to provide recommendations to users. The appearance model allows a previously discovered task-relevant object to be recognized by relying on gaze fixation, so that a video-based snippet guidance can then suggest a mode of interaction, given the current status of the object. The discovery of a TRO on the Bristol Egocentric Object Interactions Dataset (BEOID) yields an $F1$ score above 80%, while the accuracy of the modes of interaction discovery was 58%.

## Estimating Action Maps Associated With Spatial Locations

If scene identification can provide contextual information for object recognition and location-aware assistance services to the user in indoor scenarios, the spatial layout of scenes and the objects they contain encode the ability of a user to perform activities at various locations. For example, the ability to perform the activity *sit* is encoded by the presence of chairs and couches in the common area of an office space and the ability to perform the activity *walk* is encoded by the presence of free space. Learning the map between locations and activities is beneficial for indoor navigation systems. Rhinehart and Kitani [29] proposed a method that learns the relationship between actions, scenes, and objects to predict "Action Maps" that associate possible actions with every spatial location on a map over a large environment such as a house or office space. First, a 3D reconstruction of the environment is obtained from a collection of egocentric videos by using structure from motion. Second, the visual context of actions (i.e. scene appearance and object detections) is learned from a collection of egocentric human activity videos. The same videos are used to detect and spatially localize actions (see Fig. 7.2). Third, the localized action detection and visual context data are integrated using a matrix completion framework to estimate the Action Map matrix $\mathbf{R} \in \mathcal{R}_+^{M,A}$. Each element of the matrix, say $r_{m,a}$, describes to what extent the activity $a$ can be performed at location $m$. Given the sparsity of the observations, matrix $\mathbf{R}$ is initially sparse. To complete it, the authors proposed a matrix factorization framework that incorporates side information about the visual context of actions to perform a functionality estimation of unseen locations. This is achieved by minimizing an energy

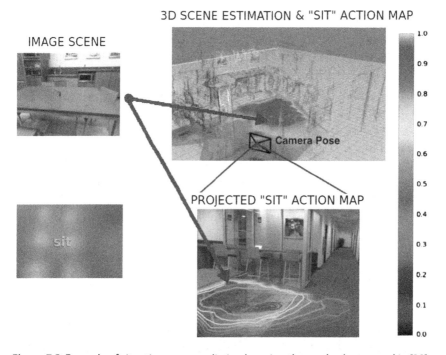

**Figure 7.2** Example of *sit* action map prediction by using the method proposed in [29]. Upper left: Sample image scene of the kitchen. Upper right: The 3D layout of the scene is estimated by using structure from motion. Bottom right: The visualization of the *sit* action map is projected onto the image scene. Image adapted from the author's project page.[2]

functional. The data fidelity term penalizes decompositions with values different from the observed values in **R**. The cross-location regularization term enforces the functional similarity of locations having similar object detections and scene classification descriptions. The cross-action regularization term enforces no penalty for differences across per location action labels. In the experiments, four distinct office buildings, three in the United States and another in Japan, and a home consisting of a kitchen, a living room, and a dining room, were considered. The average per class weighted $F1$ score is above $0.6$.

Recognizing scenes and objects, discriminating personal locations, discovering objects during a task performance, and encoding the ability to perform activities in specific locations are all valuable tasks for contextual awareness wearable computing. The automatic recognition of personal

[2] https://www.youtube.com/watch?v=IBpCRU7ovpQ.

locations, objects, and scenes are highly desired for location-aware assistance services to the user in indoor scenarios. Contextual awareness can be exploited to predict the user's intention before the activity is actually completed to provide appropriate support and to improve person–environment interaction. The next section focuses on the task of recognizing the activities the user is already engaged in.

## 7.3 FIRST-PERSON ACTIVITY RECOGNITION

Understanding human activities from a first-person perspective has a wide range of compelling assistive applications [16], ranging from lifestyle characterization [9] to telerehabilitation [30]. Nearly all of these applications demand robust methods for recognizing the camera wearer's activities. In this section, we introduce state-of-the-art computer vision methods for activity recognition in egocentric videos, including ambulatory activities such as *running* or *walking* (Section 7.3.1), person-to-object interactions such as *typing* or *drinking* (Section 7.3.2), person-to-person interactions such as *hugging* or *shaking hands* (Section 7.3.3), and more subtle activities such as *ego-engagement* (Section 7.3.4).

### 7.3.1 Ambulatory Activities

Patients suffering from obesity, diabetes, or heart disease are often required to follow a well-defined exercise routine as part of their treatment. Therefore, recognizing ambulatory activities such as walking, running, or cycling is useful in order to provide feedback to the caregiver about the patient's behavior. Ambulatory activities are characterized by the fact that they involve full body motion. Therefore, discriminant features are commonly extracted from global motion such as optical flow [31,18].

Poleg et al. [32] proposed a method for robust temporal segmentation of egocentric videos into a hierarchy of 14 motion classes, including long-term activities such as *static, standing, walking, etc.* They train a Convolutional Neural Network (CNN) directly on optical flow instead of pixel intensities. More specifically, the input to the network is sparse optical flow computed on $32 \times 32$ grid, stacked over 60 frames. The network architecture is made of two convolutional layers and two fully connected layers that precede the softmax layer. The first convolutional layer is a 3D spatiotemporal convolution that learns to discriminate long-term activities. The network achieves an average recall rate of 86% in the 14 activity class tasks. While walking, driving, or biking are recognized with very high accuracy (above 80%

F1-score), there is room for improvement when discriminating between stair climbing and sailing, boxing and cooking.

In addition to optical flow data, Abebe et al. [33] exploited magnitude, direction, and frequency (periodicity) characteristics and combined them with virtual inertial features extracted from the video. Their approach stems from the observation that some activities such as *sprinting* and *walking* have similar direction information but differ significantly in their magnitude. On the opposite side, some activities such as *sitting down* and *standing up* vary in their direction components while possessing similar magnitude values. As a consequence their approach aims to extract, for each video segment, a set of feature subgroups that encode the variations in motion magnitude, direction, and dynamics among the activities. Their starting point is to compute the Horn–Schunk optical flow [34] for each subsequent pair of frames over a grid representation to avoid redundancy. The Motion Magnitude Histogram Feature (MMHF) is derived from the histogram representation of grid optical flow magnitude. MMHF is useful to discriminate between activities involving similar direction patterns but different motion magnitudes. The Motion Direction Histogram Feature (MDHF) is the histogram representation of the motion direction. It is crucial to classify activities that might have similar motion magnitudes but different motion direction. The Motion Direction Histogram Standard deviation Feature (MDHSF) represents the standard deviation of each direction bin in MDHF across a video segment. The Fourier Transform of Motion direction Across Frame (FTMAF) is a frequency-domain feature that reflects the variation in direction bins of each MDHF associated with a video segment. Finally, the Fourier Transform of grid Motion Per Frame (FTMPF) measures the variation in grid optical flow in a frame. It helps to discriminate between complex activities with high dynamics of ego-motion (e.g. dribbling) from simple activities (e.g. standing). By computing the first and second derivative, respectively, on the sequence of Gaussian-smoothed intensity centroids, virtual inertial data in terms of centroid velocity and acceleration values are generated. Time and frequency-domain features are extracted from the virtual inertial signals. SVM is used for classification. The methods achieved nearly perfect estimations on the indoor ambulatory activity recognition (IAR) dataset and basketball activity recognition (BAR) dataset.

## 7.3.2 Person-to-Object Interactions

Activities such as *shopping* and *preparing a meal* belong to the set of Instrumental Activities of Daily Living (IADL), which include but are not limited to the activities performed on a daily basis for living at home or in a community [35]. ADL mostly involve object manipulations. Wearable cameras are particularly useful for analyzing activities involving object manipulations since the workspace containing the manipulated objects is usually visible to the camera and occlusions tend to be minimized. The objects being interacted with by the user in a given video are commonly called *active objects* and, together with hand appearance and motion, are especially helpful features for egocentric activity recognition. However, an object's appearance can change dramatically when the object is being interacted with. For example, the appearance of an empty pot may differ drastically from a full pot that is being used for cooking. To distinguish active objects from the passive ones and to learn a discriminative representation of active objects, McCandless and Grauman [36] proposed a randomized spatiotemporal pyramid framework. A space-time pyramid consists of multiple levels of bins, from coarse to fine, where each bin records the frequency with which each object category appears in particular space-time regions. The histograms of each object class are concatenated over all pyramid levels to get a single video descriptor. Instead of using uniformly placed bins, the bin locations are randomly selected following the distribution of active objects in the training set, which is biased towards the lower center field of view and nearly uniform along the temporal dimension. A boosting approach is then used to select the most discriminative spatiotemporal pyramids for the given activity recognition task and to predict its activity label. Experimental results on the ADL dataset [27] (see Fig. 7.3) corroborate the hypothesis that learned spatial cuts are essential in scenes with similar objects appearing across different actions.

In addition to objects and object attributes, hand appearance and local hand motion play an important role in characterizing first-person actions. In fact, hands appear in nearly every frame in egocentric videos since people performing a task tend to coordinate head and hand movements so that active objects lie in the visual field. Several works have assumed that the background is static and have considered the pattern of irregular optical flow as cues for detecting hands and held objects [37], without or with color features [38]. However, in realistic videos, the background can also be dynamic because of the presence of other people, whose hands may also appear in the video and should be distinguished from the hands of the camera wearer. Consequently, optical flow may be misleading in uncontrolled scenarios. Many other works [39] assume that "hands" correspond to contiguous skin regions up to the sleeves and treat the problem of hand

Doing laundry          Making tea

Washing dishes         Brushing teeth

**Figure 7.3** Sample images from the ADL dataset [27] with their corresponding activity labels.

detection as a skin detection problem, and focus on modeling local appearance and global illumination, neglecting optical flow information.

Bambach et al. [40] proposed a method for dealing with more realistic scenarios with other people in the background. They exploit the fact that the spatial location of active objects and hands is biased towards the lower center field of view for fast and accurate hand proposal generation. To this end, they estimated the probability that an object $O$ appears in a region $R$ of image $I$, say $P(O|R, I) \propto P(I|R, O)P(R|O)P(O)$, where $P(O)$, the object occurrence probability of the object $O$ is computed directly from the training data; $P(R|O)$, the prior distribution over the size, shape, and position of regions containing $O$, is estimated by fitting a four-dimensional Gaussian kernel density estimator to the training set; and $P(I|R, O)$, the appearance model, is obtained by using a nonparametric modeling of skin color in YUV color space (disregarding the luminance channel) to estimate the probability that the central pixel of $R$ is skin. Later, they used a standard CNN classification framework to classify each proposal as "hand" or "non-hand" and a semisupervised segmentation algorithm, GrabCut [29], to segment hands. Since hand segmentation conveys hand pose, the authors investigated whether activities can be detected based solely on hand pose information. To this goal they fine-tuned a CNN to four different activities from images where all non-hand background was masked using ground truth segmentation. The classifier achieved roughly 2.7 times random chance, on the test dataset with non-hand regions blanked out, confirming that there is a strong connection between activities and hand location and pose. Their ground truth experiments show that automated approaches would benefit from increased segmentation accuracy, such as that proposed by Betancourt et al. [41].

Ma et al. [42] proposed a way to recognize activities represented by an action–object pair such as *take bread*. A twin stream convolutional neural network architecture that takes into account the appearance of objects and hands, their motion, and the camera motion. In particular, one stream analyzes appearance information with the goal of identifying objects of manipulation. The other stream analyzes motion information and differentiates between action labels such as put, take, close, scoop, and spread. The appearance stream performs hand segmentation and object localization, whereas the motion stream jointly captures different local motion and temporal features by stacking optical flow of adjacent frames in a fixed length temporal window. The two networks are merged by concatenating the last fully connected layers from the two networks and adding another fully connected layer on top. Their system achieves a global accuracy of around 76%.

### 7.3.3 Person-to-Person Interactions

The ability to recognize human activities targeted to a user wearing a camera from his own viewpoint may be a crucial step in facilitating computational understanding of social communications. A system with this ability would allow a visually impaired people to be socially aware of what other people want to do, for instance give a hug, which would help them feel less alienated.

Proper understanding of person-to-person interactions such as "shaking hands with the observer", "pointing at the observer", and "punching the observer" requires encoding the link between "how the observer moves" and "what he sees" (see Fig. 7.4). For example, a third person hugging the observer moves his body in a very specific way, and in turn his movement will cause the ego-motion of the observer being hugged. Therefore motion is a crucial feature for recognizing human–human interaction activities from an egocentric perspective. Based on this observation, Ryoo and Matthies [43,44] captured the ego-motion of the observer by global motion descriptors obtained from the histograms of locations and orientations of the

Shaking hand                    Punching

Hugging                         Throwing

**Figure 7.4** Sample images of actions directed to the camera wearer, in this case a humanoid, taken from the JPL-Interaction dataset [43].

optical flow, and captured the body movements of an interacting person by local motion descriptors. The global motion descriptor is obtained as a histogram of extracted optical flows in fixed duration time intervals. The local motion descriptor is obtained by first applying a spatiotemporal saliency detector to the 3-D XYT video volume and then summarizing gradient values of the detected video patch. The histogram of the visual words obtained by clustering local and global motion descriptors separately is used to represent the video motion. Multichannel kernels that robustly combine information from global and local motion are used to compute video similarity, leading to the computation of a nonlinear decision boundary. Since there are strong cause-and-effect relations between global and local motion in person-to-person interactions, learning the structure a representation of each activity should be advantageous for recognition. Therefore, each activity is split into a continuous concatenation of its subevents and the similarity between two videos is measured by performing hierarchical segment-to-segment matching. The authors represent the "optimal kernel function" and candidate structure kernels in terms of Gram matrices and measure their similarities to evaluate structure kernels for activity recognition. Near 90% accuracy is achieved on the JPL-Interaction dataset captured by a GoPro camera mounted on the head of a humanoid.

## 7.3.4 Ego-Engagement in Browsing Scenarios

While shopping at markets or visiting a museum, people often try to locate specific objects or to inspect something more closely with the intent of getting additional information. This kind of behavior may be used as an alarm for a robotic assistant performing a physical tasks such as reading the label next to a picture in a museum or taking the item being inspected by the user and putting it into a shopping bag. Su and Grauman [45] defined the time interval where the user *is attracted by some object(s), and interrupts the ongoing flow of activity to purposefully gather more information about the object(s)* as ego-engagement. Typically, when undergoing ego-engagement, people first glance at an object, and then either turn towards the object or turn away quickly. Therefore, to capture engagement it is important to look at the sequence of actions during an interval. Following this idea, the proposed method first computes frame-wise predictions, then leverages these frame predictions to generate interval hypotheses, and finally classifies each interval. The frame descriptor captures dominant ego-motion and dynamic scene motion. It consists of temporally smoothed optical flow vectors concatenated over cells in a uniform grid, together with their mean and standard deviation. Later, positive interval proposals are obtained by generating, for a given threshold, contiguous frames whose confidence exceeds the threshold. Positive intervals generated with different thresholds are pooled together to form the final set of positive interval proposals. Each interval proposal is described by a three-level temporal pyramid representation of positive intervals and is augmented with those of its temporal neighbor intervals (i.e. before and after). The goal of the augmentation is to capture the change of motion from low engagement to high engagement and back. Finally, a random forest classifier is used over this descriptor. The proposed approach was tested on the UT Egocentric Engagement (UT EE) dataset including browsing scenarios in a mall, in a market, and in a museum, and the public UT Egocentric dataset (UT Ego). The $F1$ score accuracy of all methods is near 70%. The method generalizes to unseen browsing scenarios, suggesting that some properties of ego-engagement are independent of visual appearance and relate directly to ego-motion.

The following section focuses on predicting what the user is likely to do next, based on past and current observations.

## 7.4 FIRST-PERSON ACTIVITY FORECASTING

A system able to understand what a person is doing and to anticipate what the person is likely to do next can take preemptive actions to better meet the user's needs and/or facilitate context-aware prompting.

### Early Activity Forecasting

The problem of activity prediction or forecasting involves recognizing ongoing activities at their early stages, that is before they are fully executed. This requires real-time computations that make the prediction as early as possible. Ryoo and Matthies [44] formulated the activity prediction problem as probabilistic inference by modeling two situations: (1) subtle activities (called *onset activities*) such as standing and pointing that serve as cues to infer the next activities and (2) preactivities (called *onset signatures*) that capture how the similarity between each possible video segment and the onset activity is changing over time. Onset signatures take the form of time series and are represented as a succession of histograms of time series gradients corresponding to the different segments composing the time series. The probability that the activity $C$ is *ongoing* at frame $t$ regardless of its starting/ending time, given the video stream $V$ until frame $t$, denoted simply as $V$, can be approximated as

$$P(C^t|V) = P(C^{[t_1,t_2]}, d|V) \approx \sum_d \sum_{[t_1,t_2]} \sum_{(\mathbf{A},I)} P(C^{[t_1,t_2]}, d, \mathbf{A}, I, V),$$

where $t_1 \leq t \leq t_2$ is a time interval including $t$, $d$ is the progress level of the ongoing activity, $\mathbf{A} = \{A_1, ..., A_{|A|}\}$ is a set of all previous activity occurrences, and $I$ is the intention of the interacting person. This quantity can be further approximated by a set of binary classifiers

$$P(C^t|V) \approx \max_d \max_{[t_1,t_2]} \sum_I F_{(C,d)}(x(t), V_{[t_1,t]}) \cdot \Gamma_C([t_1, t_2], I)$$

where $x(t)$ is the vector representation of the onset signature. This classifier is learned for each pair activity $C$ and possible progress level $d$, and takes as input positive and negative samples of $V$ and $x$ together with their ground truth labels. Based on the training samples, the SVM classifier will learn to focus on particular onset signatures while ignoring irrelevant onset activities. The video features and onset representations are provided jointly as multichannel input to $F_{(C,t)}$ so that the recognition takes into account both the temporal structure and the onset of human activities. The function $\Gamma_C$ is trained in a similar way by providing $I$ and $[t_1, t_2]$ as inputs. The mean AP at the 50% observation is above 60%.

## Long-Term Activity Forecasting

Bokhari and Kitani [46] proposed a reinforcement learning approach for long-term activity forecasting, able to predict the trajectory through the environment, along with the actions performed along the trajectory, learning from only a few examples. The prior knowledge of the environment is encoded by three components: (1) the *state space*, which represents the environment, (2) the *action set*, which are the actions that can be performed in each state, and (3) the *reward function*, which describes the reward an agent receives when transitioning from one state to another upon performing a certain action. Since the goal is to forecast the agent's decisions in the environment, the total state space is expressed as the Cartesian product $S = X \times W$, where $W$ represents all possible states of an activity and $X$ represents all possible position states. The total action set is expressed as the union of $A = M \cup E$. The action set $M$ includes movement actions, such as moving in one of the four cardinal directions, and affects the location portion. The action set $E$ can be thought of as environmental interactions, such as manipulating an object, and affects the activity portion. The reward function determines which policies are encouraged and which are not, and is therefore able to encode prior knowledge about the scene such as the fact that people will avoid obstacles. Since the goal is to encourage those policies that mimic human behavior, the reward is a function of three variables: the end goal, the locations where actions may be performed, and the obstacles in the environment. Reinforcement learning leverages this prior knowledge to find the optimal policy, in other words to learn the best action to perform in each state in order to maximize the expected future reward. The optimal policy is found and evaluated through the Deep Q-Learning, an extension to classic Q-Learning that models the Q-function using a deep network. The average forecasting error expressed in terms of the Modified Hausdorff Distance is about 1.5.

Instead of forecasting the long-term trajectory of the user, Rhinehart and Kitani [47] proposed a method called "Demonstrating Agent Rewards for K-futures Online" (DARKO) able to predict the user's goal by modeling the user's relationship to the objects in their environment. Given the per person online learning nature of their approach, it requires consider-

**Figure 7.5** Illustration for activity forecasting: In the first-person frame (upper left corner), the user grabs a mug. From this frame and knowledge of the position in the home (indicated by a green ball), the system proposed in [47] forecasts that he is likely to go to the kitchen (indicated by a red ball on the map estimated by the system). Image taken from the author's project page.[3] (For interpretation of the colors in this figure, the reader is referred to the web version of this chapter.)

ably less labeled training data than off-the-shelf deep learning approaches. DARKO is based on the assumption that the user will achieve his goals in a reasonably efficient manner; for instance, if he is going to the kitchen to get a refill of coffee, he will likely take along his empty mug (see Fig. 7.5). The near-optimal user's behavior is modeled by a Markov Decision Process (MDP) that, given demonstrations from the user, is able to recover a reward function that penalizes nonoptimal behaviors. The MDP model forecasts continuously based on the current state of the user, which includes a spatial and a semantic component. The spatial component is given by the location of the user in the environment, whereas the semantic component is given by what objects a person is carrying. This information allows the objects a person will acquire and release in the future to be inferred. The Maximal Entropy inverse optimal control framework is used as the inference mechanism. The detection accuracy for the user's goal varies from 23% to 93.1% on a dataset containing large environments, such as a home and an office space.

The next section is devoted to the detection and characterization of the social interactions of the user.

---

[3] https://www.youtube.com/embed/rvVoW3iuq-s.

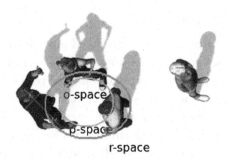

**Figure 7.6** People involved in a social interaction tend to follow a geometric spatial pattern, called F-formation, made of a convex space accessible to all participants (o-space), a narrow space surrounding it occupied by the participants themselves (p-space), and the outside world (r-space).

## 7.5 FIRST-PERSON SOCIAL INTERACTION ANALYSIS

Socializing is one of the most fundamental of human activities, one that impacts thoughts, emotions, decisions, and the overall well-being of individuals. In this regard, monitoring and facilitating social interactions constitutes an important factor for health sciences.

Humans spontaneously orient their gaze to the target of their attention. This is particularly true in the context of social interactions, since we interact with others via joint attention, i.e. coordinated attention to each other and/or to a third object or event. For example, a person engaged in a social interaction tends to look at the people he is interacting with, and vice versa, standing in a position that avoids occlusions. Therefore, in contrast to a third-person perspective, a wearable camera, especially when worn on the head, is particularly suitable for capturing many of the nonverbal communication cues that characterize human interaction such as gaze, smiling, winking, frowning, and gesturing. Early works on social interaction analysis from first-person cameras focused on the detection problem in different scenarios. The common pipeline is to first detect and track faces, then to project them into the 3D space, and finally to infer patterns of joint attention. The idea of estimating a bird's-eye view model of the scene to characterize social interactions is based on the sociological concept of F-formation [48], according to which two or more interacting people form a spatial pattern made of three social spaces: o-space, p-space, and r-space. The o-space is a convex empty space where the interaction takes place and is characterized by the fact that every participant looks inward into it, the p-space is a narrow strip surrounding the o-space that contains the bodies of interacting people, while the r-space is the area beyond the p-space (see Fig. 7.6).

## Characterizing Single Interaction Classes

In their seminal work, Fathi et al. [22] proposed a project to detect and characterize social interactions of 25 people wearing a GoPro camera during a visit to an amusement park. After detecting and tracking faces using state-of-the-art techniques, the faces were projected into the 3D space by relying on the estimated camera intrinsic parameters. The 3D location and orientation of a face provided the observation variables to infer the 3D locations the face is looking at in a Markov Random Field (MRF) model. In particular, the unary potentials of the MRF capture the likelihood of looking at a grid cell in which the space has been discretized, given the observations. The pairwise potentials model the likelihood of looking at a grid cell based on where other faces are looking, so that if many people are looking at the same grid cell, the likelihood that another person is looking at the same cell grid will be high. Social interactions were further classified into three categories, namely *discussion* (active interaction among multiple people), *dialogue* (with a single subject), and *monologue* (largely one-sided activity). In addition to patterns of attention of individuals, patterns of first-person head movement and location of faces around the first-person were used as feature vectors from a small subwindow around each frame. A hidden conditional MRF was used to connect the hidden variables corresponding to feature vectors over time and learning the latent variable corresponding to the interaction label. Extracting roles in social interactions captured by a wearable camera characterizes people's patterns of communication without imposing any restrictions on the user's interactions or environment.

## Detecting Multiple Social Groups

While the previous approach focuses on recognizing single interaction classes and does not take into account group dynamics and social relations, Alletto et al. [23] proposed a method for detecting groups in an unconstrained ego-vision scenario. They provided 3D localization without relying on camera calibration, hence generalizing the task to different wearable cameras. To avoid losing the target because of fast head movement, they adapted the TLD tracker [49] by allowing it to neglect the frames with high blur level, detected by an ad hoc algorithm. Head estimation is performed by using landmarks whenever they can be reliably computed, otherwise by a method that uses HOG features and a classification framework composed of SVM followed by HMM. Their approach for detecting social groups uses a correlation clustering that, given a set of people in the video sequence, models their pairwise relations with an affinity matrix $W$, whose elements $w_{ij}$ are positive if the people $i$ and $j$ are in the same group with confidence $w_{ij}$. The correlation clustering of a set of people is obtained as the partition that maximizes the sum of affinities for item pairs in the same cluster. The MITRE loss function is used as scoring measure in the training phase, thus spanning trees can be used to represent clusters. The proposed approach achieves an average accuracy of over 90% on the EGO-GROUP dataset.

## Detecting Social Saliency

More recently, Park and Shi [50] introduced the concept of *social saliency*, which is a measure of social significance of a 3D point and typically refers to those regions where people look. Inspired by the study of electric fields, their predictive model to detect social saliency exploits *social charges* and encodes the consensus of multiple individuals over the attention of members in a social group. Instead of relying on directional measurements such as gaze directions to detect social saliency, a gradient field induced by the social charges is used to define the relationship between gaze behaviors. The modes in a social saliency field correspond to the social charges are computed via a mean shift algorithm that automatically determines the amount of joint attention. Quantitative evaluation over real-world examples demonstrates robustness of the method for the considered purpose (77% accuracy).

## Quantifying Social Interactions

With the goal of quantifying social interactions and hence characterizing social patterns, Aghaei et al. [51,24] proposed a method for identifying people who the camera wearer interacts with using images captured over a full day. After segmenting the stream of images into events by applying a semantic regularized clustering method [19], a multiface tracking algorithm tolerant to disappearing/reappearing targets is applied to each event containing trackable people [51] (see Fig. 7.7). For a tracked face, its orientation and distance from the camera wearer are estimated. Each face is then represented by a two-dimensional time series corresponding to the temporal evolution of the distance and orientation features over time. A Long Short-Term Memory (LSTM) Recurrent Neural Network is employed to classify each time series into "interacting" or "not interacting" person. The

**Figure 7.7** Illustration for tracking the appearance of individuals involved in social interaction: Localization and tracking of face occurrences of individual 1 (top row) and individual 2 (bottom row) using the method proposed in [51].

system achieves an accuracy of above 70% on a dataset acquired under free-living conditions. In [52], interacting faces are grouped, across different events and days, into the individual identities present in these data. To this end, they leverage inner-class and inter-class constraints derived from the face tracking of people across single events. The tracking provides a set of correct variations of the same face in one sequence, called face-sets. They first calculate the dissimilarity between all the possible pairs of face-sets, and then, based on these measurements, employ a hierarchical clustering technique to discover the most similar face-sets.

## 7.6 DISCUSSION AND CONCLUSIONS

The body of work summarized in this chapter is by no means a complete review of the advances that have been seen over the last few years in the field of egocentric vision. However, it provides an overview of those results achieved by researchers that can provide tools for developing assistive technologies. In particular, ambient assisted living (AAL) systems that help elderly people live independently for as long as possible, may largely benefit from the techniques described in this chapter. AAL environments are smart environments embedded with a variety of sensors that acquire data about the state of both the environment and the individual. Since wearable cameras provide information about what the wearers see and do, they are particularly suitable to facilitate person–environment interaction. For instance, an egocentric system based on the techniques described in Section 7.2 could convey functional information to elderly people concerning the environment, suggesting which activities to perform in any location of the room and pointing to where related objects are located, hence providing support in domestic tasks. Forecasting systems relying on the methods introduced in Section 7.4 could be used to automate the smart home in preparation for upcoming activities or to send messages to an intelligent agent so that it can take anticipatory action to help the user complete these activities.

Another category of assistive technologies that can take advantage of the first-person perspective are assistive devices for the visually impaired. Detecting ego-engagement in a browsing scenario permits shifting the focus of the assistive intelligent agent to the center of attention of the user in order to provide help in everyday scenarios, such as the supermarket, and in less standard situations, such as a museum, enlarging the radius of activity of the user. Moreover, egocentric vision technology would not only grant visually impaired people greater mobility and independence, but would also make them feel less alienated by improving their social interactions. For the blind and visually impaired, moving around in unfamiliar spaces is very challenging since they have to avoid various physical obstacles and move around in social spaces created by other interacting people. When they enter a crowded lunchroom and are looking for a seat, they could be led to the first available one without knowing if there are familiar faces on site. The ability to recognize, re-identify, and localize people or groups of people in unconstrained scenarios plays a crucial role in the social interactions of people with visual disabilities. Assistive devices based on the techniques described in Section 7.5, would aid in reestablishing these abilities, giving visually impaired people the awareness of having a correct social behavior, and consequently the possibility to experience less frustration and low self-confidence, hence improving their general well-being.

Last but not least, first-person vision has several application opportunities in the field of computer-assisted technologies for disease prevention, detection, and rehabilitation. Activity recognition systems such as those described in Section 7.3, may provide services that go beyond automatic ADL monitoring, including frailty detection [53,35]; mobility assistance [54]; motor telerehabilitation [55]; lifestyle monitoring for people affected by chronic diseases such as obesity, diabetes, and cardiovascular diseases; or for people interested in improving their habits in order to prevent noncommunicable diseases [56].

Social pattern characterization systems such as those described in Section 7.5 have several applications in detection and intervention strategies. On the one hand, since the lack of social interactions is strongly correlated with depression and poor outcomes in stroke survivors, social pattern characterization could be an important factor for the prevention/detection of relapse [57]. On the other hand, given the stronger emotional burden with respect to activities performed in isolation, social interactions are expected to have a greater potential to trigger autobiographical memory in people affected by mild cognitive impairment. Therefore, a wearable device able to automatically detect social moments would be effective in selecting the most suitable events to be used in intervention strategies aiming to stimulate cognitive memory functions. Understanding the roles played by users in social interactions may help to determine if they actively participate in conversations, which is an important factor in determining the level of subjective satisfaction in social relationships [58].

Computer vision techniques summarized in this chapter suggest that first-person vision technologies are mature enough to be used in assistive applications, or at least in pilot studies. The results of recent studies show that family caregivers are willing to have their elders use a wearable camera system [59]. We may expect that major applications of egocentric vision to assistive technologies are just around the corner. However, several research opportunities are still open. We identify three main lines of research in first-person vision that would enable further applications in assistive technologies. First, we consider multimodal approaches particularly promising. Wearable cameras can collect multimodal data such as gaze information, GPS position, IMU data, etc., that may help in overcoming robustness issues in uncontrolled scenarios [60–63]. The use of additional sensors for capturing physiological signals, such as heart rate, combined with contextual information provided by the camera, such as what the person is doing and with whom, can help infer the emotional state of the wearer and reveal correlations that could be crucial in prevention and rehabilitation. Additionally, the combination of different wearable sensor data such as speech features [64] and visual data can play an important role in sensing and modeling face-to-face interactions. Second, a system able to understand *how* a person carries out ADLs such as eating, cleaning, and cooking may be used to detect early signs of cognitive decline or support caregivers' work [65]. Third, the understanding of human social behavior through wearable cameras is still largely unexplored. Further advances in the field may provide automatic systems for detecting behavioral disorders such as autism or cognitive decline at an early stage. Additionally, the ability to capture the social meaning of nonverbal communication cues, which are unaccessible to visually impaired people, may help improve their social interactions and feel less alienated. Finally, the understanding of human social behavior may help build socially intelligent agents that can be more closely integrated in our lives and help reduce the feelings of loneliness and isolation in elders.

## REFERENCES

[1] E. Sazonov, M.R. Neuman, Wearable Sensors: Fundamentals, Implementation and Applications, Elsevier, 2014.

[2] S.D. Guler, M. Gannon, K. Sicchio, A brief history of wearables, in: Crafting Wearables, Springer, 2016, pp. 3–10.

[3] S.C. Mukhopadhyay, Wearable sensors for human activity monitoring: a review, IEEE Sensors Journal 15 (3) (2015) 1321–1330.

[4] M.M. Rodgers, V.M. Pai, R.S. Conroy, Recent advances in wearable sensors for health monitoring, IEEE Sensors Journal 15 (6) (2015) 3119–3126.

[5] A.R. Doherty, P. Kelly, J. Kerr, S. Marshall, M. Oliver, H. Badland, et al., Using wearable cameras to categorise type and context of accelerometer-identified episodes of physical activity, International Journal of Behavioral Nutrition and Physical Activity 10 (1) (2013) 22.

[6] T. Kanade, First-person, inside-out vision, in: IEEE Workshop on Egocentric Vision, CVPR, vol. 1, 2009.

[7] S. Hodges, E. Berry, K. Wood, SenseCam: a wearable camera that stimulates and rehabilitates autobiographical memory, Memory 19 (7) (2011) 685–696.

[8] M.C. Allé, L. Manning, J. Potheegadoo, R. Coutelle, J.M. Danion, F. Berna, Wearable cameras are useful tools to investigate and remediate autobiographical memory impairment: a systematic PRISMA review, Neuropsychology Review (2017) 1–19.

[9] A.R. Doherty, S.E. Hodges, A.C. King, A.F. Smeaton, E. Berry, C.J. Moulin, et al., Wearable cameras in health, American Journal of Preventive Medicine 44 (3) (2013) 320–323.

[10] Q. Xu, S.C. Chia, B. Mandal, L. Li, J.H. Lim, M.A. Mukawa, C. Tan, SocioGlass: social interaction assistance with face recognition on Google Glass, Scientific Phone Apps and Mobile Devices 2 (1) (2016) 1–4.

[11] U. Rehman, S. Cao, Augmented-reality-based indoor navigation: a comparative analysis of handheld devices versus Google Glass, IEEE Transactions on Human-Machine Systems (2016).

[12] J.R. Terven, B. Raducanu, M.E. Meza-de Luna, J. Salas, Head-gestures mirroring detection in dyadic social interactions with computer vision-based wearable devices, Neurocomputing 175 (2016) 866–876.

[13] S.R. Edmunds, A. Rozga, Y. Li, E.A. Karp, L.V. Ibanez, J.M. Rehg, W.L. Stone, Brief report: Using a point-of-view camera to measure eye gaze in young children with autism spectrum disorder during naturalistic social interactions: a pilot study, Journal of Autism and Developmental Disorders (2017) 1–7.

[14] M. Devyver, A. Tsukada, T. Kanade, A wearable device for first person vision, in: 3rd International Symposium on Quality of Life Technology, 2011.

[15] M. Leo, G. Medioni, M. Trivedi, T. Kanade, G. Farinella, Computer vision for assistive technologies, Computer Vision and Image Understanding 154 (2017) 1–15.

[16] M. Bolanos, M. Dimiccoli, P. Radeva, Toward storytelling from visual lifelogging: an overview, IEEE Transactions on Human-Machine Systems (2016).

[17] A. Betancourt, P. Morerio, C.S. Regazzoni, M. Rauterberg, The evolution of first person vision methods: a survey, IEEE Transactions on Circuits and Systems for Video Technology 25 (5) (2015) 744–760.

[18] Y. Poleg, C. Arora, S. Peleg, Temporal segmentation of egocentric videos, in: Proceedings of the IEEE Conference on Computer Vision and Pattern Recognition, 2014, pp. 2537–2544.

[19] M. Dimiccoli, M. Bolaños, E. Talavera, M. Aghaei, S.G. Nikolov, P. Radeva, Sr-clustering: semantic regularized clustering for egocentric photo streams segmentation, Computer Vision and Image Understanding (2016).

[20] A.G. del Molino, C. Tan, J.H. Lim, A.H. Tan, Summarization of egocentric videos: a comprehensive survey, IEEE Transactions on Human-Machine Systems (2016).

[21] T.H. Nguyen, J.C. Nebel, F. Florez-Revuelta, Recognition of activities of daily living with egocentric vision: a review, Sensors 16 (1) (2016) 72.

[22] A. Fathi, J.K. Hodgins, J.M. Rehg, Social interactions: a first-person perspective, in: IEEE Conference on Computer Vision and Pattern Recognition (CVPR), 2012, IEEE, 2012, pp. 1226–1233.

[23] S. Alletto, G. Serra, S. Calderara, R. Cucchiara, Understanding social relationships in egocentric vision, Pattern Recognition 48 (12) (2015) 4082–4096.

[24] M. Aghaei, M. Dimiccoli, P. Radeva, With whom do I interact? Detecting social interactions in egocentric photo-streams, in: International Conference on Pattern Recognition (ICPR), 2016.

[25] A. Furnari, G.M. Farinella, S. Battiato, Recognizing personal locations from egocentric videos, IEEE Transactions on Human-Machine Systems (2016).

[26] G. Vaca-Castano, S. Das, J.P. Sousa, N.D. Lobo, M. Shah, Improved scene identification and object detection on egocentric vision of daily activities, Computer Vision and Image Understanding (2016).

[27] H. Pirsiavash, D. Ramanan, Detecting activities of daily living in first-person camera views, in: IEEE Conference on Computer Vision and Pattern Recognition (CVPR), 2012, IEEE, 2012, pp. 2847–2854.

[28] D. Damen, T. Leelasawassuk, W. Mayol-Cuevas, You-Do, I-Learn: egocentric unsupervised discovery of objects and their modes of interaction towards video-based guidance, Computer Vision and Image Understanding 149 (2016) 98–112.

[29] N. Rhinehart, K.M. Kitani, Learning action maps of large environments via first-person vision, in: Proceedings of the IEEE Conference on Computer Vision and Pattern Recognition, 2016, pp. 580–588.

[30] M. Lavallière, A.A. Burstein, P. Arezes, J.F. Coughlin, Tackling the challenges of an aging workforce with the use of wearable technologies and the quantified-self, DYNA 83 (197) (2016) 38–43.

[31] M.S. Ryoo, B. Rothrock, L. Matthies, Pooled motion features for first-person videos, in: Proceedings of the IEEE Conference on Computer Vision and Pattern Recognition, 2015, pp. 896–904.

[32] Y. Poleg, A. Ephrat, S. Peleg, C. Arora, Compact CNN for indexing egocentric videos, in: IEEE Winter Conference on Applications of Computer Vision (WACV), 2016, IEEE, 2016, pp. 1–9.

[33] G. Abebe, A. Cavallaro, X. Parra, Robust multi-dimensional motion features for first-person vision activity recognition, Computer Vision and Image Understanding 149 (2016) 229–248.

[34] B.K. Horn, B.G. Schunck, Determining optical flow, Artificial Intelligence 17 (1–3) (1981) 185–203.

[35] I. Martin-Lesende, K. Vrotsou, I. Vergara, A. Bueno, A. Diez, et al., Design and validation of the VIDA questionnaire, for assessing instrumental activities of daily living in elderly people, Journal of Gerontology & Geriatric Research 4 (214) (2015) 2.

[36] T. McCandless, K. Grauman, Object-centric spatio-temporal pyramids for egocentric activity recognition, in: BMVC, vol. 2, 2013, p. 3.

[37] X. Ren, C. Gu, Figure-ground segmentation improves handled object recognition in egocentric video, in: IEEE Conference on Computer Vision and Pattern Recognition (CVPR), 2010, IEEE, 2010, pp. 3137–3144.

[38] A. Fathi, X. Ren, J.M. Rehg, Learning to recognize objects in egocentric activities, in: IEEE Conference on Computer Vision and Pattern Recognition (CVPR), 2011, IEEE, 2011, pp. 3281–3288.

[39] C. Li, K.M. Kitani, Pixel-level hand detection in ego-centric videos, in: Proceedings of the IEEE Conference on Computer Vision and Pattern Recognition, 2013, pp. 3570–3577.

[40] S. Bambach, S. Lee, D.J. Crandall, C. Yu, Lending a hand: detecting hands and recognizing activities in complex egocentric interactions, in: Proceedings of the IEEE International Conference on Computer Vision, 2015, pp. 1949–1957.

[41] A. Betancourt, P. Morerio, E. Barakova, L. Marcenaro, M. Rauterberg, C. Regazzoni, Left/right hand segmentation in egocentric videos, Computer Vision and Image Understanding 154 (2017) 73–81.

[42] M. Ma, H. Fan, K.M. Kitani, Going deeper into first-person activity recognition, in: The IEEE Conference on Computer Vision and Pattern Recognition (CVPR), 2016.

[43] M.S. Ryoo, L. Matthies, First-person activity recognition: what are they doing to me?, in: Proceedings of the IEEE Conference on Computer Vision and Pattern Recognition, 2013, pp. 2730–2737.

[44] M. Ryoo, L. Matthies, First-person activity recognition: feature, temporal structure, and prediction, International Journal of Computer Vision 119 (3) (2016) 307–328.

[45] Y.C. Su, K. Grauman, Detecting engagement in egocentric video, in: European Conference on Computer Vision, Springer, 2016, pp. 454–471.

[46] S.Z. Bokhari, K.M. Kitani, Long-term activity forecasting using first-person vision, in: Asian Conference on Computer Vision, Springer, 2016, pp. 346–360.

[47] N. Rhinehart, K.M. Kitani, Online semantic activity forecasting with DARKO, arXiv preprint, arXiv:1612.07796, 2016.

[48] A. Kendon, Studies in the Behavior of Social Interaction, vol. 6, Humanities Press Intl., 1977.

[49] Z. Kalal, K. Mikolajczyk, J. Matas, Tracking-learning-detection, IEEE Transactions on Pattern Analysis and Machine Intelligence 34 (7) (2012) 1409–1422.

[50] S.H. Park, J. Shi, Social saliency prediction, in: Proceedings of the IEEE Conference on Computer Vision and Pattern Recognition, 2015, pp. 4777–4785.

[51] M. Aghaei, M. Dimiccoli, P. Radeva, Multi-face tracking by extended bag-of-tracklets in egocentric photo-streams, Computer Vision and Image Understanding 149 (2016) 146–156.

[52] M. Aghaei, M. Dimiccoli, P. Radeva, All the people around me: face discovery in egocentric photo-streams, arXiv preprint, arXiv:1703.01790, 2017.

[53] S.M. Schüssler-Fiorenza Rose, M.G. Stineman, Q. Pan, H. Bogner, J.E. Kurichi, J.E. Streim, D. Xie, Potentially avoidable hospitalizations among people at different activity of daily living limitation stages, Health Services Research (2016).

[54] B. Kopp, A. Kunkel, H. Flor, T. Platz, U. Rose, K.H. Mauritz, et al., The Arm Motor Ability Test: reliability, validity, and sensitivity to change of an instrument for assessing disabilities in activities of daily living, Archives of Physical Medicine and Rehabilitation 78 (6) (1997) 615–620.

[55] R. Planinc, A.A. Chaaraoui, M. Kampel, F. Florez-Revuelta, Computer vision for active and assisted living, Active and Assisted Living: Technologies and Applications (2016) 57.

[56] E.G. Spanakis, S. Santana, M. Tsiknakis, K. Marias, V. Sakkalis, A. Teixeira, et al., Technology-based innovations to foster personalized healthy lifestyles and well-being: a targeted review, Journal of Medical Internet Research 18 (6) (2016).

[57] A. Dhand, A.E. Dalton, D.A. Luke, B.F. Gage, J.M. Lee, Accuracy of wearable cameras to track social interactions in stroke survivors, Journal of Stroke and Cerebrovascular Diseases (2016).

[58] D. Fiorillo, F. Sabatini, Quality and quantity: the role of social interactions in self-reported individual health, Social Science & Medicine 73 (11) (2011) 1644–1652.

[59] J.T. Matthews, J.H. Lingler, G.B. Campbell, A.E. Hunsaker, L. Hu, B.R. Pires, et al., Usability of a wearable camera system for dementia family caregivers, Journal of Healthcare Engineering 6 (2) (2015) 213–238.

[60] S. Song, V. Chandrasekhar, B. Mandal, L. Li, J.H. Lim, G.S. Babu, et al., Multimodal multi-stream deep learning for egocentric activity recognition, in: Proceedings of the IEEE Conference on Computer Vision and Pattern Recognition Workshops 2016, 2016, pp. 24–31.

[61] S. Song, N.M. Cheung, V. Chandrasekhar, B. Mandal, J. Liri, Egocentric activity recognition with multimodal fisher vector, in: International Conference on Acoustics, Speech and Signal Processing (ICASSP), IEEE, 2016, pp. 2717–2721.

[62] K. Nakamura, A. Alahi, S. Yeung, L. Fei-Fei, Egocentric multi-modal dataset with visual and physiological signals, http://www.cbi.gatech.edu/fpv2016/abstracts/egocentric_multimodal.pdf, 2016.

[63] D. Maunder, J. Epps, E. Ambikairajah, B. Celler, Robust sounds of activities of daily living classification in two-channel audio-based telemonitoring, International Journal of Telemedicine and Applications 2013 (2013) 3.

[64] T. Choudhury, A. Pentland, Characterizing social interactions using the sociometer, in: Proceedings of NAACOS, Citeseer, 2004.

[65] P. Rashidi, A. Mihailidis, A survey on ambient-assisted living tools for older adults, IEEE Journal of Biomedical and Health Informatics 17 (3) (2013) 579–590.

# CHAPTER 8

# Computer Vision for Augmentative and Alternative Communication

**Sethuraman Panchanathan, Meredith Moore,**
**Hemanth Venkateswara, Shayok Chakraborty, Troy McDaniel**
Center for Cognitive Ubiquitous Computing (CUbiC), Arizona State University, AZ, United States

## Contents

Computer Vision for Assistive Healthcare.
DOI: https://doi.org/10.1016/B978-0-12-813445-0.00008-3
Copyright © 2018 Elsevier Ltd. All rights reserved.

## Abstract

Augmentative and Alternative Communication (AAC) includes all forms of communication that are used to supplement speech for those with impairments in the production or comprehension of spoken language. AAC technology has mostly been used by people with severe speech or language impairments. Individuals with visual impairments also face a fundamental limitation in communicating with their sighted peers, as about 65% of the information during social interactions is conveyed using nonverbal cues. In this chapter, we present our computer vision research contributions in the design and development of a Social Interaction Assistant (SIA), which is an AAC technology that can enrich the communication experience of individuals with visual impairments. Moreover, individuals with visual impairments often have specific requirements that necessitate a personalized, adaptive approach to multimedia computing, rather than a "one-size-fits-all" approach. To address this challenge, our proposed solutions place emphasis on understanding the individual user's needs, expectations, and adaptations toward designing, developing, and deploying effective multimedia solutions. Our empirical results demonstrate the significant potential in using person-centered AAC technology to enrich the communication experience of individuals with visual impairments.

## Keywords

Augmentative and alternative communication (AAC), Speech and language disorders [or impairments], Nonverbal communication, Dysarthria, Person-centered technologies, Batch mode active learning, Conformal predictions, Facial expression recognition

## Chapter Points

- AAC has mostly been used by individuals with severe speech and language impairments.
- We propose a person-centered, adaptive approach to AAC for individuals with visual impairments that caters to the needs of individual users rather than a "one-size-fits-all" solution.

- We depict the power and potential of our framework using three applications: person recognition through batch mode active learning, reliable multimodal person recognition through the conformal predictions framework, and facial expression recognition through topic models.

## 8.1 INTRODUCTION AND BACKGROUND

In this section, we present a brief background and survey of Augmentative and Alternative Communication (AAC) and discuss open research problems and challenges. In the field of AAC, communication is defined by the joint establishment of meaning involving at least two people. This co-production of meaning is essential to nearly all aspects of life, and is incredibly important for quality of life. An individual's ability to communicate effectively affects the ability to build and maintain relationships, make choices, and participate in everyday life [1].

### 8.1.1 The Communication Process

In order to fully participate in the communication process, an individual must be able to receive, send, process, and comprehend concepts of verbal, nonverbal, or graphic symbol systems. This process makes use of the systems of speech, language, and hearing.[1]

The communication process begins by an individual forming the intention to communicate, then deciding what meaning should be communicated, and finally putting that meaning into some sort of language. After deciding the concept of the words or movements to communicate that meaning, the individual then needs to send the physiological signals to the muscles necessary to coordinate the movements needed to either speak the words or make the gestures. Once the message is sent, the communication partner must receive this information. If the message was conveyed via speech, the communication partner must hear the sound, process them as words, and attribute meaning to them. If the meaning is conveyed via gestures, this means that the communication partner needs to be able to see or feel the movements made, and be able to attribute meaning to these gestures. Then the communication partner can develop the intention to communicate back to the individual following a similar process. Each of these subprocesses within communication are delicate and complicated processes.

[1] Augmentative and Alternative Communication (AAC), n.d.Am. Speech-Lang.-Hear. As-soc. http://www.asha.org/public/speech/disorders/AAC/ (accessed 5.6.16).

## 8.1.2 Diversity of Communication

When we think of communication, as humans our first impulse is to think of speech. At its simplest, speech is a series of alternating high- and low-pressure waves that are interpreted as sound. For such a simple signal, speech is very information dense. In speech alone we have the capability to convey nearly limitless ideas. However, speech is not the only modality of communication. Communication is often broken down into verbal and nonverbal communication. Verbal communication refers to communication in the form of speech or words. Both written and oral communication are included in the verbal-communication category. Sign language, although it is a series of body gestures, is also considered verbal communication as each sign has an associated semantic meaning (i.e. a word). Nonverbal communication largely involves different types of gestures including overall body language, hand/arm gestures, eye contact, facial expressions, and prosodic voice cues such as intonation, pitch, rhythm, or volume. We convey more information nonverbally than is evident.

In today's technology centric world, communication takes place on many different platforms. On a daily basis, most people communicate via phone calls, text messages, emails, video-chat platforms, etc. It is important in the design of communication tools to consider the diverse forms of communication that are used, and to make as many forms of communication accessible as possible.

## 8.1.3 Complex Communication Needs

A communication disorder is any impairment of the ability to partake in this process. As the communication process is so complicated, it is easy to see that there is a wide variety of disorders that may lead to complex communication needs (CCN). Individuals who have complex communication needs may not have the communication skills necessary to fully participate in education, employment, family, and community life [2]. AAC systems have demonstrated positive outcomes for many different disorders including autism, amyotrophic lateral sclerosis (ALS), cerebral palsy (CP), intellectual developmental disabilities, severe chronic aphasia, and motor speech disorders [2]. Individuals with CCN face barriers when it comes to employment and active participation in the community [2–9].

## 8.1.4 Introduction to Augmentative Alternative Communication

Augmentative and alternative communication (AAC) devices are used by anyone who is unable to meet their communication needs. Communication needs are defined as any skills necessary to fully participate in education, employment, family, and community life. As the needs of individuals with CCN vary dramatically, so do the AAC devices that are available. There are many AAC devices to choose from, and these devices are continually developing and changing as new technologies are introduced [10].

### 8.1.4.1 Goals of AAC

In the research and development process of AAC technologies it is easy to get caught up in the measurable gains shown by one AAC device in comparison to another. These measurable quantities include the rate of communication, automaticity, learning demands, ease of use, cognitive load, and cognitive/linguistic factors such as memory, learning, vigilance, and language development. While these measures do promote some forward movement within the field of AAC, it is important for researchers not to forget the bigger picture. In Light and McNaughton's discussion of Designing AAC Research and Interventions, they argue the necessity for a holistic approach to communication, where communication is seen only as a tool, not as the end goal. Designing AAC around the idea that it should be used to further the user's participation in education, employment, family, and community life will lead to technologies that will optimize the benefits to the user.

### 8.1.4.2 Importance of a Person-Centric Approach to AAC

Because of the wide variety of complex communication needs, the field of AAC greatly benefits from designing AAC devices from a person–centric approach. Each user's abilities and needs are highly individualized, and it is very difficult to design for a generic user profile, as is common practice in software design. In the case of AAC, it is necessary to create highly adaptable systems that are flexible enough to adjust to each individual user's needs. If AAC systems are designed in this way, they are more likely to enable the users to communicate to the best of their abilities.

### 8.1.4.3 Classification of AAC Devices

With the dramatic increase in development of AAC devices and strategies, a system for classifying these technologies has been developed. There are two main distinctions that are made in classifying different types of AAC devices, whether the AAC device is aided or unaided, and whether the device is high-tech or low-tech [11].

Aided communication devices use external technologies to meet the communication needs of the user. Common aided AAC devices include picture exchange systems (PECS), speech generating devices (SGDs), and communication boards. These devices use symbols such as photographs, drawings, animations, letters, and words. Unaided AAC strategies do not require external devices and involve the use of symbols such as manual signs, pantomimes, and gestures [11].

The level of technological involvement is also a common way to classify AAC devices. AAC interventions that do not involve computers are generally referred to as low-tech solutions, while interventions that require some form of computation are referred to as high-tech solutions. For example, a printed communication board would be considered low-tech, while a word prediction speech generating device would be considered a high-tech AAC device.

### 8.1.4.4 Brief History of AAC

In the last several decades, the field of AAC has changed significantly. Individuals with complex communication needs largely resided in mental institutions completely segregated from the rest of society. In the past forty or so years, the field of AAC has grown considerably in several different directions. There has been an increase in the number of people using AAC, a wider scope of communication needs met by AAC, a wider variety of AAC devices available, and an overall increase in the acceptance and participation of individuals with CCN in the community [10].

Originally, AAC technologies were largely low-tech nonelectronic partner communication strategies. Some of these strategies include the eye-gaze vest, the laptray, the transparent display, and the mirror/prism communicator [12]. For individuals with residual speech, these partner communication strategies included partner listening strategies [13]. Another low-tech solution often used was communication assistants (individuals who were skilled at communicating with the individual with CCN and could translate for any unfamiliar communication partner), and more specifically sign language interpreters.

Other AAC techniques developed over time, and people began using speech generating devices, and picture exchange systems. These systems were developed by a small number of manufacturers, and as such there were only a few options to choose from, and the systems were relatively cost prohibitive. Recently, however, there has been a significant increase in the variety of AAC technologies due to advances in mobile technology such as smartphones and tablets. These touchscreen interfaces provide the perfect pervasive platform for the implementation of AAC systems [14]. These advances have led to a wide choice of AAC systems available on mainstream devices. Now with the standardization of computer vision techniques, there is the same possibility for advancement of AAC technologies using computer vision. So far there have been several applications of computer vision to AAC; however, there is still significant room for advancement in the realm of assistive computer vision for AAC.

### 8.1.4.5 AAC: Open Problems and Challenges Related to Computer Vision

Computer vision has made a significant impact on the field of AAC; however, there are still several open questions, many of which have been mentioned above.

**Low-Cost Pervasive Eye Tracking.** With the improvement of eye tracking technologies, researchers are now able to use the embedded cameras on mobile devices to track a user's eye movements with reasonable accuracy. Previously, one of the greatest limitations of eye tracking technologies was that they are cost prohibitive and bulky. Using the embedded camera in mobile devices to allow the user to interface with their phone or tablet with their eyes shows a huge potential for the application of this technology to the field of AAC. If an individual with CCN is able to control a phone or tablet via the eyes rather than bulky external hardware, it very well may make social interactions easier, and therefore aid the process of communication. It also would improve access to common technologies that are sometimes inaccessible to users with movement limitations.

**Multimodal Dysarthric Speech Recognition.** Many individuals with motor speech disorders have some residual speech; however, this speech is often not easily understood by unfamiliar communication partners. While there have been many attempts to build speech recognition systems that recognize dysarthric speech, the range of differences in speech found in speech dysarthria make this problem extremely challenging. Recognition system accuracies are limited by the vast difference between individuals, and therefore to be successful require an extensive amount of person dependent training. There seems to be a strong potential in the field of dysarthric speech recognition for multimodal disambiguation of data. The fusion of audio and visual data has shown improved accuracies in "normal" speech, but there has been very little audio-visual speech recognition research within the application of dysarthric speech recognition.

**Multimodal Sign Language Recognition.** Sign Language Recognition (SLR) is another area where multimodal fusion of data shows promise in reducing recognition errors. While computer vision methodologies have made significant strides in SLR, in certain planes of view it becomes very difficult to recognize certain hand gestures. Using solely visual data also prevents these users from communicating when they cannot be seen by the camera. Again, there is a potential for pairing another modality of data to help disambiguate the cases where visual recognition fails.

**Context-Aware AAC.** The nature of communication is context-dependent, which means we can use data from the environment to derive context, and use that context to help better predict what the user wants to say. Systems have been laid out to take advantage of context-aware computing in the realm of AAC; however, these systems have yet to be implemented. Computer vision can be used in context-aware AAC to determine who the user is speaking with (and based on history what the user may want to talk to them about), as well as what objects of interest are in the environment.

**Long-Term Outcomes.** As AAC technologies are an applied technology that has a significant impact on the user's everyday life, there needs to be a stronger focus in the field of AAC research on long-term outcomes. Many recent AAC systems have been tested by a small number of users for a few hours in a nonnatural communication situation. When designing AAC systems researchers should be aware of how the device would be used naturally, and should focus on a greater number of long-term studies that demonstrate how well the tool is able to help the individual participate fully in their home, school, work, and community environments [2].

The rest of the chapter is organized as follows. Section 8.2 presents existing research on computer vision for AAC. We introduce our person-centered multimedia computing paradigm in Section 8.3 and the social interaction assistant in Section 8.4, which is an AAC technology to enrich the communication experience of individuals with visual impairments. We then present our computer vision research contributions in batch mode active learning, conformal predictions, and latent facial topics in Sections 8.5,

8.6, and 8.7 respectively. Finally, we conclude with discussions in Section 8.8.

## 8.2 COMPUTER VISION FOR AAC

### 8.2.1 Gesture Recognition

Most computer vision applications for AAC can be simplified into gesture recognition problems. Gesture recognition can be broadly defined as the mathematical interpretation of a human motion by a computing device. Within the field of gesture recognition lie problems of facial recognition, emotion recognition, voice recognition, eye tracking, lip movement recognition, and sign language recognition. Gestures can originate from any bodily motion; however, gestures most commonly originate from the face or the hand. As applied to AAC, computer vision makes its greatest contributions when it comes to eye tracking, blink recognition, head tracking, facial detection, and sign language recognition.

#### 8.2.1.1 Computer Vision for Alternate Access Interfaces

Aided AAC technologies generally involve a user navigating an interface to select the symbols that he or she wants to use to communicate. In order to navigate these interfaces two main functionalities are necessary: the ability to move the cursor and the ability to select whatever the cursor points to. This interaction paradigm can be simplified as the "point and click model". These two functionalities are all an individual really needs to get the full use out of a computer-based system. If needed, this interaction paradigm can be simplified further into just the operation of a binary switch. For this paradigm to work, the software needs to scroll through the options, allowing the user to select whenever the desired option is highlighted by the software. This methodology is referred to as switch scanning [15]. Nonetheless, the "point and click" methodology is the most popular interaction method for assistive technologies. For individuals who do not have the ability to operate a keyboard, navigate with a mouse, or interact with a touch screen, there are alternate access methodologies. Many of these alternate access methodologies employ computer vision to allow the user to "point and click". This is generally accomplished through some form of body part tracking.

For individuals with minimal movement, computer vision has been successfully used to improve the physical accessibility of AAC technologies. These users generally rely on head and/or eye movement to access their

AAC interface. Eye tracking systems work by reflecting infrared (IR) light off the surface of the eye causing a reflection off of the user's pupil that the computer can track. Using math, the software then correlates the reflected infrared light and the computer cursor through a calibration process. Eye tracking has shown positive outcomes when used as an alternate access methodology for individuals with CCN [16–19]. Eye tracking has also been used to inform the design of AAC technologies for children with autism [20].

Eye tracking has developed over the past 20 years, and is just now coming into standard use. Originally eye tracking was cost prohibitive and bulky as it required external hardware such as glasses. Proof of concept studies using these bulky technologies have been successful as an alternate method of facilitating the use of communication tools for individuals with CCN [16,19].

However, within the last year or two eye tracking has become much more accessible to the masses. A recent paper proposes using pervasive hardware such as mobile phones and tablets for eye tracking. This chapter discusses the use of convolutional neural networks, which can eliminate the need for bulky glasses, and how it is possible to use the built-in camera on a device or cellular telephone [21]. This technology has yet to be implemented in the application domain of assistive technology, but this application shows promise in increasing the accessibility of eye tracking as an alternate access method for individuals with minimal movement.

Another recent paper that discusses shedding the limitations of conventional eye tracking presents a smartphone-based gaze gesture communication tool for individuals with motor disabilities [22]. In the app proposed in this paper, GazeSpeak, a user is able to communicate through six simple eye gestures (look up, left, right, down, center, and closed). Each direction corresponds to a subset of the alphabet, and a language model predicts what words the individual wants to communicate. This application was tested with a small number of subjects who gave positive feedback on the value of a portable, low-cost technology to supplement IR forms of eye tracking.

Other forms of alternate access methodologies using computer vision include blink recognition and head pointing. These two input methodologies are very similar to eye tracking, but they track the user's head and recognize when the user blinks [17,23]. Blink recognition can be used as an input for the "click" methodology, while the head pointing can be used as an input for the "point" input. Varying the duration of the pointing methodologies can also serve as a selection input.

### 8.2.1.2 Context-Aware AAC

Communication is context dependent; people talk about things that are rooted in their environments and as such, the things we talk about are very seldom random. Context-aware computing refers to systems that examine and react to an individual's changing context [24]. One of the goals of AAC is to allow the user to communicate as efficiently as possible. To enable efficient communication, most speech generating devices (SGDs) of PEC systems are equipped with a word prediction engine that can make use of the contextual clues from the environment to make better predictions. Kane et al. explored this idea by designing a context-aware communication tool for people with aphasia [25]. In this paper, Kane et al. discuss the possible contextual cues that the system can take advantage of in a series of proposed use cases for context-aware AAC. These use cases take advantage of the context-dependent nature of life and improve the ability of the device to predict words that are relevant to the situation. For example, if an individual is at a coffee shop, the AAC device may pull up the words necessary to complete the order.

While Kane et al. contributed a great conceptual framework for a context-aware AAC device and some preliminary participatory design results, they did not fully implement the system. Their paper presents a summary of most of the research in the area of context-aware computing, specifically through the use of computer vision, and as such it remains an open question. As shown in Table 1, computer vision can be employed to recognize a subset of the contexts used in the word prediction process, mainly in recognizing the communication partner through facial recognition (although this could arguably be completed using only audio through speaker recognition [26]), or recognizing objects in the environment surrounding the user. The idea is that when we know more contextual information about the situation, it follows that we will know more about the conversation. The more that is known about the conversation, the better we are able to predict what words the user may want to use. This concept has yet to be fully tested, but seems to warrant further investigation. This idea also walks the thin line of helping the user too much. The communication partner may not appreciate what feels like talking to a computer, and would prefer to talk to the user of the device. If the computer interprets too much and helps the user respond in a way that is not natural, will this have a negative effect on the social acceptance of the device? These questions require further investigation.

### 8.2.1.3 Computer Vision for Unaided AAC

By definition, unaided AAC is a group of AAC techniques that do not rely on any additional technology. However, unaided AAC techniques are plagued with the problem of unintelligibility. For example, an individual who uses sign language to communicate is able to communicate with close family and friends, but communicating with anyone who does not know sign language will need the help of a translator. In order for individuals to fully participate in their community, they need to be able to communicate with more than a small circle of communication partners. For individuals who use unaided AAC techniques, intelligibility becomes a barrier to community participation. However, computer vision can help break down this barrier to communication by translating for the users with CCN. There are two main ways that individuals with CCN communicate using unaided AAC techniques: through speech and through gestures.

It is worth noting that while the original AAC strategy here is unaided, the addition of computer vision would change these AAC strategies from the unaided category into the aided category.

## 8.2.2 Dysarthric Speech Recognition

Many individuals with CCN still have some residual speech and as such, are still able to make repeatable utterances that are understood by their caregivers, but not as easily by unfamiliar communication partners. This is most common with individuals who have motor speech disorders. Any speech impairment that is a result of a motor-speech disorder is known as dysarthric speech. There is a large body of research that looks into automatic speech recognition for dysarthric speech, but the accuracies achieved by these studies are highly constrained [27,28]. Dysarthric speech is widely varied in all of the aspects of speech depending on the individual. The prosodic features (the elements of speech that are not individual phonetic segments, i.e. intonation, tone, stress, rhythm) in dysarthric speech are highly variable [27]. The different phonemes are also difficult to discern in dysarthric speech. Multimodal recognition methods have been shown to disambiguate noisy data and improve recognition rates of difficult recognition tasks [29]. Speech recognition from visual data, also known as lip reading, has been explored with relatively positive results considering the sparse information carried in the visual representation of speech [30–32]. To supplement audio-based speech recognition, specifically in regard to noisy audio signals, there has been a relatively large body of research into audio-visual speech recognition. The area of audio-visual speech recognition has been explored with promising recognition results [33–35]. While the Universal Access database of dysarthric speech has both audio and visual modalities and makes a point of mentioning the possibility of audio-visual speech recognition, this area remains relatively unexplored. Salama et al. [36] proposed an audiovisual dysarthric speech recognition framework. While audio-visual speech recognition for dysarthric speech has yet to be thoroughly explored, it seems like this is an area worthy of further exploration as the visual signal could be used to achieve better recognition rates for dysarthric speech recognition.

### 8.2.3 Sign Language Recognition

Individuals who are deaf or hard of hearing (DHH) often communicate via sign language. For signing communities, sign language is an ideal way to communicate; however, there is a language translation problem when communicating with someone who does not know sign language. For other spoken languages, if we wanted to translate what someone was saying the answer would be in our pockets. We could pull out a cell phone and use a translation service. This technology does not yet exist for sign language. However, there has been significant research in the area of sign language recognition (SLR) [37–45]. The majority of the research that has been done in SLR has used computer vision [46]. There have been a variety of computer vision techniques for SLR, and high levels of accuracy have been achieved. However, vision based SLR has some significant limitations. These limitations include insufficient data to recognize the gestures in certain planes of movement, and a lack of portability as there needs to be a camera somewhere in the scene to detect the sign language. Sign language recognition lends itself well to the previously discussed idea of multimodal disambiguation of data. There has been some research into multimodal SLR [47]; however, this area remains relatively unexplored.

### 8.3 AAC FOR INDIVIDUALS WITH VISUAL IMPAIRMENTS

AAC has been predominantly used by individuals with severe speech and language impairments. However, individuals with visual impairments represent another significant percentage of the population who face fundamental challenges in social interactions. In everyday interactions, people communicate so effortlessly through verbal and nonverbal cues that they

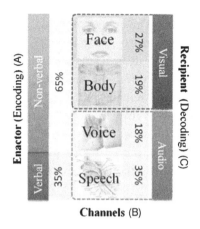

**Figure 8.1** Relative importance of (A) verbal vs. nonverbal cues, (B) four channels of nonverbal cues, and (C) visual vs. audio encoding and decoding of bilateral human interpersonal communicative cues. Based on the meta-analysis in [49].

are not aware of the complex interplay of the voice, face, and body in establishing a smooth communication channel. While spoken language plays an important role in communication, speech accounts for only 35% of the interpersonal exchanges. As depicted in Fig. 8.1, nearly 65% of all information communication happens through nonverbal cues [48]. Since most of the nonverbal cues (eye gaze, head nod, body mannerisms, facial expressions) are perceived visually, people with visual impairments are largely deprived of these vital communicative cues and face significant challenges in interacting with their sighted peers. Limited access to nonverbal cues can create miscommunications, leading to embarrassing social situations, increasing the likelihood of social isolation. To address this problem, we have focused our research efforts on the development of a Social Interaction Assistant, whose objective is to enrich the communication experience of individuals with visual impairments.

## 8.3.1 The Person-Centered Multimedia Computing Paradigm

Today's multimedia technologies are largely geared toward addressing the needs of the "able" population. Accessibility features of commercial products are often an afterthought rather than an integral part of the design from inception to completion. Special purpose assistive technologies are expensive due to smaller market segments and potential additional regulatory costs. This poses a serious challenge to the 10% of the world's population (approximately 650 million people) who live with some form of disability. In the United States, 12% (roughly 36 million people) live with at least one disability. Since January 2011, 10,000 baby boomers turn 65 every day, each of whom will experience functional limitations in their daily life activities.[2] Thus, there is a pronounced need for the development and deployment of multimedia solutions for the large population of individuals with disabilities in addition to the general population.

Human-centered computing (HCC) has emerged as a major subfield of computational science which emphasizes the understanding of human behavior, needs, and expectations in the design and development of technologies [50]. An important aspect of HCC design theory is to incorporate societal and cultural differences in system design for increased naturalness and acceptance [51]. While human–computer interaction (HCI) focuses solely on interface design and usability, HCC system design is motivated by human understanding and draws inspirations from social and behavioral sciences. Human-centered multimedia computing (HCMC) [51,52] applies the concepts and theories of HCC to address challenges in multimedia data analysis. Three key design factors have been proposed for HCMC: multimodal interaction for natural, effective use; consideration of cultural and societal differences to facilitate adoption; and accessibility beyond the desktop toward ubiquity. While HCC principles and methods have served researchers well, the increasing need for individualized solutions warrants a person-centered approach. Our work with the target population of individuals with disabilities has unraveled the impending need for coadaptive, ego-centric technologies that closely align with the needs and abilities of individual users. In general, users with disabilities are increasingly uninterested in a one-size-fits-all approach as they are reluctant to force-fit themselves to available solutions. More broadly, the abilities of the general population change over time due to factors such as age, context, and geographical locations, which necessitates technologies that adapt to people rather than people adapting to technologies.

To address this fundamental challenge, we have recently introduced the Person-Centered Multimedia Computing (PCMC) paradigm [53], which emphasizes individual user needs, preferences, and behavior toward the design and development of multimedia computing solutions. An important factor in PCMC is coadaptation [54,55], the bidirectional interaction where both the user and the system learn and adapt together over time through

---

[2] http://www.disabilitystatistics.org/.

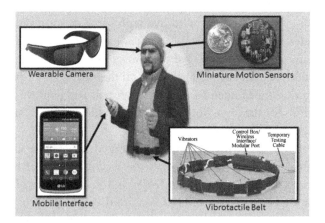

**Figure 8.2** The CUbiC Social Interaction Assistant.

continual use. Through coadaptation, ego-centric designs can be realized that still maintain applicability to a broad range of users. Since the human and the machine work in a symbiotic relationship, coadaptation leverages human cognitive capabilities to efficiently solve real-world issues, which are significantly more difficult to address using only machine intelligence.

## 8.4 THE SOCIAL INTERACTION ASSISTANT (SIA)

### 8.4.1 System Description

The Social Interaction Assistant (SIA) system is depicted in Fig. 8.2. It consists of a pair of glasses with a small camera embedded in the bridge. The video stream captured by the camera is analyzed in real time using machine learning and computer vision algorithms. Relevant information is then delivered to the user without obstructing the ongoing social interaction. To this end, sensory substitution is performed to convert visual nonverbal cues into haptic representations for delivery to the skin.

### 8.4.2 Person-Centeredness in the SIA

To fully understand the impediments faced by individuals with visual impairments in social interactions, we conducted two focus groups, made up of disability specialists, individuals who are visually impaired, and their family and friends. During the sessions, participants were asked to freely express their desires and challenges encountered during their daily activities. From these discussions, eight specific requirements to enrich daily interactions

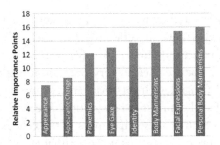

**Figure 8.3** Results from an online survey. Person recognition and facial expression recognition were ranked highest. Image source [56].

were identified through an online survey [56]. Twenty-seven people participated in the survey anonymously, of whom 16 were blind, 9 had low vision, and 2 were sighted specialists in visual impairments. The results are presented in Fig. 8.3.

The most useful cues that facilitate effective social interactions are the identity of an interaction partner, and his or her emotional state. We have therefore chosen to tackle these issues through our recent research on robust face and expression recognition. Recognition of body mannerisms, eye gaze direction, proxemics, and other nonverbal cues are part of our ongoing research.

It is important to note that there are different strata of individuals within the entire population of the visually impaired. Hence, it is imperative to design assistive technologies to accommodate the adaptation of features based on an individual user's specific abilities, needs, and preferences. There are four levels of visual function according to the International Classification of Diseases[3]: normal vision, moderate visual impairment, severe visual impairment, and blindness. Moderate and severe visual impairments are usually grouped under the term "low vision". Blindness can therefore be conceptualized as a spectrum, with each individual residing at a unique location on the scale. Consequently, their needs and expectations from an assistive technology will be different and are expected to change over time. Moreover, some people are blind from birth (early blind), while some become blind later in life (late blind) and studies have shown that they vary considerably in terms of visual experiences [57].

As evident from user feedback, the key requirements in the SIA are robust person and facial expression recognition, which pose three funda-

[3] http://www.who.int/mediacentre/factsheets/fs282/en/.

mental challenges: (i) identifying the salient and exemplar samples from large amounts of unlabeled data in order to train a classification model with a very reduced human annotation effort; (ii) computing a measure of confidence or certainty associated with every prediction made by the system, and (iii) identifying facial descriptors that can capture a richer range of emotions for improved facial expression recognition. We now elaborate our contributions addressing each of these challenges in the context of the SIA. Furthermore, from the above discussions it is clear that the SIA technology should be personalized and adaptive, and a generalized design will not cater to the needs of all individuals with visual impairments. This has essentially been the motivation for making person-centeredness and coadaptation an integral part of our solutions.

## 8.5 BATCH MODE ACTIVE LEARNING FOR PERSON RECOGNITION

Recognizing the identity of an interaction partner is perhaps the most important aspect of social interaction. The camera mounted on the pair of glasses, as depicted in Fig. 8.2, has a high frame rate (25–30 fps) and thus a large number of images will be captured in a very short time span. These face images need to be labeled offline by a human expert to train the underlying classification models so the same subjects will be accurately recognized in the future. Manual annotation of such a large amount of data is an expensive process in terms of time, labor, and human expertise. This necessitates the use of batch mode active learning (BMAL) frameworks to address this challenge. These algorithms automatically identify the salient and exemplar instances from large amounts of unlabeled data and tremendously reduce human annotation efforts in training classification/regression models.

### 8.5.1 Batch Mode Active Learning: An Introduction

In batch mode active learning, the learner is exposed to a pool of unlabeled instances and it iteratively selects a batch of samples for manual annotation. Brinker [58] proposed a diversity-based BMAL technique based on SVMs which queried a diverse batch of samples for annotation. Hoi et al. [59] used Fisher information as a measure of model uncertainty and proposed querying a batch of samples which maximally reduced the Fisher information. The authors applied the same BMAL concept to the problems of content-based image retrieval [60] and medical image classification [61]. Guo and Schuurmans proposed an optimization-based framework for batch selection by maximizing the log likelihood of the selected instances with respect to the already selected data and minimizing the uncertainty of the unselected unlabeled instances [62]. Guo also proposed a matrix partition based framework for BMAL, which was independent of the underlying classification model [63]. In our previous work, we proposed an adaptive BMAL algorithm which automatically computed the batch size based on the complexity of the unlabeled data in question [64]. We also developed a BMAL framework based on the convex relaxation of an NP-hard integer quadratic programming problem, with guaranteed bounds on the solution quality [65].

### 8.5.1.1 *Proposed Framework*

Consider a scenario where a video stream has been captured by a user who is blind and where a BMAL algorithm needs to be applied to select a batch $B$ of salient and exemplar images for manual annotation. The batch size $k$ (number of samples to be selected) is assumed to be known in advance (based on available resources). We now formulate an objective function to address this problem. The set of unlabeled samples is denoted $U_t$. Let $w^t$ denote the classification model trained on the available labeled examples $L_t$. In order to have good generalization performance, the model $w^{t+1}$ trained on $L_t \cup B$ should have high classification confidence on the set of unselected images $U_t - B$. Entropy is taken as a metric to quantify uncertainty, with high entropy values signifying higher uncertainty. We introduce a term in the objective function, which ensures that the entropy of the updated model (model trained on the initial training set together with the newly selected batch) on the remaining unselected images is minimal. However, this may result in the selection of images only from the high-density regions of the unlabeled set. This occurs because, from a data geometry point of view, the set of images that is not selected may be dominated by images from high-density regions, which constitute a large portion of the data. To address this issue, we append a term that selects images specifically from low-density regions in the data space, i.e. images that have a high measure of distance from the remaining set. More formally, the entropy of the model $w^{t+1}$ on an image $x_j$ is computed as

$$S(y|x_j, w^{t+1}) = -\sum_{y \in C} P(y|x_j, w^{t+1}) \log P(y|x_j, w^{t+1}), \qquad (8.1)$$

where $C$ is the total number of classes and $y$ is a class label. Also, let $\rho_j$ denote the average distance of an unlabeled image $x_j$ from other images in the unlabeled video $U_t$. Greater values of $\rho_j$ can be assumed to denote that the image is located in a low–density region. In order to ensure that the objective function is differentiable, we use the Euclidean distance in this work (since the Euclidean norm ($L_2$) is differentiable). Any other differentiable distance metric can also be used. The two conditions described previously can thus be satisfied by defining a performance score function $f(B)$ in the following manner

$$f(B) = \sum_{i \in B} \rho_i - \lambda \sum_{j \in U_t - B} S\big(y|x_j, w^{t+1}\big), \tag{8.2}$$

where $\lambda$ is a trade-off parameter controlling the relative importance of the two terms. The problem therefore reduces to selecting a batch $B$ of $k$ unlabeled images that produces the maximum score $f(B)$. Since the search space is exponentially large, exhaustive search methods are not feasible. We therefore use numerical optimization techniques to solve this problem. Specifically, we define a binary vector $M$ of size equal to the number of elements in the unlabeled pool $U_t$, $M \in \Re^{|U_t| \times 1}$. Each entry denotes whether the corresponding unlabeled sample will be selected for annotation ($M_i = 1$) or not ($M_i = 0$). The batch selection problem in Eq. (8.2) therefore reduces to

$$\max_{M} \sum_{j \in U_t} \rho_j M_j - \lambda \sum_{j \in U_t} (1 - M_j) S\big(y|x_j, w^{t+1}\big) \tag{8.3}$$

s.t.:

$$M_i \in \{0, 1\}, \forall i \quad \text{and} \quad \sum_{i=1}^{|U_t|} M_i = k. \tag{8.4}$$

The above optimization is an integer programming problem and is NP-hard. We therefore relax the constraints to make it a continuous optimization problem:

$$\max_{M} \sum_{j \in U_t} \rho_j M_j - \lambda \sum_{j \in U_t} (1 - M_j) S\big(y|x_j, w^{t+1}\big) \tag{8.5}$$

s.t.:

$$0 \leq M_i \leq 1, \forall i \quad \text{and} \quad \sum_{i=1}^{|U_t|} M_i = k. \tag{8.6}$$

We solve $M$ from this formulation and set the top largest $k$ entries as 1 to approximate the integer solution.

### 8.5.1.2 Solving the Optimization Problem

The objective function is written in terms of $M$ as

$$f(M) = \sum_{j \in U_t} \rho_j M_j - \lambda \sum_{j \in U_t} (1 - M_j) S(y|x_j, w^{t+1}). \qquad (8.7)$$

To solve the optimization problem, we use the Quasi-Newton method, which assumes that the function can be approximated as a quadratic in the neighborhood of the optimum point and iteratively updates the variable $M$ to guide the functional value toward this local optima. The first derivative of the function and the Hessian matrix of second derivatives need to be computed as parts of the solution procedure. Assuming $w^{t+1}$ remains constant with small iterative updates of $M$, the first order derivative vector is obtained by taking the partial of the objective with respect to $M$:

$$\nabla f(M_j) = \rho_j + \lambda_1 S(y|x_j, w^{t+1}). \qquad (8.8)$$

The Hessian starts as an identity matrix and is updated according to the BFGS method [66]. In each iteration a quadratic programming problem is solved, which yields an update direction for $M$. The step size is obtained using a backtrack line search method based on the Armijo–Goldstein equation and that guarantees monotonic convergence of the function to the local optimum. The iterations are continued until the change in the value of the objective function is negligible. The final value of $M$ is used to govern the specific points to be selected for the given data stream (by greedily setting the top $k$ entries in $M$ as 1 to recover the integer solution). For further details about the Quasi-Newton method, please refer to [66]. We note that the objective function is defined in terms of the future classifier $w^{t+1}$, which is unknown. In the Quasi-Newton iterations, $w^{t+1}$ is approximated as the classifier trained on the current training set $L_t$ together with the set of unlabeled points selected in the current iteration (through $M$), where the label of each selected unlabeled point is assumed to be the same as that of the closest training point in $L_t$. The pseudocode is outlined in Algorithm 1.

## 8.5.2 BMAL for Person Recognition in the SIA

To empirically validate our algorithm in the context of the social interaction assistant, we used the VidTIMIT [68] and the NIST Multiple Biometric Grand Challenge (MBGC) [69] datasets, both of which are publicly

---

**Algorithm 1** Proposed Batch Mode Active Learning Algorithm

---

**Require:** Training set $L_t$, Unlabeled set $U_t$, parameters $\lambda$ and batch size $k$, an initial random guess for $M$, a stopping threshold $\alpha$

1: Initialize the Hessian matrix $H$ as the identity matrix $I$.
2: Evaluate the objective function $f(M)$ (Eq. (8.7)) and the derivative vector $\nabla f(M)$ (Eq. (8.8)).
3: **repeat**
4:     Solve the QP problem as required by Quasi-Newton: $QP(H, \nabla f(M), M)$ and let the solution be $M^*$.
5:     Compute the step size $s$ from the Armijo–Goldstein Equations.
6:     Update $M$ as $M_{new} = M + s(M^* - M)$.
7:     Evaluate the new objective $f(M_{new})$ and the new derivative vector $\nabla f(M_{new})$ using $M_{new}$.
8:     Calculate the difference in objective value: $diff = abs(f(M) - f(M_{new}))$.
9:     Update the Hessian $H$ using the BFGS Equations [66].
10:    Update the objective value: $f(M) = f(M_{new})$.
11:    Update the derivative vector: $\nabla f(M) = \nabla f(M_{new})$.
12:    Update the vector $M$: $M = M_{new}$.
13: **until** $diff \leq \alpha$
14: Greedily set the top $k$ entries in $M$ as 1 to recover the integer solution.
15: Select $k$ points accordingly.

---

available. These datasets contain recordings of subjects under natural conditions where there is a redundancy of information and were hence chosen to study the performance of BMAL algorithms. The discrete cosine transform (DCT) feature was used in our experiments and the Gaussian Mixture Model (GMM) was used as the underlying classifier. Twenty-five subjects were selected for this study. The initial training set contained 250 images, the unlabeled set had 2000 images, and the test set was formed of 4500 images spanning all the subjects. The batch size $k$ was taken as 10 and we studied the performance on the test set as more unlabeled samples were iteratively labeled and appended to the training set. The proposed approach was compared with three other BMAL schemes based on heuristic scores: (i) Random Sampling, where a batch of points was randomly queried from the unlabeled pool; (ii) SVM Active Learning with Angular Diversity, where a batch of points was incrementally sampled such that at each step the hyperplane induced by the selected point maximized the angle with all the hyperplanes of the already selected points, as proposed by Brinker [58]; and (iii) Uncertainty Based Ranked Selection, where the top

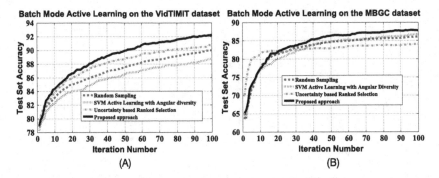

**Figure 8.4** Batch Mode Active Learning on the VidTIMIT (A) and MBGC (B) datasets. The proposed method outperforms the baseline algorithms. Image source [67].

$k$ uncertain points are queried from the unlabeled pool, $k$ being the batch size.

The results are depicted in Fig. 8.4, where the $x$-axis denotes the iteration number and the $y$-axis denotes the accuracy on the unseen test set [67]. We note that the proposed BMAL framework performs much better than the other methods as its accuracy on the test set grows at the fastest rate. The label complexity (defined as the number of queries needed to achieve a certain accuracy) is lowest in the case of the proposed technique. Thus, the proposed framework succeeds in selecting the most informative unlabeled samples for manual annotation and is tremendously useful in learning a good classification model with minimal human annotation effort.

### 8.5.3 Person-Centered BMAL for Face Recognition

The performance of the person recognition system can be further improved by incorporating person-centeredness in the BMAL formulation. Every person has a unique daily activity schedule and expects to encounter a specific set of subjects at specific locations. For instance, in an office location, there is a higher chance of meeting work colleagues than family members; at home, the situation is reversed. Thus, person-centeredness can be integrated into the system through *contextual information*. Following [70], we define context as the *location of a user*. We assume that at any given location, the user is aware of the subjects to be expected in that location (for example, work acquaintances in an office setting or family members in a home setting). This was used to construct a prior probability vector depicting the chances of seeing each subject at a given location. In this situation, the query function can be modified to ensure that the images remaining in

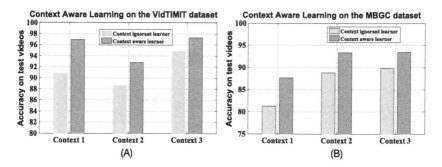

**Figure 8.5** Context-Aware Learning on the VidTIMIT (A) and MBGC (B) datasets. The context-aware learner performs better than the context-ignorant learner. Image source [67].

the unlabeled video after batch selection have low entropy with respect to the subjects expected in the given context. Thus, the entropy is computed only on the subjects that are present in a given video stream:

$$f(B) = \sum_{i \in B} \rho_i - \lambda \sum_{j \in U_t - B} S^{context}\left(\gamma | x_j, w^{t+1}\right). \qquad (8.9)$$

Here, $S^{context}$ is the context-aware entropy term. For each unlabeled image, this term was computed from the posterior probabilities, which in turn were obtained by multiplying the likelihoods returned by the trained GMM classifier with the context-aware prior. Thus, subjects not expected in a given context will have low priors and consequently, the corresponding posteriors will not contribute much in computing $S^{context}$. To simulate this situation, three contexts were arbitrarily defined and 8 random subjects (chosen from the set of 25) were assigned to each context. BMAL was used to select batches of samples from unlabeled video streams in each context. The updated classifiers were then tested on videos in the respective context. The context-ignorant learner was implemented using equal class priors in the entropy term.

Fig. 8.5 shows the accuracies obtained on the VidTIMIT and MBGC test videos (averaged over three trials in each context) [67]. It is noted that in each context, the context-aware learner produces better accuracy on test videos than the context ignorant learner. Thus, incorporation of context in the formulation further helps in querying salient images for manual annotation.

In this application, we note that the user and the system work together and learn over time. Initially, the user supplies the chances of meeting subjects at specific locations (in the form of "yes", "no", and "maybe" inputs) for better performance of the underlying algorithm. Over time the system adapts to the lifestyle and routine of the user, and thus the computation of the prior probability vector can be automated. This is the core idea of person-centeredness and coadaptation.

## 8.6 CONFORMAL PREDICTIONS FOR MULTIMODAL PERSON RECOGNITION

As detailed in Section 8.5, person recognition is an integral component of the SIA. An additional important component is the computation of reliable confidence measures, which depict the system's level of certainty in its predictions. In the SIA, an incorrect system prediction of the identity of an individual can lead to an embarrassing social situation; thus, knowledge of the machine's confidence associated with each of its predictions can enable the user to react appropriately and avoid such situations. Moreover, there has been extensive research in machine learning from multiple sources and information fusion to improve the robustness of a recognition system. In the SIA, two sources of information, audio and video, can be combined for accurate person recognition. In this section, we detail the Conformal Predictions (CP) framework for the computation of reliable confidence metrics and its application to person recognition from face and voice modalities.

### 8.6.1 Conformal Predictions: An Introduction

The theory of Conformal Predictions was developed by Vovk, Shafer, and Gammerman [71,72] based on the principles of algorithmic randomness, transductive inference, and hypothesis testing. The theory is based on the relationship derived between transductive inference and Kolmogorov complexity [73] of an identically independently distributed (i.i.d.) sequence of data instances. The merit of this framework is that the results are well-calibrated in an online setting; that is, the frequency of errors, $\epsilon$, made by the system is exactly bounded according to the confidence level, $1 - \epsilon$, defined by the user.

#### 8.6.1.1 Conformal Predictions in Classification

For a given test sample, say $x_{n+1}$, a null hypothesis is assumed that $x_{n+1}$ belongs to the class label, say $y_p$. Using this assignment, the nonconformity measures [74] of all the data points in the system so far are recomputed assuming the null hypothesis is true. Sample nonconformity measures for various classification algorithms can be found in [72]. A p-value function is defined as

$$p\left(\alpha_{n+1}^{y_p}\right) = \frac{count\left\{i : \alpha_i^{y_p} \geq \alpha_{n+1}^{y_p}\right\}}{n+1}, \tag{8.10}$$

where $\alpha_{n+1}^{y_p}$ is the nonconformity measure of $x_{n+1}$, assuming it is assigned the class label $y_p$. It is evident that the p-value is highest when all nonconformity measures of training data belonging to class $y_p$ are higher than that of the new test point, $x_{n+1}$, which shows that $x_{n+1}$ is *most conformal* to the class $y_p$. This process is repeated with the null hypothesis supporting each of the class labels, and the highest of the p-values is used to decide the actual class label assigned to $x_{n+1}$, thus providing a transductive inferential procedure for classification. If $p_j$ and $p_k$ are the two highest p-values obtained (in respective order), then $p_j$ is called the *credibility* of the decision, and $1 - p_k$ is the *confidence* of the classifier in the decision. The conformal prediction regions are presented as regions representing a specified confidence level, $\Gamma_\epsilon$, which contains all the class labels with a p-value greater than $1 - \epsilon$. These regions are *conformal*, i.e. the confidence threshold, $1 - \epsilon$, directly translates to the frequency of errors, $\epsilon$, in the online setting [71]. The framework is summarized in Algorithm 2.

---

**Algorithm 2** Conformal Predictors for Classification

---

**Require:** Training set $T = \left\{(x_1, y_1), ..., (x_n, y_n)\right\}$, $x_i \in X$, number of classes $M$,
    $y_i \in Y = y_1, y_2, ..., y_M$, classifier $C$

1: Get new unlabeled example $x_{n+1}$.
2: **for** all class labels, $y_j$, where $j = 1, ..., M$ **do**
3:    Assign label $y_j$ to $x_{n+1}$.
4:    Update the classifier $C$ with $T \cup \left\{x_{n+1}, y_j\right\}$.
5:    Compute nonconformity measure value, $\alpha_i^{y_j} \forall i = 1, ..., n+1$ to compute the p-value, $P_j$, with respect to class $y_j$ (Eq. (8.10)) using the conformal predictions framework.
6: **end for**
7: Output the conformal prediction regions $\Gamma_{1-\epsilon} = \left\{y_j : P_j > 1 - \epsilon, y_j \in Y\right\}$, where $1 - \epsilon$ is the confidence level.

---

### 8.6.1.2 Conformal Predictors for Information Fusion

Our proposed methodology for obtaining conformal predictions in information fusion settings is fundamentally premised on multiple hypothesis testing [75]. We propose that each modality (or data feature) considered for fusion can be formulated as an independent hypothesis test, and the p-values obtained from each hypothesis test can be combined using established statistical methods [76]. Given a new test data instance, the CP framework outputs a p-value for every class label. When there are multiple data sources describing a single class label entity (e.g. different modalities like face and speech, or different image feature spaces obtained from a single face image for person recognition), we use a classifier for each individual data source with appropriate nonconformity measures, and obtain p-values for each class label uniquely for each data source. Thus, for every class label $y_j$, $j \in \{1, \ldots, M\}$, we have an individual null hypothesis for each data source, $H_{01}, H_{02}, \ldots, H_{0N}$, and an individual alternate hypothesis, $H_{A1}, H_{A2}, \ldots, H_{AN}$, where $M$ is the number of class labels and $N$ is the number of data sources. Thus, for every class label $y_j$, we obtain $N$ p-values, $p_i$, $i = 1, \ldots, N$, one for each modality. These p-values are then combined into a new test statistic $\phi = \phi(p_1, \ldots, p_N)$, which is used to test the combined null hypothesis $H_0$ for class label $y_j$. The conformal prediction region at a specified confidence level, $\Gamma_\epsilon$, is then presented as a set containing all the class labels with a p-value greater than $1 - \epsilon$. The pseudo-code is depicted in Algorithm 3.

---

**Algorithm 3** Conformal Predictors for Information Fusion

---

**Require:** Number of data sources $N$; Training sets for each data source $T_1 = \{(x_{11}, y_1), \ldots, (x_{1n}, y_n)\}, \ldots, T_N = \{(x_{N1}, y_1), \ldots, (x_{Nn}, y_n)\}$; Number of classes $M$, $y_i \in Y = y_1, y_2, \ldots, y_M$; classifiers $C_1, \ldots, C_N$ for each data source

1: Get the new unlabeled example with respect to each data source $x_{1,n+1}, \ldots, x_{N,n+1}$.
2: Using Algorithm 2 and classifiers $C_1, \ldots, C_N$ corresponding to each data source, compute p-values $p_{ij}$, where $i = 1, \ldots, N$ and $j = 1, \ldots, M$.
3: **for** each class label, $y_j, j = 1, \ldots, M$ **do**
4:     Compute p-value $\hat{p}_j$ of combined hypothesis using $N$ modalities.
5: **end for**
6: Output the conformal prediction regions $\Gamma_{1-\epsilon} = \{y_j : \hat{p}_j > 1 - \epsilon, y_j \in Y\}$, where $1 - \epsilon$ is the confidence level.

---

Multiple hypothesis testing has been extensively studied and a variety of methods have been proposed to combine p-values from multiple hypothesis tests. We use the Quantile combination method, where a relevant parametric cumulative distribution function (CDF), $F$, is selected, and

**Table 8.1** Standard Normal Fusion results on the VidTIMIT and MOBIO datasets. The error rate is calibrated at each of the confidence levels for both datasets. Table source [81]

| Dataset | Percentage of errors at confidence level | | | | | | |
|---|---|---|---|---|---|---|---|
| | 50% | 60% | 70% | 80% | 90% | 95% | 99% |
| VidTIMIT | 44.46% | 35.37% | 25.79% | 14.91% | 2.59% | 0.82% | 0.80% |
| MOBIO | 46.05% | 37.73% | 28.92% | 20.49% | 7.92% | 2.18% | 0.91% |

the p-values $p_i$s, are transformed into distributional quantiles, $q_i = F^{-1}(p_i)$, where $i = 1, 2, \ldots, k$ for each of the class labels. These $q_i$s are subsequently combined as $\phi = \sum_i q_i$, and the p-value of the combined test $H_0$ is computed from the sampling distribution of $\phi$. Examples of CDFs used in these methods include chi-square [77], standard normal [78], uniform [79], and logistic [80]. We use the standard normal fusion (SNF) technique because it is widely used and for its promising empirical performance.

## 8.6.2 Conformal Predictions for Person Recognition in the SIA

We used the VidTIMIT [68] and the MOBIO [82] biometric datasets to validate our algorithm in the context of the SIA. Both these datasets contain frontal images and speech data of subjects captured under challenging real-world conditions. We used the DCT feature for the video modality and SVM was used as the underlying classification model. The Lagrangian multipliers obtained while training an SVM were used as the nonconformity measures, as suggested in [72]. The speech signal was downsampled to 8 kHz and a short-time 256 pt Fourier analysis is performed on a 25 ms Hamming window (10 ms frame rate). The magnitude spectrum was transformed to a vector of Mel-Frequency Cepstral Coefficients (MFCCs). A gender-dependent 512-mixture GMM Universal Background Model was initialized using a $k$-means algorithm and then trained by estimating the GMM parameters via the Expectation Maximization algorithm. To adapt this to the CP framework, the negative of the likelihood values generated by the GMM were used as the nonconformity scores, as suggested by Vovk et al. [72].

The fusion results are reported in Table 8.1 [81]. It depicts the percentage of errors made by the system at different values of the confidence level. It is evident that the results are statistically calibrated across the two datasets; that is, the error rate, $\epsilon$, is consistently close to the confidence level, $1 - \epsilon$. This corroborates the usefulness of the CP framework for reliable person recognition in the SIA application.

### 8.6.3 Person-Centered Recognition Using the CP Framework

To maintain the calibration property, the CP framework may sometimes produce multiple predictions for a given test sample. This occurs mostly for high-confidence predictions where the system has to deliver accurate outputs to ensure the calibration property. For low-confidence predictions, the number of multiple predictions is much lower (which has been validated by our experiments). The concept of person-centeredness can be judiciously used to handle the multiple prediction problem in the context of the SIA.

As noted before, individuals who are blind or visually impaired fall on a continuum where each person resides at a specific location based on the level of blindness. Consider the case of a low-vision individual who can rely on his or her sensory abilities to some extent. This person can use a moderate to high threshold on the percentage of errors (50–60%) and thus will get relatively imperfect results but with a much lower number of multiple predictions. The user's limited vision together with the system results will lead to a decision regarding the identity of the interaction partner. Over time, however, it is possible that the user's condition may deteriorate toward complete blindness. In this case, the user will become increasingly reliant on the machine and will need to use a much lower threshold on the percentage of errors (5–10%) to ensure perfect results from the system. Multiple predictions can be handled based on the context (a particular subject may have a very low probability of being present at a given location, e.g. a home acquaintance in an office setting). The machine also adapts to the user's condition over time and can automatically decide the error threshold needed to avoid embarrassing social situations.

As another example, consider the case of a blind individual. The error threshold can be decided based on this user's location at a given point of time. For instance, when in an office meeting, a low error threshold should be selected as it is imperative to accurately identify all the subjects in the meeting. When the user is at home, a relatively higher error threshold can be selected (minimizing the number of multiple predictions), as incorrect recognition does not have severe consequences in a home setting. The threshold selection can be automated over time, based on the user's daily activity schedule. In both of these instances, the user and the system work together and adapt over time, another example of person-centeredness and coadaptation.

## 8.7 TOPIC MODELS FOR FACIAL EXPRESSION RECOGNITION

Facial expression recognition is an inherent part of the SIA; understanding the emotional state of an interaction partner is crucial in driving a fruitful conversation. Human emotional states can be described using seven discrete states (anger, disgust, contempt, fear, joy, sadness, and surprise). There has been an extensive body of work explaining the seven discrete states in terms of facial Action Units (AUs), as presented in the Facial Action Coding System (FACS) of Ekman and Friensen [83]. There has been extensive research in recognizing these seven emotional states. In our research, we have used Latent Facial Topics (LFTs) to explore a more complex range of emotions. Our extensive empirical studies emphasize the usefulness of LFTs for discrete expression recognition and also establish that facial topics derived using this method have semantic validity and are visualizable.

Topic models have been used successfully in a variety of computer vision tasks including object recognition, scene segmentation, and image annotation [84]. The fundamental idea of topic models is that every document contains latent concepts called topics. In text mining, a topic is defined as a collection of words that can cooccur in a given corpus of documents. In our research, we adapt topic models to facial video processing, where we can either model an entire video or an individual video frame as a document and explain the activity within each facial video or image document using LFTs. Our methodology in this work is illustrated in Fig. 8.6. The facial documents are created by quantizing facial features (e.g. shape or appearance features, which we call base features) obtained from face images. We used two popular probabilistic topic models, Latent Dirichlet Allocation (LDA) [85], which is unsupervised, and Supervised Latent Dirichlet Allocation (SLDA) [86], which is supervised, to extract LFTs from these facial documents.

### 8.7.1  Facial Expression Recognition in the SIA

We used the Cohn–Kanade Plus (CK+) database [88], which contains 327 image sequences, annotated with 7 facial expressions and 34 AUs from 118 subjects, to study the usefulness of LFTs for emotion recognition. We used a subject-based leave-one-out strategy to evaluate the performance. Using the LFT distributions as features, two classifiers were trained: Support Vector Machines using Linear (LFT-SVML) and RBF kernels (LFT-SVMR). The proposed approach was compared against the similarity-normalized shape (SPTS) features (SPTS-SVML) which were originally proposed for

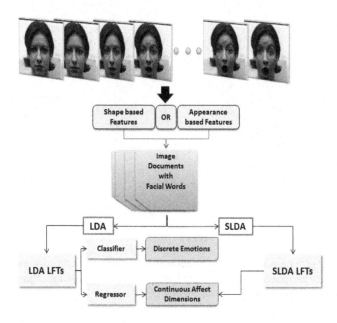

**Figure 8.6** Emotion Recognition using Latent Facial Topics (LFTs). Image source [87].

**Table 8.2** Accuracies for Discrete emotion recognition on the CK+ dataset. LFTs outperform SPTS features by a large margin. Table source [87]

| Classifier | SPTS-SVML | LFT-SVML | LFT-SVMR |
|---|---|---|---|
| Accuracy | 66.68% | 85.62% | 84.4% |

this database [88]. The results are given in Table 8.2 and corroborate that LFTs lead to improved recognition performance over SPTS features [87].

Latent facial topics discovered automatically using LDAs are shown in Fig. 8.7 [87]. LFTs are probability distributions over facial words, where the probability values are presented by the length of the blue lines; we note that LFTs are implicitly meaningful, as is visually evident in Fig. 8.7. A deeper analysis also reveals a relationship with facial AUs, as defined by Ekman and Friensen [83]. For example, LFT 8 and 18 together correspond to AU 1 (Inner Brow Raiser) and LFT 49 corresponds to AU 12 (Lip corner pull). These results demonstrate the efficacy of LFTs as salient features in human emotion recognition and their semantic interpretation and visualizibility.

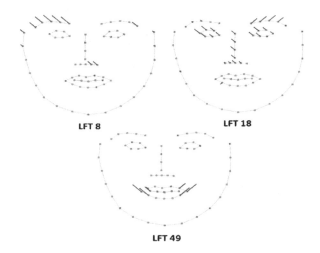

**Figure 8.7** Plots of LDA based LFTs from CK+ dataset. LFTs have semantic meaning and are visualizable. Image source [87]. (For interpretation of the colors in this figure, the reader is referred to the web version of this chapter.)

## 8.7.2 Person-Centered Facial Expression Recognition

A system recognizing emotions using LFTs can be used to convey the results at different granularities; it can either convey the discrete emotion or more subtle information such as the movement of the facial muscles (an eyebrow raise, lip-corner pull, etc.). Depending on the context in question, one particular output may be preferred over another. Consider the application of an individual who is visually impaired, attending an interview or in a meeting. In this case, it is imperative to gauge the mind-set of the interaction partner to better understand his or her reactions. A system delivering the minute details about the facial muscle movements is immensely useful in such a scenario. On the other hand, in a more casual setting like home, conveying the intricate details may result in an unnecessary overload of information; a device furnishing merely the discrete emotion is more convenient. These inputs can be passed manually to the system initially; over time, the system adapts to the schedule and habits of a particular user and can operate autonomously to deliver appropriate outputs. Thus, the machine and the user adapt to each other over time to solve a challenging computer vision task. This aligns with the core principles of person-centered multimedia computing.

## 8.8 CONCLUSION AND DISCUSSION

In this chapter, we presented our computer vision research contributions toward the design and development of a person-centered social interaction assistant, which is an AAC technology to enrich the communication experience of individuals with visual impairments. This is the first research effort toward the development of adaptive solutions that can cater to the needs and preferences of individual users instead of a one-size-fits-all approach. Through coadaptation, the user and the system work together to solve challenging real-world problems and adapt to each other over time. Our user studies have shown that integration of person-centeredness in the design and development of solutions is pivotal to addressing individual users' needs, expectations, and adaptations.

While the SIA was originally designed for individuals with visual impairment, this effort naturally generalizes to the broader population. In general, the explicit needs of individuals with disabilities may be the unspecified implicit needs of the general population. For example, the SIA could be helpful for sighted individuals in remote audio communication, where the visual modality may not be available or accessible. Providing access to nonverbal visual social cues can be quite valuable in these situations. We view the spectrum of disabilities as a range of abilities; an ability is a continuum from "disability" to "ability" to "super-ability". Hence, the focus on a person-centered approach provides valuable insights that not only address the needs of individuals with disabilities, but also empower individuals throughout the spectrum of abilities. We believe that the PCMC approach will increasingly become the methodology of choice in the design of new technologies and applications for the general population.

## ACKNOWLEDGMENTS

The authors would like to thank Arizona State University and the National Science Foundation for their funding support. This material is partially based upon work supported by the National Science Foundation under Grant Nos. 1069125 and 1116360.

## REFERENCES

[1] S. Blackstone, M. Williams, D. Wilkins, Key principles underlying research and practice in AAC, Augmentative and Alternative Communication (2007).
[2] J. Light, D. McNaughton, Designing AAC research and intervention to improve outcomes for individuals with complex communication needs, Augmentative and Alternative Communication (2015).

[3] S. Baxter, P. Enderby, P. Evans, S. Judge, Barriers and facilitators to the use of high-technology augmentative and alternative communication devices: a systematic review and qualitative synthesis, International Journal of Language and Communication Disorders (2012).

[4] K. Button, F. Rossera, Barriers to communication, The Annals of Regional Science (1990).

[5] K. Chew, T. Iacono, J. Tracy, Overcoming communication barriers: working with patients with intellectual disabilities, Australian Family Physician (2009).

[6] D. McNaughton, J. Light, L. Groszyk, "Don't give up": employment experiences of individuals with amyotrophic lateral sclerosis who use augmentative and alternative communication, Augmentative and Alternative Communication (2001).

[7] D. McNaughton, D. Bryen, AAC technologies to enhance participation and access to meaningful societal roles for adolescents and adults with developmental disabilities who require AAC, Augmentative and Alternative Communication (2007).

[8] D. McNaughton, D. Bryen, Enhancing participation in employment through AAC technologies, Assistive Technology (2002).

[9] J. Swain, S. French, C. Barnes, C. Thomas, Disabling Barriers, Enabling Environments, SAGE Publications, 2004.

[10] J. Light, D. McNaughton, The changing face of augmentative and alternative communication: past, present, and future challenges, Augmentative and Alternative Communication (2012).

[11] P. Mirenda, Toward functional augmentative and alternative communication for students with autism: manual signs, graphic symbols, and voice output communication aids, Language, Speech & Hearing Services in Schools (2003).

[12] C. Goossens, S. Crain, Overview of nonelectronic eye-gaze communication techniques, Augmentative and Alternative Communication (1987).

[13] K. Hustad, C. Dardis, A. Kramper, Use of listening strategies for the speech of individuals with dysarthria and cerebral palsy, Augmentative and Alternative Communication (2011).

[14] D. McNaughton, J. Light, The iPad and mobile technology revolution: benefits and challenges for individuals who require augmentative and alternative communication, Augmentative and Alternative Communication (2013).

[15] H. Koester, S. Levine, Learning and performance of able-bodied individuals using scanning systems with and without word prediction, Assistive Technology (1994).

[16] Z. Al-Kassim, Q. Memon, Designing a low-cost eyeball tracking keyboard for paralyzed people, Computers & Electrical Engineering (2017).

[17] A. Al-Rahayfeh, M. Faezipour, Eye tracking and head movement detection: a state-of-art survey, IEEE Journal of Translational Engineering in Health and Medicine (2013).

[18] C. Jen, Y. Chen, Y. Lin, C. Lee, A. Tsai, M. Li, Vision based wearable eye-gaze tracking system, in: IEEE International Conference on Consumer Electronics, 2016.

[19] V. Raudonis, R. Simutis, G. Narvydas, Discrete eye tracking for medical applications, in: International Symposium on Applied Sciences in Biomedical and Communication Technologies, 2009.

[20] K. Gillespie-Smith, S. Fletcher-Watson, Designing AAC systems for children with autism: evidence from eye tracking research, Augmentative and Alternative Communication (2014).

[21] K. Krafka, A. Khosla, P. Kellnhofer, H. Kannan, S. Bhandarkar, W. Matusik, A. Torralba, Eye tracking for everyone, in: IEEE Conference on Computer Vision and Pattern Recognition, 2016.

[22] X. Zhang, H. Kulkarni, M. Morris, Smartphone-based gaze gesture communication for people with motor disabilities, in: ACM CHI, 2017.

[23] K. Grauman, M. Betke, J. Gips, G. Bradski, Communication via eye blinks – detection and duration analysis in real time, in: IEEE Conference on Computer Vision and Pattern Recognition, 2001.

[24] B. Schilit, N. Adams, R. Want, Context-aware computing applications, in: Workshop on Mobile Computing Systems and Applications, 1994.

[25] S. Kane, B. Linam-Church, K. Althoff, D. McCall, What we talk about: designing a context-aware communication tool for people with aphasia, in: ACM SIGACCESS Conference on Computers and Accessibility, 2012.

[26] J. Campbell, Speaker recognition: a tutorial, Proceedings of the IEEE (1997).

[27] V. Young, A. Mihailidis, Difficulties in automatic speech recognition of dysarthric speakers and implications for speech-based applications used by the elderly: a literature review, Assistive Technology (2010).

[28] K. Rosen, S. Yampolsky, Automatic speech recognition and a review of its functioning with dysarthric speech, Augmentative and Alternative Communication (2000).

[29] S. Oviatt, Taming recognition errors with a multimodal interface, Communications of the ACM (2000).

[30] A. Hassanat, Visual Speech Recognition, Technical Report, Mu'tah University, Jordan, 2014.

[31] D. Howell, S. Cox, B. Theobald, Visual units and confusion modelling for automatic lip-reading, Image and Vision Computing (2016).

[32] Y. Lan, R. Harvey, B. Theobald, Insights into machine lip reading, in: IEEE International Conference on Acoustics, Speech and Signal Processing, 2012.

[33] S. Chu, T. Huang, Audio-visual speech modeling using coupled hidden Markov models, in: IEEE International Conference on Acoustics, Speech, and Signal Processing, 2002.

[34] S. Dupont, J. Luettin, Audio-visual speech modeling for continuous speech recognition, IEEE Transactions on Multimedia (2000).

[35] A. Nefian, L. Liang, X. Pi, X. Liu, K. Murphy, Dynamic Bayesian networks for audio-visual speech recognition, EURASIP Journal on Advances in Signal Processing (2002).

[36] E. Salama, R. El-khoribi, M. Shoman, Audio-visual speech recognition for people with speech disorders, International Journal of Computer Applications (2014).

[37] R. Akmeliawati, M. Ooi, Y. Kuang, Real-time Malaysian sign language translation using colour segmentation and neural network, in: IEEE Instrumentation Measurement Technology Conference, 2007.

[38] P. Dreuw, D. Stein, T. Deselaers, D. Rybach, M. Zahedi, J. Bungeroth, H. Ney, Spoken language processing techniques for sign language recognition and translation, Technology and Disability (2008).

[39] G. Fang, W. Gao, D. Zhao, Large vocabulary sign language recognition based on fuzzy decision trees, IEEE Transactions on Systems, Man and Cybernetics (2004).

[40] Z. Halim, G. Abbas, A Kinect-based sign language hand gesture recognition system for hearing- and speech-impaired: a pilot study of Pakistani sign language, Assistive Technology (2015).

[41] J. Hernandez-Rebollar, N. Kyriakopoulos, R. Lindeman, A new instrumented approach for translating American sign language into sound and text, in: IEEE International Conference on Automatic Face and Gesture Recognition, 2004.

[42] Y. Li, X. Chen, J. Tian, X. Zhang, K. Wang, J. Yang, Automatic recognition of sign language subwords based on portable accelerometer and EMG sensors, in: International Conference on Multimodal Interfaces: Workshop on Machine Learning for Multimodal Interaction, 2010.

[43] R. McGuire, J. Hernandez-Rebollar, T. Starner, V. Henderson, H. Brashear, D. Ross, Towards a one-way American sign language translator, in: IEEE International Conference on Automatic Face and Gesture Recognition, 2004.

[44] R. Su, X. Chen, S. Cao, X. Zhang, Random forest-based recognition of isolated sign language subwords using data from accelerometers and surface electromyographic sensors, Sensors (2016).

[45] S. Wei, X. Chen, X. Yang, S. Cao, X. Zhang, A component-based vocabulary-extensible sign language gesture recognition framework, Sensors (2016).

[46] Y. Wu, T. Huang, Vision-based gesture recognition: a review, in: Lecture Notes in Computer Science, Springer, Berlin, Heidelberg, 1999.

[47] H. Brashear, T. Starner, P. Lukowicz, H. Junker, Using multiple sensors for mobile sign language recognition, in: IEEE International Symposium on Wearable Computers, 2003.

[48] M. Knapp, Nonverbal Communication in Human Interaction, Holt, Rinehart and Winston, New York, 1978.

[49] N. Ambady, R. Rosenthal, Thin slices of expressive behavior as predictors of interpersonal consequences: a meta-analysis, Psychological Bulletin (1992).

[50] M. Guzdial, Human-centered computing: a new degree for Licklider's world, Communications of the ACM (2013).

[51] A. Jaimes, N. Dimitrova, Human-centered multimedia: culture, deployment and access, IEEE Multimedia (2006).

[52] A. Jaimes, N. Sebe, D. Gatica-Perez, Human-centered computing: a multimedia perspective, in: ACM Multimedia Conference, 2006.

[53] S. Panchanathan, T. McDaniel, V. Balasubramanian, Person-centered accessible technologies: improved usability and adaptation through inspirations from disability research, in: ACM Workshop on User Experience in e-Learning and Augmented Technologies in Education, 2012.

[54] S. Panchanathan, T. McDaniel, V. Balasubramanian, An interdisciplinary approach to the design, development and deployment of person-centered accessible technologies, in: International Conference on Recent Trends in Information Technology (ICRTIT), 2013.

[55] S. Panchanathan, T. McDaniel, Person-centered accessible technologies and computing solutions through interdisciplinary and integrated perspectives from disability research, Universal Access in the Information Society (2014).

[56] S. Krishna, D. Colbry, J. Black, V. Balasubramanian, S. Panchanathan, A systematic requirements analysis and development of an assistive device to enhance the social interaction of people who are blind or visually impaired, in: Workshop on Computer Vision Applications for the Visually Impaired, 2008.

[57] H. Burton, A. Snyder, T. Conturo, E. Akbudak, J. Ollinger, M. Raichle, Adaptive changes in early and late blind: a fMRI study of braille reading, Journal of Neurophysiology (2002).

[58] K. Brinker, Incorporating diversity in active learning with support vector machines, in: International Conference on Machine Learning (ICML), 2003.

[59] S. Hoi, R. Jin, M. Lyu, Large-scale text categorization by batch mode active learning, in: International Conference on World Wide Web (WWW), 2006.

[60] S. Hoi, R. Jin, J. Zhu, M. Lyu, Semi-supervised SVM batch mode active learning with applications to image retrieval, ACM Transactions on Information Systems (2009).

[61] S. Hoi, R. Jin, J. Zhu, M. Lyu, Batch mode active learning and its application to medical image classification, in: International Conference on Machine Learning (ICML), 2006.

[62] Y. Guo, D. Schuurmans, Discriminative batch mode active learning, in: Neural Information Processing Systems (NIPS), 2007.

[63] Y. Guo, Active instance sampling via matrix partition, in: Neural Information Processing Systems (NIPS), 2010.

[64] S. Chakraborty, V. Balasubramanian, S. Panchanathan, Dynamic batch mode active learning, in: IEEE Conference on Computer Vision and Pattern Recognition (CVPR), 2011.

[65] S. Chakraborty, V. Balasubramanian, Q. Sun, S. Panchanathan, J. Ye, Active batch selection via convex relaxations with guaranteed solution bounds, IEEE Transactions on Pattern Analysis and Machine Intelligence (2014).

[66] J. Nocedal, S.J. Wright, Numerical Optimization, Springer, 1999, 636 pp.

[67] S. Chakraborty, V. Balasubramanian, S. Panchanathan, Generalized batch mode active learning for face-based biometric recognition, Pattern Recognition (2013).

[68] C. Sanderson, Biometric Person Recognition: Face, Speech and Fusion, VDM Verlag, 2008.

[69] M. Tistarelli, M. Nixon, Advances in biometrics, in: Third International Conference on Biometrics, Springer-Verlag, Berlin, Heidelberg, 2009.

[70] A.K. Dey, G.D. Abowd, D. Salber, A conceptual framework and a toolkit for supporting the rapid prototyping of context-aware applications, Human-Computer Interaction (2001).

[71] G. Shafer, V. Vovk, A tutorial on conformal prediction, Journal of Machine Learning Research (2008).

[72] V. Vovk, A. Gammerman, G. Shafer, Algorithmic Learning in a Random World, Springer-Verlag, New York, 2005.

[73] M. Li, P. Vitanyi, An Introduction to Kolmogorov Complexity and Its Applications, Springer-Verlag, New York, 1997.

[74] K. Proedrou, I. Nouretdinov, V. Vovk, A. Gammerman, Transductive confidence machines for pattern recognition, in: European Conference on Machine Learning (ECML), 2002.

[75] J. Shaffer, Multiple hypothesis testing, Annual Review of Psychology (1995).

[76] T. Loughin, A systematic comparison of methods for combining p-values from independent tests, Computational Statistics & Data Analysis (2004).

[77] R. Fisher, Statistical Methods for Research Workers, Macmillan Pub. Co., 1970.

[78] T. Liptak, On the combination of independent tests, A Magyar Tudományos Akadémia Matematikai Kutató Intézetének Közleményei (1958).

[79] E. Edgington, An additive method for combining probability values from independent experiments, The Journal of Psychology: Interdisciplinary and Applied (1972).

[80] G. Mudholkar, E. George, The logit method for combining probabilities, The Journal of Psychology: Interdisciplinary and Applied (1979).

[81] V. Balasubramanian, S. Chakraborty, S. Panchanathan, Conformal predictions for information fusion: a comparative study of p-value combination methods, Annals of Mathematics and Artificial Intelligence (AMAI) (2013).

[82] S. Marcel, C. McCool, S. Chakraborty, V. Balasubramanian, Mobile biometry (MO-BIO) face and speaker verification evaluation, in: International Conference on Pattern Recognition (ICPR), 2010.

[83] P. Ekman, W. Friensen, The Facial Action Coding System (FACS): A Technique for the Measurement of Facial Action, Consulting Psychologists Press, 1978.

[84] J. Varadarajan, J. Odobez, Topic models for scene analysis and abnormality detection, in: IEEE International Conference on Computer Vision (ICCV), 2009.

[85] D. Blei, A. Ng, M. Jordan, Latent Dirichlet allocation, Journal of Machine Learning Research (2003).

[86] W. Chong, D. Blei, F. Li, Simultaneous image classification and annotation, in: IEEE Conference on Computer Vision and Pattern Recognition (CVPR), 2009.

[87] P. Lade, V. Balasubramanian, S. Panchanathan, Latent facial topics for affect analysis, in: IEEE International Conference on Multimedia and Expo (ICME), 2013.

[88] P. Lucey, J. Cohn, T. Kanade, J. Saragih, Z. Ambadar, I. Matthews, The extended Cohn–Kanade dataset (CK+): a complete dataset for action unit and emotion-specified expression, in: IEEE Conference on Computer Vision and Pattern Recognition Workshops (CVPRW), 2010.

# CHAPTER 9

# Computer Vision for Lifelogging
## Characterizing Everyday Activities Based on Visual Semantics

**Peng Wang\*, Lifeng Sun\*, Alan F. Smeaton†, Cathal Gurrin†, Shiqiang Yang\***

\*National Laboratory for Information Science and Technology, Tsinghua University, China
†Insight Centre for Data Analytics, Dublin City University, Dublin, Ireland

## Contents

## Abstract

The rapid development of mobile devices capable of sensing our interactions with the environment has made it possible to assist humans in daily living, for example helping patients with cognitive impairment or providing customized food intake plans for patients with obesity, etc. All of this can be achieved through the passive gathering of detailed records of everyday behavior known as lifelogging. For example, the widely adopted smart mobile phones and newly emerging consumer wearable devices like Google Glass, Baidu Eye, and Narrative Clip are usually embedded with rich sensing capabilities including a camera, accelerometer, GPS, digital compass, etc., which can help to capture daily activity unobtrusively. Among these heterogeneous sensor readings, visual media contain more semantics to assist in characterizing everyday activities, and visual lifelogging is a class of personal sensing that employs wearable cameras to capture images or video sequences of everyday activities. This chapter will focus on the

Computer Vision for Assistive Healthcare.
DOI: https://doi.org/10.1016/B978-0-12-813445-0.00009-5

most recent research methods used to understand visual lifelogs, including semantic annotations of visual concepts, use of contextual semantics, recognition of activities, visualization of activities, etc. We also discuss some research challenges that indicate potential directions for future research. This chapter is intended to support readers in the area of assistive living using wearable sensing and computer vision for lifelogging, and human behavior researchers aiming at behavioral analysis based on visual understanding.

## Keywords

Visual lifelogging, Context sensing, Semantic indexing, Automatic tagging, Occurrence patterns, Attribute-based activity recognition, Event-based browsers, Ethics and privacy

## Chapter Points

- Comprehensive description of lifelogging and visual lifelogging in assistive living.
- State-of-the-art computer vision processing for visual understanding of everyday contexts.
- A full understanding of everyday activities from static to dynamic points of view.
- Practical experience and guidance in visual lifelogging interpretation.

## 9.1  INTRODUCTION AND BACKGROUND

The proliferation of modern sensing devices has opened the possibility of technically assisted living that benefits the well-being of dependents such as the elderly, patients in need of special care, and independently healthy people. To this end, it is imperative to understand the everyday activities of the individual in order to provide customized services or treatments. For example, the accurate sensing of the Activities of Daily Living (ADL) has many benefits in analyzing human lifestyle, diet monitoring, occupational therapy, aiding human memory, active rehabilitation, etc. This has led to the phenomenon of *lifelogging*, which is used to describe the process of automatically and digitally recording our own daily activities for our own personal purposes, using a variety of sensor types.

### 9.1.1  Lifelogging in General

Lifelogging is a very broad topic both in terms of the technologies that can be used and the applications for the lifelogged data. Compared to traditional digital monitoring where users are monitored by others, lifelogging introduces a new form of "sousveillance", i.e. capturing data about oneself for use by oneself [1,2]. This is in contrast to having somebody else record what we are doing and using the logged data for some public or shared purpose [3]. Among various applications of lifelogging, assistive living accounts for an important part of the research aiming at improving people's health and well-being both mentally (e.g. helping with memory recall) and physically (e.g. in anomaly detection) thanks to the advantages of lifelogging in measuring activities longitudinally in fine granularity.

### 9.1.1.1 Context Sensing for Lifelogging

As an integrated part of our lives, our context is changing dynamically and if we can capture some parts of this context, then they can be used as cues that reflect our activities. By "context" we mean the features of where we are, who we are with, what we are doing, and when we are doing it. Since the context includes various aspects of the environment in which the user interacts with digital devices, the plurality of context can be applied intelligently to detect meaningful changes in the environment for assistive living for example. The increasing adoption of sensors makes it possible to gather more context information on mobile phones or wearable devices which are important data sources for activity recognition and understanding. These applications of heterogeneous sensors in context sensing is known as multimodel context awareness. Based on the collection of low-level sensor information we can infer cues about the host and the environment.

The earliest motivation behind automatic context recording and the generation of personal digital archives can be traced back to 1945 when Bush expressed his vision [4] that our lives can be recorded with the help of technology and access can be made easier to these "digital memories". This idea is unprecedentedly acceptable in the era of mobile sensor networks where the cost of integrated sensors, storage, and computational power is much lower. However, there is still a lack of consensus for a definition of lifelogging. In this chapter, we borrow the definition from [5]:

---

**BOX 9.1 Definition of Lifelogging**

*Lifelogging* is the process of passively gathering, processing, and reflecting on life experience data collected by a variety of sensors, and is carried out by an individual, the lifelogger. The corresponding data gathered in lifelogging is known as a *lifelog*, which could be in hetero-

> geneous formats such as videos, pictures, or sensor streams (like GPS locations or accelerometer traces).

From this definition, we can find that the accurate quantification of human activities using lifelogging can help to measure our diets, entertainments, leisure and sports, etc., more effectively. The longitudinal profiles in terms of digital media can provide better ways to record, analyze, understand, and improve ourselves. These digitally recorded contexts usually compensate the subjectivity of human feelings, which tend to be limited by human intuition. For example, subjects with obesity often underreport their food intake and this significantly limits the food recording method [6,7].

Context metadata like date, time, and location may be sufficient for many lifelogging applications, but there are others that require searching through lifelogs based on visual content, and for this to happen the automatic recording of visual inputs needs to be introduced. When visual sensing devices such as digital cameras or camera-enabled mobile devices are involved in the recording, we refer to the process as visual lifelogging, as defined in Box 9.2. Because visual information contains more semantics of events that can be used to infer other contextual information like "who", "what", "where", and "when", visual lifelogging can usually act as the surrogate of the lifelogger's experience, and this forms a stream of lifelogging research for assistive living.

---

**BOX 9.2 Visual Lifelogging**

*Visual lifelogging* represents a branch of lifelogging research in which digital cameras or camera-enabled mobile devices are employed as sensors to gather visual media to reflect the interaction between the lifelogger and his/her environment.

---

As the tools that permit visual lifelogging, camera-enabled sensors are used in wearable devices to record still images [40] or video sequences [3,13,32] taken from a first-person view, i.e. representing the subject's view of everyday activities. Visual lifelogging has already been widely applied in assistive living applications including aiding human memory recall, monitoring diet, diagnosing chronic disease, recording activities of daily living, and so on. Example visual lifelogging projects include Steve Mann's WearCam [31,32], the DietSense project at UCLA [38], the WayMarkr project at New York University [4], the InSense system at MIT [3], and the IMMED system [33]. Microsoft Research catalyzed research in this area with the development of the SenseCam [12,40], which was made available to other research groups in the late 2000s.

### 9.1.1.2 Visual Lifelogging Categories

In terms of sensing devices, assistive visual lifelogging can be roughly categorized into two groups, in situ lifelogging and wearable lifelogging.

#### In Situ Visual Lifelogging

In situ lifelogging can be described simply as sensing in instrumented environments such as homes or workplaces. This means that human activities can be captured through sensors such as video cameras installed in the local infrastructure; therefore, the recording is highly dependent on instrumented environments, such as PlaceLab (MIT) [8]. Typical use of video sensors for in situ sensing also includes the works reported in [9–12] and [13]. Jalal et al. [11] proposed a depth video-based activity recognition system for smart spaces based on feature transformation and HMM recognition. Similar technologies that can recognize human activities from body depth silhouettes are applied in other work by the same authors in [9]. In related work by Song et al. [12], depth data is utilized to represent the external surface of the human body. By proposing the body surface context features, human action recognition is robust to translations and rotations. As with Jalal's work in [10], Song's work [12] still depends on static scenes with an embedded sensing infrastructure. Current activity recognitions in such lifelogging settings usually assume there is only one actor in the scene, and determining how these solutions can work in more realistic and challenging settings such as the outdoors is a difficult task.

#### Wearable Visual Lifelogging

Because frequent in situ observations for activity measuring are limited to the instrumented environments, scaling the system up to more realistic and challenging settings such as the outdoors is difficult. The wide adoption of mobile or wearable devices makes it feasible to measure everyday activities digitally with their built-in sensing and computational capabilities, which helps to alleviate the challenges of in situ sensing. Meanwhile, activity recognition within such non-instrumented environments using wearable visual sensing is also a focus of assistive sensing and independent living. In wearable assistive living, the sensing devices are portable and worn directly by the subjects and can include head-mounted cameras in works by Hori and Aizawa [14] and Mann et al. [15] or cameras mounted on the users' chests in works by Blum et al. [16] and by Sellen et al. [17]. These papers all reflect the common phenomenon of continuous recording of everyday activity details based on wearable sensors. In this chapter, unless explicitly specified, we refer to wearable lifelogging as "lifelogging" for purposes of simplicity.

## 9.1.2 Typical Applications in Assistive Living

As mentioned previously, the development of sensors and low-cost data storage have make continuous or long-term lifelogging possible. However, this is only the prerequisite of lifelogging because the necessary condition of lifelogging includes various applications that motivate the lifeloggers to do it. Gordon Bell, a scientist at Microsoft, is a senior lifelogger who attempted to digitize his lifetime archives including articles, letters, photos, and medical records in the MyLifeBits project [18]. In his coauthored book Total Recall [19], he envisioned the roles of lifelogging in changing various aspects of human daily life like study, work, domesticity, etc. In this book, he also mentioned Cathal Gurrin, one author of this chapter, who started to wear a SenseCam in mid-2006 to gather an extensive visual archive of his everyday activities. According to Gurrin's experience, the passively captured visual lifelogs constructed a surrogate memory for reexperiencing his episodes of interest [19], such as when and where he met with an important person for the first time.

This overseen or experienced superiority of visual lifelogging benefits from its detailed sensing, high storage, and multimedia presentation capabilities. These advantages mean that visual lifelogging can be embraced in assistive living to satisfy the needs of different groups. The typical applications can be summarized as memory aid, diet monitoring, ADL analysis, disease diagnosis, and so on, though new application areas are emerging.

**Memory Aid.** Memory aid is a potential medical benefit which can be supported by lifelogging technologies. By recording various aspects of our recent daily activities, visual lifelogging will offer an approach for wearers to reexperience, recall, or look back through recent past events. In [20], a user study with a patient suffering from amnesia is conducted with SenseCam images and highlights the usefulness of these images in reminiscing about recent events by the patient. In [17], evidence is found that SenseCam images do facilitate people's ability to connect to their recent past. Similar applications of turning lifelogging into a short-term memory aid can also be found in [21] and [22]. Another good example of this is the work by Browne et al. [23] who used the visual lifelog from a SenseCam to stimulate autobiographical recollection, promoting consolidation and retrieval of memories of significant events. All these clinical explorations seem to agree that visual lifelogging provides a "powerful boost to autobiographical recall, with secondary benefits for quality of life" [23,3].

**Diet Monitoring.** Diet monitoring is another application of visual lifelogging for medical purposes. Though dietary patterns have been proved to be a critical contributing factor to many chronic diseases [24], traditional strategies based on self-reported information do not fulfill the task of accurate diet reporting. More usable and accurate ways to analyze the dietary information of an individual's daily food intake are badly needed. Visual media like images and videos provide hugely increased sources of sensory observations about human activities among which food intake can be monitored for diet analysis. The application of visual lifelogging in diet monitoring can support both patients with obesity and health care professionals analyzing diets. DietSense [24] is an example of such a lifelogging software system using mobile devices to support automatic multimedia documentation of dietary choices. The captured images can be *ex post facto* audited by users and researchers with easy authoring and dissemination of data collection protocols [24]. Professional researchers can also benefit in performing diet intake studies with the help of lifelog browsing and annotation tools. Both audio recorders and cameras are combined in [25]. According to [26], individuals' self-reported energy intake frequently and substantially underestimates true energy intake, while a Microsoft Sense-Cam wearable camera can help more accurately report dietary intake within various sporting populations.

**Disease Diagnosis.** Project IMMED [27] is a typical application of lifelogging to ADL, the goal of which is assessing the cognitive decline caused by dementia. Audio and video data of the instrumented activities of a patient are both recorded in [27] and indexed for later analysis by medical specialists. In [28], a wearable camera is used to capture videos of patients' daily activities. A method for indexing human activities is presented for studies of progression of the dementia diseases. The indexes can then be used by doctors to navigate throughout the individual video recordings in order to find early signs of dementia in everyday activities. The same rationale is also reported in [29]. Most recently, by combining biosensor information with frequent medical measurements, wearable devices proved useful in the identification of the early signs of Lyme disease and inflammatory responses [30].

**ADL Analysis.** More concern is now being shown in modern society about the individual health and well-being of everyday life. However, any long-term investigation into daily life comes across many difficulties in research and in the medical treatment area. Occupational therapy aims to analyze the correlation between time spent on our daily activities and our actual health, and there is a growing body of evidence indicating the relationship [31,32]. Observational assessment tools are needed to correctly establish care needs and identify potential risks. Long-term daily routines and activity engagement assessments are necessary to evaluate the impact on activities of daily living caused by diseases or old age in order to provide a proper program that meets the needs of each patient. While traditional self-reporting or observational measures are time consuming and have limited granularity, lifelogging can provide an efficient approach to providing broader insights into activity engagement. [33] has shown that wearable cameras represent the best objective method currently available to categorize the social and environmental context of accelerometer-defined episodes of activity in free-living conditions. In [34], the feasibility of using visual wearable cameras to reconstruct time use through image prompted interview is demonstrated.

## 9.2 SEMANTIC INDEXING OF VISUAL LIFELOGS: A STATIC VIEW

While this chapter focuses on computer vision for lifelogging applications, this takes place within the context of a huge growth in the amount of and use of generated multimedia content. Nowadays, we are not just consumers of multimedia through the traditional channels of broadcast TV, movies, etc., we are also generators of multimedia, especially visual multimedia, though the widespread availability of smartphones and the strong support for sharing of our images and videos through our online social networks like Facebook and Twitter. While the ubiquitous mobile smartphone device has provided access to technology for creating and sharing visual media and has catalyzed the growth of such media creation so that we can easily create permanent memory records of our lives, this then creates huge challenges to analyze, index, and retrieve information from this visual media.

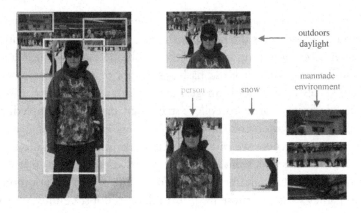

**Figure 9.1** Using metadata and lightweight content analysis to tag an image.

Perhaps the easiest way to provide access to visual media, from whatever source, is to use automatically created metadata. In fact this is true of all retrievable artifacts, whether analog, digitized, or born digital, and to support this there is a legacy of decades of work in developing standards for metadata creation and access. This varies from the Dublin Core [35], which is general purpose, to EXIF metadata for images taken with almost all cameras. From such simple metadata like date, time, and location we can actually go quite far and support access to images based on grouping images from the same or different users into "events", and we can then augment the description of images and videos for these events using external data. For the specific case of lifelogging, there is past work that showcases such lifelog event augmentation [36] using tags from similar images taken at the same location and at the same or different times.

While this is inventive and satisfies some information seeking needs, and is completely automated, by adding even a small amount of *content* processing we can go much further. Consider the image shown in Fig. 9.1. Knowing the date, time, and GPS coordinates from EXIF metadata, resolving this using a gazetteer, drawing some further linked data from Wikipedia or another open source, analyzing the content to perform face detection and recognition against a database of friends' faces and resolving who the person is and detecting some simple setting characteristics, we can tag this image with the following . . . *1 person, date = "22 Feb 2007", time = "2pm", setting = "outdoors, daylight", location = "Auron, Southern France", weather = "sunny", setting = "ski resort", altitude = "1,622m", setting = "snow", setting = "manmade environment".*

The next obvious enhancement to describing image content is manual annotation. Annotation is the process of generating high-level semantic metadata to describe something and has become the de facto norm in applications like Twitter where we use hashtags, in annotating the presence of our friends' faces in Facebook images, and there are many other cases where we use dedicated forms to collect structured data, like completing a web form for a quotation for car insurance. This is both boring, and not scalable, and even where we outsource a task like image annotation to a crowd as in Amazon's Mechanical Turk or microworkers, or even if we gamify the annotation through micro-games for online players [37] or more sophisticated games with a purpose [38], this still does not scale upwards to very large numbers. In addition, there is a latency between the time of image capture and time of annotation that could be important if we want to do anything in real time with a lifelog.

In reality, the only way to achieve real time analysis and content indexing of lifelog imagery is to automate the process, and initial efforts in this area were to leverage the developments in semantic annotation of images by predefined tags. Automatic annotation of images has been a long-standing application of machine learning and for many years researchers tried to use low-level image (and video) features as a basis for building classifiers for individual content tags. These were based on extracting features such as color histograms, dominant colors, textures, and shapes from entire images and also from localized areas within images. In recent years additional features such as SIFT/SURF characteristics have been included and indeed the MPEG-7 standard, which was explicitly defined to encode image and video metadata, supported this. Extracting low-level features from an image is computationally fast and can be used to build a compact representation and hence a small feature space for images. Once a ground truth of images annotated to the presence or absence of a semantic concept or tag is available, conventional machine learning tools can be used to learn the differences between those that have and those that do not have the particular tag, all done within the feature space of the low-level features.

For many years, within the context of the TRECVid video benchmarking activity, researchers struggled to achieve a high enough accuracy for the classifiers, as well as a large enough numbers of tags, in order to be usable [39]. Then, in 2012, things changed with the significant improvements in recognition accuracy obtainable when deep learning networks were applied to this computer vision problem for classifying and tagging images,

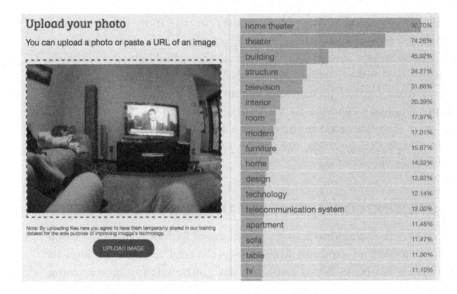

**Figure 9.2** Automatically tagging a lifelog image using www.imagga.com.

all led by the work of Geoffrey Hinton's team [40]. Suddenly there was a perfect storm of conditions for effective and usable automatic image tagging: accuracy levels improved almost to human levels of agreement, large data repositories of tagged images became available through the manual annotation of user generated content, and the large computational requirements needed to train (and run) deep networks could be addressed by using relatively cheap GPUs.

The level of accuracy and the scale and size of these (independent) taggers is leveling off and there are now emerging cloud services that can automatically tag. The output of one of these, www.imagga.com, is shown in Fig. 9.2. These services can easily be hosted on cloud services and we can expect them to be built into consumer photography and social media image processing. Indeed, Google Photos and Facebook now tag uploaded images in this way, initially presented as an aid to those with visual impairment.

While these are extremely useful, there is still a long way to go with this technology because, for example, semantic concepts are treated as independent of each other and there is no consolidation across the set of assigned tags (although we are making progress in this area, as we show in Section 9.3).

When we examine the developments in semantic image tagging through the lens of lifelogging, a simple set of semantic tags is not what we really want for a lifelog. This has recently been highlighted in work on the Kids'Cam project, which annotated almost 1.5M images taken from cameras worn by 169 preteen children in New Zealand in a project to measure children's exposure to fast food advertising [41]. Here we found that where there are errors in the machine learning recognition of semantic tags in a lifelog image, these need to be *pruned* in a much more dynamic manner from the set of tags.

## 9.3 UTILIZING CONTEXTUAL SEMANTICS: A DYNAMIC VIEW

In Section 9.2, the automatic indexing of visual lifelogs is discussed based on the detection of concepts from low-level visual features. Though its effectiveness has been shown using the state-of-the-art machine learning methods, current concept detectors are mostly the one-per-class classifiers that ignore the contextual correlation between concepts. However, the different concepts do not appear independently from each other. Instead, the cooccurrence and reoccurrence patterns of various concepts imply that concepts interact during the evolution of everyday activity engagements. This section will elaborate the use of concept contextual semantics, which benefits both the indexing of visual lifelogs and the characterization of everyday activities.

### 9.3.1 Modeling Global and Local Occurrence Patterns

One day of continuous visual lifelogs in the form of image streams can usually be segmented by using the technique introduced in [42] into dozens of events; this eases the representation and interpretation of everyday engagements. A lifelog event corresponds to a single activity in the wearer's day such as watching TV, commuting, or eating a meal, and the durations vary widely due to the engagement natures of the various activities.

After applying the automatic image indexing methods introduced in Section 9.2, we can characterize each lifelog event as a series of concept detection results performed on each of the images representing it. Assume that event $E_i$ consists of successive images $I^{(i)} = \{Im_1^{(i)}, Im_2^{(i)}, \ldots, Im_k^{(i)}\}$ and we obtained $M$ automatic concept annotators in Section 9.2. By representing $Im_j^{(i)}$ with a vector of concept appearance $C_j^{(i)} = \{c_{j1}^{(i)}, c_{j2}^{(i)} \ldots c_{jM}^{(i)}\}$, the total event set can be described as a confidence matrix $C_{N \times M}$, where $N = \sum_{i=1}^{n} k_i$ and $k_i$ is the number of images in each event $E_i$.

Compared to low-level features like color, shape, or texture features, the results of concept detection provide a more natural way to describe and index vision content that is close to human expectations. However, the initial detection results such as $C_{N \times M}$ are still noisy because the current machine learning methods are far from perfect due to the dependence on the large volume of training corpora. Though breakthroughs have been achieved using deep learning [40], the effective transfer of learned models to different application domains such as visual lifelogs is still questionable.

Detection refinement or adjustment methods [43–47] represent a stream of postprocessing methods that can enhance detection scores obtained from individual detectors, allowing independent and specialized classification techniques to be leveraged for each concept. In this section, we introduce an approach that can exploit the interconcept relationships implicitly from concept detection results of $C_{N \times M}$ in order to provide better quality semantic indexing. This is in contrast to current refinement methods that learn interconcept relationships explicitly from training corpora and then apply them to test sets. Because acceptable detection results can be obtained for concepts with enough training samples, as witnessed by TRECVid benchmark [39] and ImageNet competition [48], it is feasible to utilize detections with high accuracies to enhance overall multiconcept detections since the concepts are highly correlated.

### 9.3.1.1 Factorizing Indexing Results

The framework of result factorization is to exploit the global pattern in concept occurrence contexts. The intuition behind the factorization method is that the high-probable correct detection results are selected to construct an incomplete but more reliable matrix, which can then be completed by a factorization method. Depending on the organization of concept detection result $C$, we can apply different forms of factorizations such as matrix or tensor factorizations in order to overlay a consistency on the underlying contextual pattern of concept occurrences. Cooccurrence and reoccurrence patterns for concepts are a reflection of the contextual semantics of concepts since everyday concepts usually cooccur within images rather than in isolation. In some potentially long-duration activities like "using a computer", "driving", etc., the indicated concepts may appear frequently and repeatedly in the first-person view.

As we can see from Fig. 9.3, the concept detection result $C$ can also be represented as a three-way tensor $T$. The rationale behind this is to

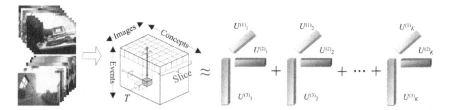

**Figure 9.3** NTF-based concept detection enhancement framework. Reprinted from [49], Copyright (2016), with permission from Elsevier.

avoid information loss from event segmentation and to utilize the temporal features reflected in different events. As illustrated in Fig. 9.3, a tensor is advantageous in representing the structure of multidimensional data more naturally. The procedure for concept tensor construction and factorization is shown in Fig. 9.3. As illustrated, each slice is a segmented part of an event and is represented by a confidence matrix. The slices are then stacked one below another to construct a three-dimensional tensor that preserves the two-dimensional characters of each segment while keeping temporal features along the event dimension and avoids significant loss of contextual information.

In tensor factorization, the Canonical Decomposition (CD) [50] model simplifies the approximation of tensor $C$ as a sum of three-fold outer products with rank-$K$ decomposition $\hat{T} = \sum_{f=1}^{K} U_f^{(1)} \otimes U_f^{(2)} \otimes U_f^{(3)}$, which means each element $\hat{T}_{ijk} = \sum_{f=1}^{K} U_{if}^{(1)} U_{jf}^{(2)} U_{kf}^{(3)}$. This form of factorization means that each factor matrix has the same number of columns, i.e. the length of latent features has the fixed value of $K$. When $n = 2$, $T$ is simply a two-mode tensor, which is indeed the initial matrix $C$, and the factorization degenerates as $\hat{T} = \sum_{f=1}^{K} U_f^{(1)} \otimes U_f^{(2)} = U^{(1)} U^{(2)T}$.

The CD approximation factorization defined above can be solved by optimizing the cost function defined to quantify the quality of the approximation. For an arbitrary $n$-order tensor, the cumulative approximation error can be used to define the cost function, which has the form $F = \frac{1}{2} \| T - \hat{T} \|_F^2 = \frac{1}{2} \| T - \sum_{f=1}^{K} \otimes_{i=1}^{n} U_f^{(i)} \|_F^2$. In factorizing the confidence tensor, the weighted measure is more suitable since detection performance is different due to the characteristics of concepts and quality of the training set. Because each value $c_{ij}$ in $C$ denotes the probability of the occurrence of concept $v_j$ in sample $s_i$, the estimation of the existence of $v_j$ is more likely to be correct when $c_{ij}$ is high, which is also adopted by [43,51] under the same assumption that the initial detectors are reasonably reliable if the returned confidences are higher than a threshold. To distinguish the contribution of different concept detectors to the cost function, the weighted cost function is employed as

$$F = \frac{1}{2}\|T - \hat{T}\|_W^2 = \frac{1}{2}\left\|\sqrt[2]{W} \circ (T - \hat{T})\right\|_F^2$$

$$= \frac{1}{2}\sum_{i_1, i_2, \ldots, i_n} w_{i_1, i_2, \ldots, i_n}\left(T_{i_1, i_2, \ldots, i_n} - \sum_{f=1}^{K} \otimes_{i=1}^{n} U_{\cdot f}^{(i)}\right)^2$$

$$\text{subject to } U^{(1)}, U^{(2)}, \ldots U^{(n)} \geq 0, \tag{9.1}$$

where $\circ$ denotes element-wise multiplication, order-$n$ ($n$-dimensional) tensor $W$, and $T \in \mathcal{R}^{I_1 \times I_2 \times \cdots \times I_n}$ ($I_1, I_2, \ldots, I_n$ denotes the size of each of the tensor's dimensions), $W$ denotes the weight tensor whose elements are higher for reliable detections and lower for less reliable detections, and $\|\cdot\|_F^2$ denotes the Frobenius norm, i.e. the sum of the squares of all entries in the tensor. The nonnegative constraint guarantees that all the components described by $U^{(i)}$ are additively combined. The discussion of the solution of the above formalized problem is beyond the scope of this chapter; we provide the iterative updating rule using the multiplicative method of [52,53] as

$$U_{i,f}^{(t)} \leftarrow U_{i,f}^{(t)} \frac{\sum_{I-i_t}(W \circ T)_{i_t} \prod_{r \neq t} U_{i_r,f}^{(r)}}{\sum_{I-i_t}(W \circ \hat{T})_{i_t} \prod_{r \neq t} U_{i_r,f}^{(r)}}, \quad 1 \leq t \leq n, \tag{9.2}$$

where $I = i_1, i_2, \ldots, i_n$ is an $n$-tuple index whose value is in the range of $i_t \in [1, I_t]$, $1 \leq t \leq n$, $I - i_t$ denotes the $n - 1$ subset with $i_t$ removed from $I$, and $(\cdot)_{i_t}$ denotes the $t$th dimension assigned with value $i_t$ for a given tensor. Taking the three-way tensor as an instance, the refinement can be expressed as a fusion of confidence tensors after factorization:

$$T' = \alpha T + (1 - \alpha)\hat{T} = \alpha T + (1 - \alpha)\sum_{f=1}^{K} U_f^{(1)} \otimes U_f^{(2)} \otimes U_f^{(3)}. \tag{9.3}$$

### 9.3.1.2 Temporal Neighborhood-Based Propagation

For much of the visual media we use for assistive living purposes, there is a temporal aspect. For example the image streams captured continuously in visual lifelogging is inherently temporal as they capture imagery over time, and thus they may have related content taken from the same scene or have the same characteristics of related activities. For such "connected" visual media it makes sense to try to exploit any temporal relationships when postprocessing initial concept detection, and to use the "neighborhood" aspect of visual media.

Following refinements based on global context using high-order tensor or degenerated matrix factorization, detection results will have been adjusted in a way consistent with the latent factors modeled in the factorization. While this procedure exploits general contextual patterns that are modeled globally by factorization, the similarity propagation method can further refine the result by exploiting any local relationships between samples.

A new confidence matrix $C'$ can be recovered from the refined tensor $T'$, and $C'$ has rows and columns representing image samples and concepts, respectively. As a refined result of $C_{N\times M}$, $C'$ can provide better measures to localize closely related neighbors for similarity-based propagation. The similarity between samples $s_i$ and $s_j$ can be calculated by the Pearson Correlation, formulized as

$$P_{i,j} = \frac{\sum_{k=1}^{M}(c'_{ik} - \bar{c}'_i)(c'_{jk} - \bar{c}'_j)}{\sqrt{\sum_{k=1}^{M}(c'_{ik} - \bar{c}'_i)^2}\sqrt{\sum_{k=1}^{M}(c'_{jk} - \bar{c}'_j)^2}},$$

where $c'_i = (c'_{ik})_{1 \leq k \leq M}$ is the $i$th row of $C'$, and $\bar{c}'_i$ is the average weight for $c'_i$. To normalize the similarity, we employ the Gaussian formula and denote the similarity as $P'_{i,j} = e^{-(1-P_{i,j})^2/2\delta^2}$, where $\delta$ is a scaling parameter for sample-wise distance. Based on this we can localize the $k$ nearest neighbors of any target sample $s_i$.

The localized $k$ nearest neighbors can be connected with the target sample using an undirected graph for further propagation. For this purpose, the label propagation algorithm [54] is derived to predict more accurate concept detection results based on this fully connected graph whose edge weights are calculated by the similarity metric as calculated by $P'_{ij}$. Mathematically, this graph can be represented with a sample-wise similarity matrix as $G = (P'_{i,j})_{(k+1)\times(k+1)}$, where the first $k$ rows and columns stand for the $k$ nearest neighbors of a target sample to be refined, which is denoted as the last row and column in the matrix. The propagation probability matrix $G'$ is then constructed by normalizing $G$ at each column as

$$g'_{i,j} = \frac{P'_{i,j}}{\sum_{l=1}^{k+1} P'_{l,j}},$$

which guarantees the probability interpretation at columns of $G'$. By denoting the row index of $k$ nearest neighbors of a sample $s_i$ to be refined as $n_i$ $(1 \leq i \leq k)$ in $C'$ and stacking the corresponding rows one below another, the neighborhood confidence matrix can be constructed as $C_n = (c'_{n_1}; c'_{n_2}; \ldots; c'_{n_k}; c'_i)$. The propagation algorithm is carried out iteratively by updating

$$C_n^t \leftarrow G' C_n^{t-1}, \qquad (9.4)$$

where the first $k$ rows in $C_n$ stand for the $k$ neighborhood samples in $C'$ indexed by subscript $n_i$ and the last row corresponds to the confidence vector of the target sample $s_i$. Since $C_n$ is a subset of $C'$, the graph $G$ constructed on $C_n$ is indeed a subgraph of the global graph constructed on $C'$. During each iteration, the neighborhood concept vector $c'_{n_i}$ needs to be clamped to avoid fading away. After a number of iterations, the algorithm converges to a solution in which the last row of $C_n$ is a prediction based on similarity propagation. In this way the local relationships between neighbors can be used for a more comprehensive refinement.

## 9.3.2 Attribute-Based Everyday Activity Recognition

Though concept detection and refinement can provide a semantic representation for visual assistive living, many assistive living applications require the ability to recognize semantic concepts that have a temporal aspect corresponding to activities or events, i.e. characterizing the whole time series rather than merely interpreting single images. While low-level feature-based methods have been shown to be ill-suited for multimedia semantic indexing due to the lack of semantics for user interpretation, high dimensionality, etc., high-level concept attributes are widely employed in the analysis of complex semantics corresponding to things like events and activities. Since such semantic structures can be represented as typical time series, the recognition of events or activities can be regarded as dynamics-based recognition using concept detection results. This is usually carried out by representing the time series as a sequence of units such as video clips or image frames. After concatenating the results of concept detectors on each unit, time series can then be represented by a temporally ordered sequence of vectors, as shown in Fig. 9.4.

Fig. 9.4 shows the paradigm of utilizing concept temporal dynamics for high-level activity detection in which typical indoor activities, like "cooking", "watching TV", and the corresponding trajectories are demonstrated.

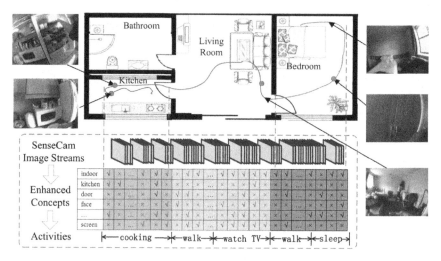

**Figure 9.4** The dynamics of concept attributes returned by concept detections. Reprinted from [49], Copyright (2016), with permission from Elsevier.

The concept detection results temporally aligned with these activities are marked as "✓" or "×" in the diagram to represent the presence or absence of concepts, respectively. It is important to note that the concept detection shown in Fig. 9.4 does have errors, and this can affect further analysis to various degrees. Therefore, in order not to propagate these errors into the subsequent analysis for activity and behavior characterization, the original concept detections can be enhanced using the refinement methods introduced in Section 9.3.1.

As shown in Fig. 9.4, everyday activities can be regarded as stochastic temporal processes consisting of various lengths of concept vectors. In this way, the dynamic evolution of concept vector occurrences can characterize a deeper meaning of underlying or derived human activities if the evolution patterns can be modeled. Attribute-based events and activity detection have attracted much research attention. More importantly, it has been found that although state-of-the-art concept detections are far from perfect, they still provide useful clues for event classification [55]. [56] also revealed that this representation outperforms—and is complementary to—other low-level visual descriptors for event modeling. This is also tested in a more comprehensive visual lifelogging experiment in [57]. Similar work is also carried out using concept detections to characterize everyday activities, as reported in [2,3], where activity recognition is built on the basis of concept detection. In this chapter, we investigate a selection of attribute-based recognition methods which are suitable for activity characterization in assistive living. These time series recognition methods investigated include temporal and non-temporal features, generative and discriminative graph models, holistic and pyramid representations, signatures of dynamic systems, etc. [57]

**Non-Temporal Features.** As one of the fusion operations for concept detection results, Max Pooling [58] has been demonstrated to give better performance than other fusions for most complex events. In max pooling [59,60], the maximum confidence is chosen from all key frame images (or video subclips) for each concept to generate a fixed-dimensional vector for an event or activity sample. Since by definition the maximum value cannot characterize a temporal evolution of concepts within a time series, this method can be regarded as nontemporal. Similar to max pooling, the Bag of Features approach is a way of aggregating concept detection results by averaging the confidences over a time window [58,61,56]. Because bag of features and max pooling reflect the statistical features within the holistic time series, they both ignore the temporal evolution of concept detection results.

**Temporal Features.** Motivated by the spatial pyramid method, the temporal pyramids proposed in [61] approximate temporal correspondence with a temporally binned model [62]. In this method, the histogram over the whole time series represents the top level, while the next level can be represented by concatenating two histograms of temporally segmented sequences. More fine-grained representations can be formalized in the same manner. By applying the multiscale pyramid to approximate a coarse-to-fine temporal correspondence, this method generates fixed-length features with temporal embeddings. Compared to non-temporal features, which are usually represented as holistic features, the temporal feature in [61] is modeled as a pyramid representation.

**Graph Models.** A generative method based on Hidden Markov Model (HMM) used in [3] is employed in this representation. HMMs are first trained for each activity class and the log-likelihood representations of per class posteriors can be concatenated into a vector. While HMM assumes all of the observations in a sequence are independent and conditional on the hidden states, Conditional Random Field (CRF) has no such constraints; it allows for the existence of non-local dependencies between hidden states and observations. Because the dependence is allowed in a wider range, a CRF model can flexibly adapt to dynamic sequences in which high correlations might exist between different regions. This is especially useful for modeling the kind of activities recorded by wearable lifelog cameras. In [49], the hidden CRF is applied to characterize everyday activities as a discriminative model. The Fisher vector-based method [63] is also a discriminative method which can be derived from HMM. The principle of the Fisher kernel is that similar samples should have similar dependence on the generative model, i.e. the gradients of the parameters [64]. Instead of directly using the output of generative models, such as in the HMM method, using a Fisher kernel tries to generate a feature vector that describes how the parameters of the activity model should be modified in order to adapt to different samples [57].

**Liner Dynamic System (LDS).** As a natural way of modeling temporal interaction within time series, Liner Dynamic Systems [60,65] can characterize temporal structure with attributes extracted from within a sliding window. The time series can be arranged in a block Hankel matrix $H$ whose elements in a column have the length of a sliding window (denoted $r$) and successive columns are shifted by one time step. According to [60], singular value decomposition of $H \cdot H^T$ can achieve comparable performance to more complex representations.

The application of the above methods in characterizing everyday activities is based on the assumption that the concepts contained in these images can reflect significant temporal patterns that characterize the semantics of the activities they represent. In this case, the temporal consistency of certain types of concepts like "indoor", "screen", and "hands" can be viewed as cues for concepts like "using a computer". Even though some activities require the user to repeatedly change locations like "walking" and "doing housework", the dynamics of concepts present in the activity will still show some patterns, such as the frequent appearance of "road" for the "walking" activity in an urban environment, or the transitions between "kitchen" and "bathroom" for the "housework" activity. Therefore, appropriate concept selection is necessary to characterize the dynamic evolution of time series-based concepts. According to [57,66], the recognition performance of activities can be enhanced if more appropriate concepts are utilized. This is more obvious when the original concept detections are less satisfactory [57]. In practice, the assistive living developers who want to apply attribute-based activity recognition should focus on three affecting factors including concept selection, concept detection, and activity recognition methods.

## 9.4 INTERACTING WITH VISUAL LIFELOGS

When considering the various approaches that have been used in lifelog data visualization and user interfaces, an understanding of the application use-cases and types of data involved is required. In a simple implementation of assistive living personal data access, and one that many readers will be familiar with, the personal quantified-self data gathered by wearable fitness trackers and related sensors is typically aggregated into meaningful infographics and charts, summarizing the activities or performance of the individual across a number of useful dimensions. However, given our definition of lifelogging that includes wearable camera data along with other potential sources of data, then this simple data aggregation summarization approach will not be appropriate and more advanced forms of visualization are required [5].

There has been some initial work on design considerations for lifelog content. Whittaker et al. [67] proposed a set of design principles for developing lifelogging systems that support human memory, and Hopfgartner et al. [68] presented a set of user interaction templates for the design of a lifelogging system. Additionally, Byrne et al. [69] presented a set of guidelines for the presentation and visualization of lifelog content, including the need for a simple and intuitive interface, segmenting the content into comprehensible units, aiding human memory, and support for exploration and comparison, all of which offer meaningful insights into how to design the user interface and user experience for lifelog applications.

However, the first applications of wearable camera data for assistive living were concerned with utilizing lifelog technologies to support memory studies of the individual. These studies were concerned with helping the individual to recollect past experiences, either as a form of scientific experimentation or as an assistive technology for individuals with memory impairments. Van den Hoven et al. [70] provide a set of design considerations that support using lifelogs to help people remember in everyday situations. These are based around the concept that when recollecting or reliving past experiences, we know that visual media, especially images captured from the first-person viewpoint, provide very powerful memory cues and can lead to what are referred to as Proustian moments of recall [71]. Hence, the early lifelog interaction scenarios to support human memory have focused on assisting the user to browse through lifelog archives. The initial work in the area from Microsoft Research presented the SenseCam image browser, which the facilitated rapid playback of image sequences, along with a manual event segmentation tool [23]. This interface presents

**Figure 9.5** An event-based Lifelog Browser. Reused from [42], with permission of Springer.

a stop-motion style rapid playback through visual lifelogs and provides a baseline visualization of visual lifelog data, though one that quickly presents challenges to the user due to the lack of data organization when the archive of data grows to weeks, months, or even years [5].

A first effort at organizing lifelog data was the event-based SenseCam image browser [72] which automatically segmented lifelog data into events and then allowed users to manually annotate those events, thereby supporting a basic level of structured browsing and linkage of lifelog data. The key organizational metaphor for the data here was a diary with a chronological ordering of events on any given day. It is worth pointing out that an event-level browser or lifelog data can reduce many thousands of images per day to a more manageable set of events (typically about 30–40). These interfaces usually include drill-down functionality to enable the user to examine some or all images that occur within one selected event. An example of an event-level browser [72] is shown in Fig. 9.5. Another related approach was ShareDay [73] which allows both browsing through lifelog events, and event sharing with family and friends.

One of the key challenges facing any lifelogger or lifelog practitioner is the volume of data to be organized. Event browsers are a step in the right direction in that they convert a temporal stream of lifelog data into a chronologically ordered sequence of events. However, even with segmenting a lifelog into events, it is still challenging for the lifelogger or practitioner to actually find anything in a large archive. We know from prior research into personal photograph collections that a rich annotation mechanism enhances user access [74]. The inclusions of annotations, at the image or event level, helps a user search through lifelog archives. One example is the food search and browse engine from Kitamura et al. [75], which encourages users to manually annotate food events within a lifelog archive. However, relying on human annotations is inherently problematic as users are unlikely to expend a lot of effort in manually tagging large volumes of lifelog data. Hence, it is necessary to incorporate automatic annotation of lifelog data, using approaches such as those outlined in this chapter. In this way, users can expect to be able to perform automated searches through the lifelog archives, and data abstraction and linkage models could be developed to support user access. Early work in this area includes the ethnography browser, which automatically annotates lifelog events for lifestyle traits and real-world logos and objects [76]; it was developed as a proof-of-concept lifelog search engine allowing cross-demographic analytics of lifelogs.

While there has been some effort in developing richer annotation tools and even in developing automated search tools for lifelog data ([77–80]), there is no standard approach and progress on interactive lifelog interfaces has been even slower. In 2015, the first collaborative effort to develop high-quality search engines, some with visual interfaces [81], was developed as part of the NTCIR-Lifelog collaborative benchmarking exercise [82]. Over the coming years, initiatives such as NTCIR-Lifelog and ImageClef-Lifelog are expected to expedite the progress in developing new interfaces and visualizations of lifelog data.

There have also been a number of interfaces developed that specifically provide assistive technologies to assist users in visualizing, comprehending, and accessing lifelog data. Caprani et al. [83] developed a touchscreen browser that incorporated event segmentation to allow computer illiterate older adults to browse and share lifelog data. Gurrin et al. developed a gesture-based event browser for use on the living room TV [84] that utilized the Nintendo Wii platform to browse through lifelog data via a gesture-based interface. Finally, a color-based data abstraction [85] in which temporal lifelog visual data was represented by a "color-of-life" wheel on the desktop and touchscreen devices.

These examples represent a summary of the different user interfaces and access methodologies that have been developed to search and browse through (mostly visual) lifelog data; however, there has been no clear evaluation of the effectiveness of these approaches in comparison to each other. The best solution may represent a combination of some of the approaches, but this requires future research. Initiatives such as NTCIR-Lifelog are providing the impetus to encourage progress in this area. Extending the existing annotation approaches to incorporating user activities and visual concepts will provide a richer annotation of the data, resulting in the assumption of better quality search and browsing support for lifelog data. However, this is still a work in progress, and it is likely to be a number of years before lifelogs get their equivalent of the top ten ranked list in web search.

## 9.5 CONCLUSION AND FUTURE ISSUES

While the sensing facilities are relatively mature thanks to the quick emergence of smartphones and wearable devices that are lightweight and embedded with powerful sensing capabilities and large storage, the understanding of visual lifelogging for the purpose of assistive living is in its infancy, due to the challenges of the wearer's movement, visual diversity, etc. Meanwhile, how to design novel and effective interactive applications is another important factor for changing visual lifelogging from write-only memory surrogates. These two aspects of understanding and interactive design of visual lifelogging were discussed in order to make visual lifelogging more adoptable in assistive living, including the understanding of concepts and activities and the visualization and interactions for possible lifelogging applications, which can be summarized as follows:

- **Understanding of Visual Lifelogging.** The semantic indexing of visual lifelogging is based on the automatic mapping, from visual features to a set of concepts. By utilizing the contextual correlations of concepts, the refinement of concept indexing results can help to improve the understanding of visual lifelogging by effectively modeling the global and local occurrence patterns. Finally, a segment of activity can be recognized by characterizing the time series representation with attribute-based everyday activity recognition.

- **Interactive Visual Lifelogging.** We have seen that preliminary research has been carried out on the development of access methodologies for visual lifelogging, yet there are still significant challenges to overcome. For example, there is not yet a clear understanding of how best to present visual lifelogs on a user interface; it is likely that some query-dependent visualization and access methodology will be required. Exactly how to develop these lifelog search engines is also unknown. What we can say is that initiatives like NTCIR-Lifelog are supporting the first comparative efforts in the development of visual lifelog search engines and analytics interfaces.

  Though preliminary research has been carried out towards the understanding and interaction of visual lifelogging which can benefit assistive living in various aspects. Visual lifelogging is still faced with the challenges of satisfying the needs of assistive living. Some future directions can be summarized:

- **Real-Time Concerns.** Recognition accuracy is currently the main research concern in understanding everyday activities in assistive visual lifelogging. While some analysis can be carried out offline (life pattern analysis, diet monitoring, memory aid, etc.), some other applications for assistive living requires further adoption of such technical schemes with real-time undersanding (context-aware personalized service, smart home, behavior monitoring, etc.). Similar to the way that humans learn about the content in the image at a glance, the reframing of object detection can be performed more efficiently in a deep neural network [86] with recent progresses. Another possible solution for tackling efficiency in understanding assistive visual lifelogging is the distributed data processing strategy. For example, in [87,88] the elastic nature of cloud infrastructure can be used as a trade-off between performance and cost. In this case, the server side responsible for high computational complexity processing can enable quick recognition of everyday concepts and activities. In addition, the fusion of heterogeneous senor readings can also tackle the scalability and performance challenges of real-time analysis in assistive living, i.e. substituting the high-power/memory consuming sensor with a more efficient one when the lower power/memory consuming sensor data can be fused and still maintain comparable activity recognition accuracy [89].

- **Incorporating External Knowledge.** Compared to some other forms of visual media that can be publicly collected from the Internet, it is more difficult to gather large collections of visual lifelogs for research or experimental purposes as this requires enough volunteers who continuously collect their everyday activities. In addition, due to the concerns of privacy issues, making these collections available to others is unrealistic so far. Transfer learning [90] is a useful method for improving the learning of the target predictive function using the knowledge learned from a source domain. Transfer learning is particularly valuable when the number of labels in the target domain is much smaller than in the source domain. How to effectively transfer the learned knowledge from large volume datasets such as ImageNet [48] to the characterization of everyday activities for assistive living is definitely a research direction to be exploited. As introduced in Section 9.3.2, the appropriate selection of semantic concepts as attributes impacts the recognition accuracy of activities. Even though manual construction of semantic space, such as a topic-related concept selection user experiment [91,92], has shown its merits, this method is less flexible when dealing with various activity types that do not exist in the predefined activity set. A more feasible solution is to exploit external online resources in order to reflect human knowledge on activity-specific concept selection. [93] points out an interesting method by choosing the WikiHow online forum to extract related tags from human daily life event queries. Transferring hierarchical structure constructed in such a way to concept-based representation of activity recognition can be another direction in understanding visual lifelogging.

- **Automatic Captioning Visual Lifelog.** One of the reasons why a set of tags is insufficient as a representation of a lifelog is that a lifelog tells a story of a person's day, and a set of tags does not. Even a set of tags for a set of images, arranged into some sequence, does not tell a story. There are things in everybody's day that are routine, regular, and maybe they are important enough to be included in the story of the day. There are also unexpected things in our day, surprises, unanticipated events, and maybe they form part of the story too. So the most relevant work on automatically describing what appears in visual lifelogs is automatic captioning [94] of images or of videos, and it is only very recently that we are seeing work emerging from research groups showing this technique. The quality of this work is mostly poor and there are many problems to be overcome. For example, in image tagging we tend to tag objects that appear in the image, like bicycles and people and trees, rather than activities and actions like running or jumping or greeting somebody. Detecting activities and actions is much more complex than detecting objects, especially when the actions are spread over time, like "preparing a cup of tea" or "eating a meal". In 2016 the TRECVid activity introduced a task on automatic captioning of video [39], where the target videos to be captioned were social media videos from the Vine website of up to 8 seconds each. For some videos, captioning was comparable to manual annotation, but for many videos the automatic captions were poor because they lacked context, i.e. information on the events and activities surrounding the video clip in question.

- **Ethical and Privacy Issues.** Because this book is about computer vision, the detailed discussion of these issues is beyond the scope of this chapter. However, the social sensitivity of these issues related to new technologies and devices such as Google Glass are concerns to be taken while developing assistive lifelogging. Given current breakthroughs in computer vision and artificial intelligence, the ethical and privacy issues should no longer be the obstructive factors that might impede the development of computer vision in lifelogging. On the contrary, the maturity of assistive computer vision can help to deal with these issues more effectively. Because computational capabilities are improving, privacy can be tackled by extracting features locally on mobile devices or through edge routers at home. The detection of privacy-related images or streams will be more accurate, which can help to filter them out. In this case, both the privacy concerns of lifelogger and bystanders who might be captured by visual devices can be alleviated. Currently, we can deal with the privacy issues by integrate "privacy by design" into the development process [5,95,96], i.e. embedding the issues as core considerations throughout the entire life cycle of this technology. More specifically, seven principles [97] can be taken into account when developing a privacy-aware lifelogging systems. Similarly, some other ethical issues can be dealt with by embedding the necessary principles in the framework, as discussed in [98].

## ACKNOWLEDGMENTS

This chapter is supported by the National Natural Science Foundation of China under Grant Nos. 61502264, 61210008, 61521002, 61472204, Beijing Key Laboratory of Networked Multimedia, and Science Foundation Ireland under grant number SFI/12/RC/2289.

## REFERENCES

[1] A.L. Allen, Dredging-up the past: lifelogging, memory and surveillance, The New Yorker (2007) 38–44.

[2] A.R. Doherty, N. Caprani, C.O. Conaire, V. Kalnikaite, C. Gurrin, A.F. Smeaton, N.E. O'Connor, Passively recognising human activities through lifelogging, Computers in Human Behavior 27 (2011) 1948–1958, https://doi.org/10.1016/j.chb.2011.05.002.

[3] P. Wang, A. Smeaton, Using visual lifelogs to automatically characterise everyday activities, Information Sciences 230 (2013) 147–161.

[4] V. Bush, As we may think, The Atlantic Monthly (1945).

[5] C. Gurrin, A.F. Smeaton, A.R. Doherty, LifeLogging: personal big data, Foundations and Trends in Information Retrieval 8 (2014) 1–107, https://doi.org/10.1561/1500000033.

[6] D.H. Wang, M. Kogashiwa, S. Kira, Development of a new instrument for evaluating individuals' dietary intakes, Journal of the American Dietetic Association 106 (2006) 1588–1593.

[7] U. Alexy, W. Sichert-Hellert, M. Kersting, V. Schultze-Pawlischko, Pattern of long-term fat intake and BMI during childhood and adolescence—results of the Donald study, International Journal of Obesity and Related Metabolic Disorders 28 (2004) 1203–1209.

[8] Placelab (MIT), http://web.mit.edu/cron/group/house_n/placelab.html (Accessed March 2017).

[9] A. Jalal, S. Kamal, Real-time life logging via a depth silhouette-based human activity recognition system for smart home services, in: 11th IEEE International Conference on Advanced Video and Signal Based Surveillance (AVSS), 2014, pp. 74–80.

[10] A. Jalal, Y. Kim, D. Kim, Ridge body parts features for human pose estimation and recognition from RGB-D video data, in: International Conference on Computing, Communication and Networking Technologies (ICCCNT), 2014, pp. 1–6.

[11] A. Jalal, N. Sharif, J.T. Kim, T.S. Kim, Human activity recognition via recognized body parts of human depth silhouettes for residents monitoring services at smart homes, Indoor and Built Environment 22 (2013) 271–279.

[12] Y. Song, J. Tang, F. Liu, S. Yan, Body surface context: a new robust feature for action recognition from depth videos, IEEE Transactions on Circuits and Systems for Video Technology 24 (6) (2014) 952–964, https://doi.org/10.1109/TCSVT.2014.2302558.

[13] A. Jalal, S. Kamal, D. Kim, A depth video sensor-based life-logging human activity recognition system for elderly care in smart indoor environments, Sensors 14 (7) (2014) 11735–11759.

[14] T. Hori, K. Aizawa, Context-based video retrieval system for the life-log applications, in: Proceedings of the 5th ACM SIGMM International Workshop on Multimedia Information Retrieval, MIR '03, ACM, New York, NY, USA, ISBN 1-58113-778-8, 2003, pp. 31–38, http://doi.acm.org/10.1145/973264.973270.

[15] S. Mann, J. Fung, C. Aimone, A. Sehgal, D. Chen, Designing EyeTap digital eyeglasses for continuous lifelong capture and sharing of personal experiences, in: Proceedings of the CHI 2005 Conference on Computer–Human Interaction, ACM Press, Portland, OR, USA, 2005.

[16] M. Blum, A.S. Pentland, G. Tröster, InSense: interest-based life logging, IEEE Multimedia 13 (4) (2006) 40–48, https://doi.org/10.1109/MMUL.2006.87.

[17] A. Sellen, A. Fogg, M. Aitken, S. Hodges, C. Rother, K. Wood, Do life-logging technologies support memory for the past? An experimental study using SenseCam, in: Proceedings of the CHI 2007, ACM Press, New York, NY, USA, 2007, pp. 81–90.

[18] J. Gemmel, G. Bell, R. Lueder, MyLifeBits: a personal database for everything, Communications of the ACM 49 (1) (2006) 88–95.

[19] G. Bell, J. Gemmell, Total Recall: How the E-Memory Revolution Will Change Everything, 2009.

[20] S. Hodges, L. Williams, E. Berry, S. Izadi, J. Srinivasan, A. Butler, et al., SenseCam: a retrospective memory aid, in: Proceedings of the 8th International Conference on UbiComp, Orange County, CA, USA, 2006, pp. 177–193.

[21] E. Berry, N. Kapur, L. Williams, S. Hodges, P. Watson, G. Smyth, et al., The use of a wearable camera, SenseCam, as a pictorial diary to improve autobiographical memory in a patient with limbic encephalitis: a preliminary report, Neuropsychological Rehabilitation 17 (4–5) (2007) 582–601, https://doi.org/10.1080/09602010601029780.

[22] S. Vemuri, W. Bender, Next-generation personal memory aids, BT Technology Journal 22 (4) (2004) 125–138, https://doi.org/10.1023/B:BTTJ.0000047591.29175.89.

[23] G. Browne, E. Berry, N. Kapur, S. Hodges, G. Smyth, P. Watson, K. Wood, SenseCam improves memory for recent events and quality of life in a patient with memory retrieval difficulties, Memory 19 (7) (2011) 713–722, https://doi.org/10.1080/09658211.2011.614622.

[24] S. Reddy, A. Parker, J. Hyman, J. Burke, D. Estrin, M. Hansen, Image browsing, processing, and clustering for participatory sensing: lessons from a dietsense prototype, in: EmNets'07: Proceedings of the 4th Workshop on Embedded Networked Sensors, ACM Press, Cork, Ireland, 2007, pp. 13–17.

[25] C.H. Kaczkowski, P.J.H. Jones, J. Feng, H.S. Bayley, Four-day multimedia diet records underestimate energy needs in middle-aged and elderly women as determined by doubly-labeled water, The Journal of Nutrition 130 (4) (2000) 802–805.

[26] G. O'Loughlin, S.J. Cullen, A. McGoldrick, S. O'Connor, R. Blain, S. O'Malley, G.D. Warrington, Using a wearable camera to increase the accuracy of dietary analysis, American Journal of Preventive Medicine 44 (3) (2013) 297–301, https://doi.org/10.1016/j.amepre.2012.11.007.

[27] R. Mégret, V. Dovgalecs, H. Wannous, S. Karaman, J. Benois-Pineau, E.E. Khoury, et al., The IMMED project: wearable video monitoring of people with age dementia, in: Proceedings of the International Conference on Multimedia, MM '10, ACM, New York, NY, USA, ISBN 978-1-60558-933-6, 2010, pp. 1299–1302, http://doi.acm.org/10.1145/1873951.1874206.

[28] S. Karaman, J. Benois-Pineau, R. Megret, J. Pinquier, Y. Gaestel, J.F. Dartigues, Activities of daily living indexing by hierarchical HMM for dementia diagnostics, in: The 9th International Workshop on Content-Based Multimedia Indexing (CBMI), Madrid, Spain, 2011, pp. 79–84.

[29] R. Mégret, D. Szolgay, J. Benois-Pineau, P. Joly, J. Pinquier, J.F. Dartigues, C. Helmer, Wearable video monitoring of people with age dementia: video indexing at the service of healthcare, in: International Workshop on Content-Based Multimedia Indexing, CBMI '08, London, UK, 2008, pp. 101–108.

[30] X. Li, J. Dunn, D. Salins, G. Zhou, W. Zhou, S.M. Schu, Digital health: tracking physiomes and activity using wearable biosensors reveals useful health-related information, PLoS Biology 15 (1) (2017), https://doi.org/10.1371/journal.pbio.2001402.

[31] M. Law, S. Steinwender, L. Leclair, Occupation, health and well-being, Canadian Journal of Occupational Therapy 65 (2) (1998) 81–91.

[32] K. McKenna, K. Broome, J. Liddle, What older people do: time use and exploring the link between role participation and life satisfaction in people aged 65 years and over, Australian Occupational Therapy Journal 54 (4) (2007) 273–284.

[33] A.R. Doherty, P. Kelly, J. Kerr, S. Marshall, M. Oliver, H. Badland, et al., Using wearable cameras to categorise type and context of accelerometer-identified episodes of physical activity, International Journal of Behavioural Nutrition and Physical Activity 10 (1) (2013) 22, https://doi.org/10.1186/1479-5868-10-22.

[34] P. Kelly, E. Thomas, A. Doherty, T. Harms, O. Burke, J. Gershuny, C. Foster, Developing a method to test the validity of 24 hour time use diaries using wearable cameras: a feasibility pilot, PLoS ONE 10 (2015).

[35] S. Weibel, The Dublin Core: a simple content description model for electronic resources, Bulletin of the American Society for Information Science and Technology 24 (1) (1997) 9–11.

[36] A.R. Doherty, A.F. Smeaton, Automatically augmenting lifelog events using pervasively generated content from millions of people, Sensors 10 (3) (2010) 1423–1446.

[37] L. von Ahn, L. Dabbish, Labeling images with a computer game, in: Proceedings of the SIGCHI Conference on Human Factors in Computing Systems, CHI '04, ACM, New York, NY, USA, ISBN 1-58113-702-8, 2004, pp. 319–326, http://doi.acm.org/10.1145/985692.985733.

[38] L. Von Ahn, L. Dabbish, Designing games with a purpose, Communications of the ACM 51 (8) (2008) 58–67.

[39] G. Awad, C.G. Snoek, A.F. Smeaton, G. Quénot, TRECViD semantic indexing of video: a 6-year retrospective, ITE Transactions on Media Technology and Applications 4 (3) (2016) 187–208.

[40] A. Krizhevsky, I. Sutskever, G.E. Hinton, ImageNet classification with deep convolutional neural networks, in: Advances in Neural Information Processing Systems, 2012, pp. 1097–1105.

[41] A.F. Smeaton, K. McGuinness, C. Gurrin, J. Zhou, N.E. O'Connor, P. Wang, et al., Semantic indexing of wearable camera images: Kids'Cam concepts, in: Proceedings of the ACM Multimedia 2016 Workshop on Vision and Language Integration Meets Multimedia Fusion, Amsterdam, The Netherlands, October 16, 2016, ACM Press, 2016.

[42] H. Lee, A.F. Smeaton, N.E. O'Connor, G.J.F. Jones, M. Blighe, D. Byrne, et al., Constructing a SenseCam visual diary as a media process, Multimedia Systems 14 (6) (2008) 341–349, https://doi.org/10.1007/s00530-008-0129-x.

[43] L.S. Kennedy, S.F. Chang, A reranking approach for context-based concept fusion in video indexing and retrieval, in: CIVR'07, ISBN 978-1-59593-733-9, 2007, pp. 333–340, http://doi.acm.org/10.1145/1282280.1282331.

[44] C. Wang, F. Jing, L. Zhang, H.J. Zhang, Image annotation refinement using random walk with restarts, in: ACM MM'06, ISBN 1-59593-447-2, 2006, pp. 647–650, http://doi.acm.org/10.1145/1180639.1180774.

[45] C. Wang, F. Jing, L. Zhang, H.J. Zhang, Content-based image annotation refinement, in: CVPR'07, 2007, pp. 1–8.

[46] Y.G. Jiang, J. Wang, S.F. Chang, C.W. Ngo, Domain adaptive semantic diffusion for large scale context-based video annotation, in: ICCV'09, 2009, pp. 1420–1427.

[47] Y.G. Jiang, Q. Dai, J. Wang, C.W. Ngo, X. Xue, S.F. Chang, Fast semantic diffusion for large-scale context-based image and video annotation, IEEE Transactions on Image Processing 21 (6) (2012) 3080–3091, https://doi.org/10.1109/TIP.2012.2188038.

[48] O. Russakovsky, J. Deng, H. Su, J. Krause, S. Satheesh, S. Ma, et al., ImageNet large scale visual recognition challenge, International Journal of Computer Vision 115 (3) (2015) 211–252, https://doi.org/10.1007/s11263-015-0816-y.

[49] P. Wang, L. Sun, S. Yang, A.F. Smeaton, C. Gurrin, Characterizing everyday activities from visual lifelogs based on enhancing concept representation, Computer Vision and Image Understanding 148 (2016) 181–192, https://doi.org/10.1016/j.cviu.2015.09.014.

[50] S. Rendle, L. Schmidt-Thieme, Pairwise interaction tensor factorization for personalized tag recommendation, in: Proceedings of the Third ACM International Conference on Web Search and Data Mining, 2010, pp. 81–90.

[51] P. Wang, A.F. Smeaton, C. Gurrin, Factorizing time-aware multi-way tensors for enhancing semantic wearable sensing, in: MultiMedia Modeling, MMM 2015, 2015, pp. 571–582.

[52] D. Lee, H. Seung, Learning the parts of objects by nonnegative matrix factorization, Nature 401 (1999) 788–791.

[53] A. Shashua, T. Hazan, Non-negative tensor factorization with applications to statistics and computer vision, in: Proceedings of the International Conference on Machine Learning, ICML, 2005, pp. 792–799.

[54] D. Xu, P. Cui, W. Zhu, S. Yang, Find you from your friends: graph-based residence location prediction for users in social media, in: ICME 2014, 2014, pp. 1–6.

[55] C.C. Tan, Y.-G. Jiang, C.-W. Ngo, Towards textually describing complex video contents with audio-visual concept classifiers, in: ACM Multimedia '11, 2011, pp. 655–658.

[56] M. Merler, B. Huang, L. Xie, G. Hua, A. Natsev, Semantic model vectors for complex video event recognition, IEEE Transactions on Multimedia 14 (1) (2012) 88–101.

[57] P. Wang, L. Sun, S. Yang, A.F. Smeaton, What Are the Limits to Time Series Based Recognition of Semantic Concepts?, Springer International Publishing, Cham, ISBN 978-3-319-27674-8, 2016, pp. 277–289.

[58] J. Guo, D. Scott, F. Hopfgartner, C. Gurrin, Detecting complex events in user-generated video using concept classifiers, in: CBMI '12, 2012, pp. 1–6.

[59] J. Liu, Q. Yu, O. Javed, S. Ali, A. Tamrakar, A. Divakaran, H. Cheng, H. Sawhney, Video event recognition using concept attributes, in: WACV 2013, 2013, pp. 339–346.

[60] S. Bhattacharya, M.M. Kalayeh, R. Sukthankar, M. Shah, Recognition of complex events exploiting temporal dynamics between underlying concepts, in: CVPR 2014, 2014.

[61] H. Pirsiavash, D. Ramanan, Detecting activities of daily living in first-person camera views, in: CVPR 2012, 2012, pp. 2847–2854.

[62] I. Laptev, M. Marszalek, C. Schmid, B. Rozenfeld, Learning realistic human actions from movies, in: IEEE Conference on Computer Vision and Pattern Recognition, 2008, pp. 1–8.

[63] C. Sun, R. Nevatia, ACTIVE: activity concept transitions in video event classification, in: ICCV 2013, 2013, pp. 913–920.

[64] T.S. Jaakkola, D. Haussler, Exploiting generative models in discriminative classifiers, in: Proceedings of the 1998 Conference on Advances in Neural Information Processing Systems II, MIT Press, Cambridge, MA, USA, ISBN 0-262-11245-0, 1999, pp. 487–493, http://dl.acm.org/citation.cfm?id=340534.340715.

[65] W. Li, Q. Yu, H. Sawhney, N. Vasconcelos, Recognizing activities via bag of words for attribute dynamics, in: CVPR 2013, 2013, pp. 2587–2594.

[66] P. Mettes, D.C. Koelma, C.G. Snoek, The ImageNet shuffle: reorganized pre-training for video event detection, in: Proceedings of the 2016 ACM on International Conference on Multimedia Retrieval, ICMR '16, ACM, New York, NY, USA, ISBN 978-1-4503-4359-6, 2016, pp. 175–182, http://doi.acm.org/10.1145/2911996. 2912036.

[67] S. Whittaker, V. Kalnikaitė, D. Petrelli, A. Sellen, N. Villar, O. Bergman, et al., Socio-technical lifelogging: deriving design principles for a future proof digital past, in: Special Issue on Designing for Personal Memories, Human-Computer Interaction 27 (1–2) (2012) 37–62.

[68] F. Hopfgartner, Y. Yang, L. Zhou, C. Gurrin, User interaction templates for the design of lifelogging systems, in: Semantic Models for Adaptive Interactive Systems, Springer, London, 2013, pp. 187–204.

[69] D. Byrne, H. Lee, G.J.F. Jones, A.F. Smeaton, Guidelines for the presentation and visualisation of lifelog content, in: iHCI 2008 – Irish HCI 2008, 2008.

[70] E. van den Hoven, A future-proof past: designing for remembering experiences, Memory Studies 7 (3) (2014) 370–384, https://doi.org/10.1177/1750698014530625.

[71] G. Stix, Photographic memory: wearable cam could help patients stave off effects of impaired recall, Scientific American (2011).

[72] A. Doherty, A. Smeaton, Automatically segmenting lifelog data into events, in: Ninth International Workshop on Image Analysis for Multimedia Interactive Services, IEEE, 2008, pp. 20–23.

[73] L. Zhou, N. Caprani, C. Gurrin, N.E. O'Connor, ShareDay: a novel lifelog management system for group sharing, in: S. Li, A. Saddik, M. Wang, T. Mei, N. Sebe, S. Yan, et al. (Eds.), Advances in Multimedia Modeling, in: Lecture Notes in Computer Science, vol. 7733, Springer, Berlin, Heidelberg, ISBN 978-3-642-35727-5, 2013, pp. 490–492, https://doi.org/10.1007/978-3-642-35728-2_47.

[74] M. Naaman, S. Harada, Q. Wang, H. Garcia-Molina, A. Paepcke, Context data in geo-referenced digital photo collections, in: MULTIMEDIA '04: Proceedings of the 12th Annual ACM International Conference on Multimedia, ACM, New York, NY, USA, ISBN 1-58113-893-8, 2004, pp. 196–203, http://doi.acm.org/10.1145/1027527. 1027573.

[75] K. Kitamura, T. Yamasaki, K. Aizawa, Food log by analyzing food images, in: Proceedings of the 16th ACM International Conference on Multimedia, MM '08, ACM, New York, NY, USA, ISBN 978-1-60558-303-7, 2008, pp. 999–1000, http://doi.acm.org/ 10.1145/1459359.1459548.

[76] M. Hughes, E. Newman, A.F. Smeaton, N.E. O'Connor, A lifelogging approach to automated market research, in: Proceedings of the SenseCam Symposium 2012, 2012.

[77] L. Xia, Y. Ma, W. Fan, VTIR at the NTCIR-12 2016 lifelog semantic access task, in: Proceedings of NTCIR-12, Tokyo, Japan, 2016.

[78] H.L. Lin, T.C. Chiang, L.P. Chen, P.C. Yang, Image searching by events with deep learning for NTCIR-12 lifelog, in: Proceedings of NTCIR-12, Tokyo, Japan, 2016.

[79] B. Safadi, P. Mulhem, G. Quénot, J.P. Chevallet, MRIM-LIG at NTCIR lifelog semantic access task, in: Proceedings of NTCIR-12, Tokyo, Japan, 2016.

[80] H. Scells, G. Zuccon, K. Kitto, QUT at the NTCIR lifelog semantic access task, in: Proceedings of NTCIR-12, Tokyo, Japan, 2016.

[81] G. de Oliveira Barra, A.C. Ayala, M. Bolaños, M. Dimiccoli, X. Giro-i-Nieto, P. Radeva, LEMoRE: a lifelog engine for moments retrieval at the NTCIR-lifelog LSAT task, in: Proceedings of NTCIR-12, Tokyo, Japan, 2016.

[82] C. Gurrin, H. Joho, F. Hopfgartner, L. Zhou, R. Albatal, Overview of NTCIR-12 lifelog task, 2016, http://eprints.gla.ac.uk/131460/.

[83] N. Caprani, A.R. Doherty, H. Lee, A.F. Smeaton, N.E. O'Connor, C. Gurrin, Designing a touch-screen SenseCam browser to support an aging population, in: CHI '10 Extended Abstracts on Human Factors in Computing Systems, CHI EA '10, ACM, New York, NY, USA, ISBN 978-1-60558-930-5, 2010, pp. 4291–4296, http://doi.acm.org/10.1145/1753846.1754141.

[84] C. Gurrin, H. Lee, N. Caprani, Z. Zhang, N. O'Connor, D. Carthy, Browsing large personal multimedia archives in a lean-back environment, in: S. Boll, Q. Tian, L. Zhang, Z. Zhang, Y.-P. Phoebe Chen (Eds.), Advances in Multimedia Modeling, in: Lecture Notes in Computer Science, Springer, Berlin Heidelberg, 2010, pp. 98–109.

[85] P. Kelly, A.R. Doherty, A.F. Smeaton, C. Gurrin, N.E. O'Connor, The colour of life: novel visualisations of population lifestyles, in: Proceedings of the International Conference on Multimedia, MM '10, ACM, 2010.

[86] J. Redmon, S. Divvala, R. Girshick, A. Farhadi, You only look once: unified, real-time object detection, in: IEEE Conference on Computer Vision and Pattern Recognition (CVPR), 2016, pp. 779–788.

[87] C. Dobbins, M. Merabti, P. Fergus, D. Llewellyn-Jones, F. Bouhafs, Exploiting linked data to create rich human digital memories, Computer Communications 36 (2013) 1639–1656.

[88] Y. Li, Y. Guo, Wiki-Health: from quantified self to self-understanding, Future Generations Computer Systems 56 (2016) 333–359.

[89] P. Wang, Measuring activities of daily living: digitally, visually and semantically, in: Proceedings of Measuring Behavior 2016: 10th International Conference on Methods and Techniques in Behavioral Research, ISBN 978-1-873769-59-1, 2016, pp. 549–556.

[90] S.J. Pan, Q. Yang, A survey on transfer learning, IEEE Transactions on Knowledge and Data Engineering 22 (10) (2010) 1345–1359.

[91] P. Wang, A.F. Smeaton, Semantics-based selection of everyday concepts in visual lifelogging, International Journal of Multimedia Information Retrieval 1 (2) (2012) 87–101.

[92] P. Wang, L. Sun, S. Yang, A.F. Smeaton, Semantically Smoothed Refinement for Everyday Concept Indexing, Springer International Publishing, 2016, pp. 318–327.

[93] G. Ye, Y. Li, H. Xu, D. Liu, S.F. Chang, EventNet: a large scale structured concept library for complex event detection in video, in: Proceedings of the 23rd ACM International Conference on Multimedia, MM '15, ACM, New York, NY, USA, ISBN 978-1-4503-3459-4, 2015, pp. 471–480, http://doi.acm.org/10.1145/2733373. 2806221.

[94] O. Vinyals, A. Toshev, S. Bengio, D. Erhan, Show and tell: a neural image caption generator, in: IEEE Conference on Computer Vision and Pattern Recognition (CVPR), 2015, pp. 3156–3164.

[95] T. Ye, B. Moynagh, R. Albatal, C. Gurrin, Negative faceblurring: a privacy-by-design approach to visual lifelogging with Google Glass, in: Proceedings of the 23rd ACM International Conference on Information and Knowledge Management, CIKM '14, ACM, New York, NY, USA, ISBN 978-1-4503-2598-1, 2014, pp. 2036–2038, http://doi.acm.org/10.1145/2661829.2661841.

[96] M. Langheinrich, Privacy by design – principles of privacy-aware ubiquitous systems, in: Proceedings of the 3rd International Conference on Ubiquitous Computing, Ubi-Comp '01, Springer-Verlag, London, UK, ISBN 3-540-42614-0, 2001, pp. 273–291, http://dl.acm.org/citation.cfm?id=647987.741336.

[97] A. Cavoukian, Privacy by Design: The 7 Foundational Principles, Information and Privacy Commissioner of Ontario, Canada, 2010.

[98] P. Kelly, S.J. Marshall, H. Badland, J. Kerr, M. Oliver, A.R. Doherty, C. Foster, An ethical framework for automated, wearable cameras in health behavior research, American Journal of Preventive Medicine 44 (3) (2013) 314–319.

# CHAPTER 10

# Computational Analysis of Affect, Personality, and Engagement in Human–Robot Interactions*

**Oya Celiktutan***, **Evangelos Sariyanidi**[†], **Hatice Gunes**[‡]
*Imperial College London, Electrical Engineering, Personal Robotics Lab, United Kingdom
[†]Centre for Autism Research, Philadelphia, PA, United States
[‡]University of Cambridge, Computer Laboratory, United Kingdom

## Contents

## Abstract

This chapter focuses on recent advances in social robots that are capable of sensing their users, and support their users through social interactions, with the ultimate goal of fostering their cognitive and socio-emotional wellbeing. Designing social robots with socio-emotional skills is a challenging research topic still in its infancy. These skills are important for robots to be able to provide physical and social support to human users, and to engage in and sustain long-term interactions with them in a variety of application domains that require human–robot interaction, including healthcare, education, entertainment, manufacturing, and many others. The availability of commercial

---

\* The research reported in this chapter was completed while O. Celiktutan and E. Sariyanidi were with the Computer Laboratory, University of Cambridge, United Kingdom.

robotic platforms and developments in collaborative academic research provide us with a positive outlook; however, the capabilities of current social robots are quite limited. The main challenge is understanding the underlying mechanisms of humans in responding to and interacting with real life situations, and how to model these mechanisms for the embodiment of naturalistic, human-inspired behavior in robots. Addressing this challenge successfully requires an understanding of the essential components of social interaction, including nonverbal behavioral cues such as interpersonal distance, body position, body posture, arm and hand gestures, head and facial gestures, gaze, silences, vocal outbursts, and their dynamics. To create truly intelligent social robots, these nonverbal cues need to be interpreted to form an understanding of the higher level phenomena including first-impression formation, social roles, interpersonal relationships, focus of attention, synchrony, affective states, emotions, personality, and engagement, and in turn defining optimal protocols and behaviors to express these phenomena through robotic platforms in an appropriate and timely manner. This chapter sets out to explore the automatic analysis of social phenomena that are commonly studied in the fields of affective computing and social signal processing, together with an overview of recent vision-based approaches used by social robots. The chapter then describes two case studies to demonstrate how emotions and personality, two key phenomena for enabling effective and engaging interactions with robots, can be automatically predicted from visual cues during human–robot interactions. The chapter concludes by summarizing the open problems in the field and discussing potential future directions.

## Keywords

Social robotics, Human–robot interaction, Affective computing, Social signal processing, Personality computing, Computer vision, Machine learning

## 10.1 INTRODUCTION

Humanoid robots are being deployed in public spaces including hospitals [1], banks [2], and airports [3]. An increasing number of individuals needing companionship and psychological support push the need for socially assistive robotics. Socially assistive robotics focuses on building robots that can facilitate an effective interaction with their human users for the purpose of assisting them at the social and cognitive level, namely, aiding them to achieve their goals, managing their medical needs, or enhancing their overall well-being. In the context of heath care and therapy, there is a significant body of work on how Paro, a robotic seal, improves well-being and reduces depression and anxiety in elderly people [4]. KASPAR, Kinesics And Synchronization in Personal Assistant Robotics, is a child-sized humanoid robot designed to develop basic social interaction skills in children with autism through turn taking and imitation games [5]. SPRITE, Stewart Platform Robot for Interactive Tabletop Engagement, helps a group of people to complete a task by manipulating turn-taking patterns and the participants' attention, with the goal of increasing group cohesion [6].

In education, several studies have already shown the benefits of using robots in one-to-one tutoring sessions and classroom settings. Students performed better in mathematics when a robot tutored them [7], and were more emotionally expressive when engaged in an interactive educational task with a social robot than when performing the same task with a tablet [8]. Personalizing a robot's actions to individual differences has been shown to be compulsory for achieving good learning outcomes in several studies. Keepon, a tabletop robot, was made to provide personalized feedback using a skill assessment algorithm in [9]. To accommodate children's short attention spans, Nao, a child-sized humanoid robot, was programmed to offer breaks based on personalized timing strategies [10]. Similarly, in [11], Nao tutored language learning by adapting its feedback to the children's skills and observed behaviors.

User modeling, adaptation, and personalization are key to the effective deployment of social robots in real-world settings. The generic system of such a robot consists of three modules [12]: (1) the perception module; (2) the reasoning (intermediate) module; and (3) the action module. The perception module acquires information regarding the human user by capturing (multimodal) data through both the robot's sensors and the environmental sensors, and analyzes the human user's behaviors based on the information collected during interactions. The action module deals with the design and generation of behaviors for the robot. The reasoning (intermediate) module connects the perception and action modules to deliver robot behaviors that are shaped by the output of the perception module. In this chapter, we exclusively focus on the perception module, in particular from the perspective of affect and social signal analysis from visual cues.

Affective and social signals are integral parts of communication. Humans exchange information and convey their thoughts and feelings through gaze, facial expressions, body language, and tone of voice along with spoken words, and infer 60–65% of the meaning of the communicated messages from these nonverbal behaviors [13]. These nonverbal behaviors carry significant information regarding higher level social phenomena such as emotions, personality, and engagement. Recognizing and interpreting these signals is a natural routine for humans, and automatizing these mechanisms is necessary for robots to be successful in their interactions with humans.

The objective of this chapter is to present a survey of computational approaches to the analysis of affective and social signals, together with recent techniques used by social robots, to categorize the available algorithms and to highlight the latest trends. The chapter starts with representative techniques for the analysis of an individual's emotions, personality, and engagement state, three social phenomena that have been commonly studied in the area of affective and social signal processing (see Section 10.2). The chapter then focuses on summarizing the state of the art of robotic platforms endowed with the capability of analyzing these social phenomena. To provide concrete examples, the chapter presents two case studies to describe how a computational method can be built for predicting emotions and personality from visual cues during human–robot interactions (see Section 10.3). The chapter concludes by summarizing the open problems in the field and discusses potential solutions to these problems (see Section 10.4).

## 10.2 AFFECTIVE AND SOCIAL SIGNAL PROCESSING

In this section, we first introduce the state-of-the-art computer vision-based approaches to affective and social signal processing, and then review the prominent techniques used by the currently available social robots. We scope out and explore three social phenomena that are widely studied in this context: (i) emotion; (ii) personality; and (iii) engagement.

### 10.2.1 Emotion

Emotion (or affect) recognition has been one of the most active research areas across multiple disciplines ranging from psychology to computer science and social robotics. There have already been several extensive surveys on automatic emotion recognition from facial cues [14,15] and bodily cues [16].

Emotion recognition methods from facial cues aim at recognizing the appearance of facial actions or the expression of emotions conveyed by these actions, and usually rely on the Facial Action Coding System (FACS) [17]. FACS consists of facial Action Units (AUs), which are codes that describe certain facial muscle movements (e.g. AU 12 is lip corner puller). The temporal evolution of an expression is typically modeled with four temporal phases [17]: neutral, onset, apex, and offset. Neutral is the expressionless phase with no signs of muscular activity. Onset denotes the period during which muscular contraction begins and increases in intensity. Apex is a plateau where the intensity usually reaches a stable level. Offset is the phase of muscular action relaxation.

There have been two lines of approaches proposed in the literature that are associated with two models of emotions, namely, the categorical model and the dimensional model. The categorical model refers to the affect model developed by Ekman and his colleagues, who argued that the production and interpretation of certain expressions are hard-wired in our brains and are recognized universally (e.g. [18]). The emotions conveyed by these expressions are grouped into six classes, known as the *six basic emotions*: happiness, sadness, surprise, fear, anger, and disgust. AUs can be mapped to the six basic emotions. For example, using a simple rule-based method, happiness can be represented as a combination of AU6 (cheek raiser) and AU12 (lip corner puller) [14]. However, the categorical model is believed to be limited in its ability to represent the broad range of everyday emotions [19]. To represent a wider range of emotions, the dimensional approach is used to continuously model emotions in terms of affect dimensions [19]. The most established affect dimensions are arousal, valence, power, and expectation [19].

The categorical and dimensional models were evaluated in two prominent affect recognition challenges: Facial Expression Recognition and Analysis (FERA) [20,21] and Audio/Visual Emotion Challenges (AVEC) [22]. The FERA challenge evaluates AU detection/AU intensity estimation and discrete emotion classification for four basic emotions (anger, fear, happiness, sadness) and one nonbasic emotion (relief). The AVEC challenge evaluates dimensional emotion models along arousal and valence dimensions.

De la Torre et al. [23] addressed the AU detection problem using a personalized learning approach based on a Selective Transfer Machine (STM) that learns a classifier while simultaneously reweighting the training samples that are most relevant to the test subject. They extracted appearance features based on Scale-Invariant Feature Transform (SIFT) descriptors, from patches centered on the automatically detected facial landmarks. The proposed method achieved superior performance compared to the conventional classification methods such as Support Vector Machines (SVMs) for classifying five emotions on the FERA 2011 benchmark [20]. The recent trend for AU detection has been deep learning methods. Jaiswal and Valstar [24] simultaneously learned dynamic appearance and shape features within a time window using Convolutional Neural Networks (CNNs), and applied Bidirectional Long Short-Term Memory (BLSTM) networks on top of the time-windowed CNN features to model temporal relationships. The proposed method outperformed the previous approaches in the FERA 2015 challenge datasets [21].

Recent works adopting the dimensional model were characterized by combining visual data with different modalities, usually audio and physiological data, and employing BLSTM for predicting arousal and valence in a time-continuous manner [25,26]. For example, the winner of the AVEC 2015 challenge [27] combined two appearance features, namely, Local Gabor Binary Patterns from Three Orthogonal Planes (LGBP-TOP), which were baseline features provided by the challenge organizers, and Local Phase Quantization from Three Orthogonal Planes (LPQ-TOP) together with geometric features computed from facial landmarks. Different feature types were fused using a model-level fusion strategy, where the outputs of single modality models were smoothed and combined using a second layer of BLSTM. Chen and Jin [26] proposed a multimodal attention fusion method that automatically assigns weights to different modalities according the current modality features and history information, which outperformed the traditional fusion strategies (e.g. early-fusion, model-level fusion, late-fusion) in the detection of valence in the same database.

**Emotion Recognition in HRI.** Emotion recognition methods used by social robots were extensively surveyed by Yan et al. in [12] and McColl et al. in [28]. Here, we only considered the prominent works that performed the recognition task by automatically extracting features from visual cues, and integrated the developed method on a robotic platform.

The categorical model of emotion has been the most widely adopted approach in the literature. Cid et al. [29] developed an emotion recognition system by extracting features based on the Facial Action Coding System (FACS) [17], and implemented it on a robotic head, Muecas [30], for an imitation task. For emotion recognition, they first applied a preprocessing step to the face image taken by Muecas to normalize the illumination and remove the noise, and a Gabor filter to highlight the facial features. From the processed face, a set of edge-based features were extracted and modeled using Dynamic Bayesian Networks to detect a total of 11 AUs. The detected AUs were used to represent four basic emotions including happiness, sadness, fear, and anger according to a rule-based approach, and were mapped on the Muecas robot to display the inferred emotion in real-time. In [31], the authors used similar visual features (i.e. Gabor filter responses) to enable the robot to learn facial expressions of emotion from interactions with humans through an online learning algorithm based on neural networks. The Muecas robot was able to learn all the emotions successfully, except for sadness. This was due to the large intra-class variability for sadness, namely, each person expressed sadness in a different manner.

In [32], Leo et al. developed an automatic emotion recognition system to measure the facial emotion imitation capability of children with Autism Spectrum Disorders (ASD). The R25 robot from Robokind [33], a small cartoon character-like robot, was first made to display a facial expression, and then the child was instructed to imitate the displayed facial expression while being analyzed through the camera located in R25's right eye. The emotion recognition method was based on a generic pipeline that consisted of four components: Viola–Jones face detection, face registration, Histogram of Gradient (HoG) face representation, and classification with SVMs. The method was tested via a study involving three children with ASD, and it achieved good emotion recognition performance, especially for happiness and sadness.

Among works adopting the dimensional model of emotion, Castellano et al. [34] focused on valence of an affect, representing it with three discrete states: positive, neutral, and negative. They designed an affect-sensitive robotic game companion, with the goal of detecting these three states and selecting an empathic strategy for the robot to display. For this purpose, they combined visual features including smiling gestures and gaze patterns with contextual information such as game state and game evolution. For detecting smiles, first an off-the-shelf application was used to estimate head pose and track facial landmark points, and then a geometry-based descriptor was defined based on the spatial locations of the facial landmarks with respect to the head pose. The developed method was integrated onto the iCat platform, a desktop user-interface robot with animated facial expressions [35] to test with children during the course of a chess game. Schacter et al. [36] focused on the prediction of both arousal and valence dimensions. They extracted geometry-based features from facial landmarks that were detected using Constrained Local Models [37], and applied Support Vector Regression (SVR) for prediction. The proposed method was tested using the onboard camera of their in-house robot called Social Robot Brian.

In this chapter, we exclusively focus on facial cues. However, body postures and hand gestures are important sources of information, especially in the context of HRI, when facial cues cannot be observed reliably. Most of the emotion recognition methods from bodily cues has relied on real-time skeleton tracking algorithm of Kinect depth sensor [38]. Wang et al. [39]

aimed at modeling arousal and valence dimensions in a time-continuous manner. They captured visual recordings using a Kinect depth sensor during the course of a game of Snakes and Ladders played by a child against the Nao robot [40]. Nao's behaviors were manipulated to display either competitive or supportive behaviors in order to elicit different emotional responses from the participated children. From these recordings, they modeled bodily expressions using the 3D skeleton tracking algorithm, and skeletal representations were used to extract two types of features: (i) a set of low-level features comprising spatial distances between hands, elbows, and shoulders, and the angles between the spine and the upper arms, and the orientation of the shoulders; (ii) a set of high-level features describing body movement activity and power, body spatial extension, and head bending. These features were then used to train Online Recursive Gaussian Processes for real-time emotion recognition from bodily expressions, where they found that the valence dimension was more difficult to model than the arousal dimension.

## 10.2.2 Personality

Individuals' interactions with others are shaped by their personalities and their impressions regarding others' behaviors and personalities [41]. This has also been shown to be the case for interactions with social robots [42]. The traditional approach to describing personality is the trait theory that focuses on the measurement of general patterns of behaviors, thoughts, and emotions, which are relatively stable over time and across situational contexts [43]. The Big Five Model is currently the dominant paradigm in personality research which defines traits along five broad dimensions: *extroversion* (assertive, outgoing, energetic, friendly, socially active), *neuroticism* (a tendency to negative emotions such as anxiety, depression, or anger), *openness* (a tendency to changing experience, adventure, new ideas), *agreeableness* (cooperative, compliant, trustworthy), and *conscientiousness* (self-disciplined, organized, reliable, consistent).

There are two strategies coupled with two main problems in automatic personality analysis [44], which are personality recognition (prediction of actual personality) and personality perception (prediction of personality impressions). In both problems, the commonly used method to measure Big Five personality traits is the Big Five Inventory (BFI) [45]. In personality recognition, an individual is asked to fill in the BFI which aims to assess personal behavioral tendencies, i.e. how an individual sees herself in the way she approaches problems, likes to work, deals with feelings, and manages relationships with others. In personality prediction, external observers are asked to view a video of the individual and rate the individual along the Big Five personality dimensions based on thin slices of behavior ranging from 10 seconds to several minutes. However, employing observers to carry out this tedious task is in itself a problem. A number of researchers [46,47] obtained manual annotations through online crowd-sourcing services such as the Amazon Mechanical Turk (MTurk) service. Typically, several folds of independent ratings are run since there is rarely full agreement between the raters.

Nonverbal behaviors are significant predictors of personality. Gaze and head movement are strongly correlated with personality. For example, *dominance* and *extroversion* are found to be related to holding a direct facial posture and long durations of eye contact during interaction, whereas *shyness* and *social anxiety* are highly correlated with gaze aversion [48]. Extroverted people are found to be more energetic, leading to higher head movement frequency, more hand gestures, and more posture shifts than introverted people [49,50]. Research has demonstrated that these nonverbal behaviors can be reliably modeled from visual cues for predicting personality.

Among the works focusing on facial and head cues, Joshi et al. [51] investigated varied situational contexts using audio–visual recordings of conversations between a human and four different virtual characters using the SEMAINE corpus [52]. The SEMAINE corpus comprises audio–visual recordings of interactions between human participants and four different virtual characters. Facial cues were extracted using the pyramid of HoG, which counts the gradient orientations in the whole face and in the localized portions. The mean and the standard deviation of the histograms accumulated from all the frames were fed into SVMs for regression. The visual features used yielded the best prediction accuracy for *conscientiousness* among the Big Five personality traits.

High-level features were taken into account by Teijeiro-Mosquera et al. [46] using videos from Youtube, so-called "video blogs", with annotations generated through the MTurk service. They detected facial expression of emotions (e.g. anger, happiness, fear, sadness) on a frame-by-frame basis and extracted emotion activity cues from sequences either by thresholding or by using an HMM-based method. These features were then fed into SVMs for predicting the five traits. Their results showed that facial expressions were a strong predictor of *extroversion*.

Another line of work has focused on the fusion of facial/head cues and bodily cues at the feature level. Aran and Gatica-Perez [53] used recordings from the ELEA corpus [54] involving three or four participants performing a Mission Survival task [55]. They represented the visual cues by extracting two types of features, namely, activity features and attention features. The participants' heads and bodies were tracked in videos, and optical flow was computed from tracked head and body parts, yielding the binary occurrence of head/body activity at a specific time instant and the amount of activity. Activity features were then computed by aggregating the occurrences and amount of activity over the whole sequence, which included head/body activity length, head/body activity turns, standard deviations of head/body activity in $x$ and $y$ directions, etc. In addition to head/body activity features, simple statistics were calculated from weighted Motion Energy Images (MEI) in order to encapsulate the whole body activity over time. Attention features were extracted based on the visual focus of attention analysis during interactions, which included attention given while speaking/listening, attention received while speaking/listening, and visual dominance ratio. Ridge regression was used both for the prediction of *extroversion* level and for the binary classification of *extroversion, agreeableness,* and *openness.* For both regression and classification, the best results were achieved by combining all the features. However, the prominent visual features were attention features and MEI statistics in the classification of *extroversion.*

From human–virtual character interactions [52], Celiktutan and Gunes [56] modeled the face/head and body movements by extracting three sets of features: (i) spatial and spatio-temporal appearance features (e.g. Zernike moments, gradient and optical flow); (ii) geometric features (e.g. spatio-temporal configuration of facial landmark points, horizontal and vertical trajectories over time); and (iii) hybrid features (e.g. the fusion of local appearance and motion information around facial landmark points). These features were then used in conjunction with Long Short-Term Memory Networks for predicting personality traits continuously in space and time, which yielded the highest coefficient of determination ($R^2$) for *conscientiousness* using the face appearance features and for *neuroticism* and *openness* using the body appearance features.

**Personality Prediction in HRI.** Incorporating human personality analysis to adapt a robot's behavior for engaging a person in an activity is a fundamental component of social robots [57,47]. One prominent work by Rahbar et al. [58] focused on the prediction of the *extroversion* trait only, when a participant was interacting with the humanoid iCub [59], a robot shaped like a four-year-old child. They combined individual features and interpersonal features that were extracted from Kinect recordings. More explicitly, the individual features included the participant's quantity of motion computed from the depth images. The interpersonal features modeled synchrony and dominance between the movements of iCub and the participant, and also proxemics (i.e. the distance between iCub and the participant). They achieved the best F-measure when they fused individual and interpersonal features at the feature level using Logistic Regression.

Some works focused on the robot's personality to improve the quality of the human experience with the robot: humans tend to be attracted by characters that have either matching personality traits (similarity rule) or non-matching personality traits (complementarity rule) [60]. Salam et al. [47] investigated the impact of the participants' personalities on their engagement states from the Kinect depth sensor recordings. These recordings contained interactions between two participants and Nao [40], a small humanoid robot. They extracted two sets of features, namely, individual and interpersonal features, similarly to [58]. Individual features described the individual behaviors of each participant, e.g. body activity computed from articulated pose and motion energy images, body appearance, etc. Interpersonal features characterized the interpersonal behaviors of the participants with respect to each other and the robot. These include the visual focus of attention (VFOA), the global quantity of movement, the relative orientation of the participants, the relative distance between the participants, and the relative orientation of the participants with respect to the robot. They first applied Gaussian process regression for personality prediction. They then combined the predicted personality labels with the individual and interpersonal features to classify whether the participants were engaged or not. The best results were achieved using individual features together with personality labels.

Despite its importance, automatic personality analysis as a part of a social robot has been scarce; indeed, to the best of our knowledge, there has been no system that is integrated onto a robot, and performs real-time analysis of personality in the course of interaction. In [61], Celiktutan et al. used a real-time implementation of their method of personality prediction from nonverbal cues [56], and demonstrated this system, called MAPTRAITS, together with the Nao robot. Using a Wizard of Oz setup, Nao asked the participants a predefined set of questions about their jobs, hobbies, and memories while the MAPTRAITS system (running on a PC) analyzed the participants' personalities in real-time using a camera placed on a tripod.

The predicted personality scores were displayed to each participant instantaneously on a screen; however, no quantitative analysis was conducted.

### 10.2.3 Engagement

Engagement is the process by which interactors start, maintain, and end their perceived connection to each other during an interaction [62]. When individuals interact with each other, they display affective and social signals that give away information regarding their engagement states (i.e. intention to engage, engagement, and disengagement).

Most of the methods for predicting engagement have focused on observable visual cues including social gaze patterns, facial gestures, and body posture. Although these cues were manually annotated, Kapoor et al. [63] exploited features based on facial gestures and body posture in order to predict the level of interest of a child who was solving a puzzle. In particular, facial gestures were coded in terms of manually annotated facial action units associated with upper face muscle movements around the eyes, eyebrows, and upper cheeks, and body posture was determined using a sensor chair. Their results showed that body posture alone was more informative than facial gestures, yielding a better classification performance with Hidden Markov Models (HMMs). Oertel and Salvi [64] only relied on features extracted from manually annotated social gaze patterns to model group involvement and individual engagement in game-based group interactions. They divided the social gaze patterns into four groups, namely, looking at another participant, looking away, looking down, and eyes closed, that were converted into a matrix for each participant. They then extracted group-level features and individual-level features from these matrices. While group-level features modeled interpersonal dynamics such as mutual gaze, individual features intended to capture individual differences in gaze behaviors. Good classification results were obtained with Gaussian Mixture Models (GMMs) for detecting the high level of group involvement and group forming/getting familiar with each other, whereas the low-level group involvement was classified poorly.

Peters et al. [65] focused on automatic gaze estimation and shared attention detection from a web camera during interactions with a virtual agent. They first estimated head pose and gaze by automatically detecting and tracking facial landmark points. The user's head and gaze directions were then mapped on the computer screen in order to model the level of attention and the level of engagement. While the level of attention was measured in terms of gaze fixations onto the virtual objects on the screen (including the virtual agent itself), the scene background, or outside of the screen, the level of engagement was defined as how much the user looks at the relevant objects in the scene at the appropriate times.

There is another line of research investigating the impact of personality on engagement in human–virtual character interactions. Cerekovic et al. [66] considered two virtual agents from the SEMAINE System [52], namely, Obadiah and Poppy. While Obadiah was gloomy and neurotic with low variation in speech and a flat tone, Poppy was cheerful and extroverted with frequent gestures and head nods. They measured the engagement level of each participant along three dimensions: quality, rapport, and likeness. In order to predict the levels of these three dimensions, they took into account both audio-visual features and manually annotated personality trait labels collected from external observers. As visual features, they computed the distribution of body leans and frequency of shifts from one body posture to another using the 3D skeleton tracking information from the Kinect depth sensor. Similar features were computed for manually annotated hand gestures, and facial expressions were modeled using an off-the-shelf facial expression recognition toolbox. They achieved the best results when they combined nonverbal features with personality scores. They found that extroverted people tended to like the neurotic agent, whereas people that score high on *neuroticism* liked the cheerful agent, supporting the interpersonal complementarity rule [60].

**Engagement Prediction in HRI.** Understanding the user's engagement is important to ensure that the user maximally benefits from an activity conducted with the assistance of the robot, particularly in health-related applications and education settings. In [67], Sanghvi detected engagement states during a chess game played by a child and iCat [34]. In order to detect whether the child was engaged or not, the child's body silhouette was first extracted, and then a set of features was extracted based on the posture and body movements. These features included (i) body lean angle, a measure of the orientation of the child's upper body when playing the game with the robot; (ii) slouch factor, a measure of the curvature of the child's back; (iii) quantity of motion, a measure of the amount of detected motion from the extracted silhouette; and (iv) contradiction index, a measure of the degree of contraction and expansion of the upper body. Using the extracted features in conjunction with ADTree and OneR classifiers yielded a high accuracy for engagement classification.

In [68], Benkaouar and Vaufreydaz proposed a multimodal approach for recognizing nonverbal cues and inferring engagement in a home environment where they used a Kinect depth sensor mounted onto a mobile robot called Kompai from Robosoft [69]. They extracted three sets of visual features: (i) proxemics features such as distance to the robot, speed from the recorded depth data; (ii) face location and face size from the recorded RGB data; and (iii) positions of stance, hips, torso, and shoulders, and their relative rotations from the tracked skeletons. The most relevant features yielding the best engagement detection accuracy were selected using the Minimum Redundancy Maximum Relevance method. Their results showed that shoulder rotation, face position and size, relative distance, and speed played an important role in engagement detection.

Salam and Chetouani [70] conducted a study in a triadic HRI scenario to investigate to what extent it is possible to infer an interactor's engagement state starting from the cues of the others in the interaction. They considered two set of features from two human participants and a robot. Each participant's features were composed of manually annotated social cues including head nods, visual focus of attention (VFOA), head pose, face location, and utterances. In addition to these cues, they extracted simple features over time, e.g. VFOA shifts, sliding windows statistics of head pose and face location, etc. The robot's features comprised utterances, addressee (addressing the speech to an interactor), and the topic of the speech. These features were used, both singly and in pairwise combinations (i.e. combining features of both participants, or combining a participant's features with the robot's features), in conjunction with SVMs for engagement classification. Their results showed that in a multiparty interaction, the cues of the other interactors can be used to infer the engagement state of the individual in question, which suggests that inter-personal context plays an important role in engagement classification.

## 10.3  TWO CASE STUDIES

In this section, we describe two automatic methods for modeling emotion and personality in interactions with a robot. First, we present a novel AU detection method. AU detection has been a popular research problem in computer science; however, there are fewer works performing AU detection in the context of HRI. Differently from [30], for more robust AU detection, our method combines shape and appearance information, and exploits differential features with respect to an individual's neutral face. Then, we introduce how this method can be implemented on the humanoid robot Nao in real-time and can be used in live public demonstrations.

Second, we describe a pipeline for automatic prediction of an individual's personality in the course of their interactions with Nao, from experimental study design to data collection and feature extraction. Despite its importance, there are only a few works performing automatic personality prediction in the context of HRI. Additionally, most of these works investigate the relationship between the personality traits and engagement state based on self reports, which might not be available in real-life applications. Here, we show that personality can be predicted from a set of low-level features extracted from videos captured from a first-person perspective.

## 10.3.1 Automatic Emotion Recognition

In this chapter, we introduce a novel method for detecting Action Units (AUs) in video sequences, and present comparative figures on a state-of-the-art database. We also demonstrate how this approach can be used for public engagement at various events (e.g. science festivals).

### 10.3.1.1 Action Unit Detection Methodology

There has been a significant body of work in the area of automatic AU detection. Recently, Sariyanidi et al. [15] highlighted the importance of two practices: (i) combining shape and appearance features, which yields better performance because they carry complementary information, and (ii) using differential features that describe information with respect to a reference image (i.e. a neutral face in the case of emotion recognition). The main advantage of the differential features is to place greater emphasis on the facial action by reducing person-specific appearance cues.

**Feature Extraction.** In the light of these insights, we extracted four types of features, namely, shape, appearance, differential-appearance (hereafter $\delta$-appearance) and differential-shape (hereafter $\delta$-shape) as follows. Shape features were obtained by concatenating the vertical and horizontal coordinates of the facial landmarks that were estimated using the Supervised Descent Method (SDM) in [71]; $\delta$-shape features were computed by subtracting the shape representation of a given facial image from the shape representation that was computed from a facial image, of the same subject, with a neutral expression.

Appearance features were extracted using the Quantized Local Zernike Moments (QLZM) method [72]. The use of this method was previously demonstrated for affect recognition based on both the categorical and dimensional models of emotion [72]. The QLZM method consists of two

steps: (i) computing local Zernike moments to describe image disconti-
nuities at various scales and orientations, and (ii) performing non-linear
encoding and pooling to improve the robustness against image noise and
translation. Here we computed the appearance features in a part-based
manner. Using the estimated facial landmarks, we first cropped three square
patches that contained the left eye, right eye, and mouth, and then computed the QLZM histograms from each part.

We computed $\delta$-appearance features using the Gabor motion energy
filters [73], where we adopted a part-based representation similarly to the
appearance features. We used Gabor motion energy filters to describe the
motion between a given face image of a subject and the subject's neutral
face image. One advantage of using the Gabor representation over using
simpler representations (e.g. difference between neutral and apex phases)
is its robustness to illumination variations. During the on-the-fly tests, we
ensured that we had the neutral face of human subjects by asking them to
stand still and make a neutral face in front of the camera prior to beginning
a test session.

Note that each AU can occur either in the upper part or in the lower
part of the face. For example, AU1 (inner brow raiser) occurs in the upper part and AU25 (lips part) occurs in the lower part. Therefore, when
detecting an AU, we took into account the above-mentioned four features extracted either from the upper part or from the lower part of face.
For shape and $\delta$-shape features, this resulted in a 60-length feature vector,
corresponding to the landmarks associated with eyes and eyebrows, and a
38-length feature vector, corresponding to the landmarks associated with
mouth. For the appearance and $\delta$-appearance features, this resulted in a
800-length and a 512-length feature vector, respectively, computed from
the left and right eye patches, and a 400-length and a 256-length feature
vector, respectively, computed from the mouth patch.

**Decision Fusion.** We trained four binary SVM classifiers, each in conjunction with one of the above-mentioned feature types, for each AU. The
final AU detection decision was given by fusing the outputs of the four
individual classifiers. Specifically, we adopted the *consensus fusion* approach,
where an AU was detected based on the condition that all four classifiers
were in full agreement. The advantage of the consensus fusion approach is
that it yields a low False Alarm Rate (FAR). The downside is that it can
also lead to a low True Positive Rate (TPR) because the consensus cannot
be reached even when one of the classifiers misses an AU. To address this
issue, we decreased the AU detection threshold for each classifier, where

we empirically set the threshold to 0.95 TPR on the training dataset. This also increased the False Positive Rate (FPR) of the individual classifiers, but the overall FPR of the consensus fusion approach was low, as shown in the next section.

### 10.3.1.2 Experimental Results

In this work, we focused on a total of seven AUs, namely, inner brow raiser (AU1), outer brow raiser (AU2), brow lowerer (AU4), cheek raiser (AU6), lip corner puller (AU12), lips parted (AU25), and jaw drop (AU26). For these AUs, we evaluated the performance of the proposed AU detection pipeline using the MMI Facial Expression dataset [74], one of the most widely used benchmark datasets in the field.

**Experimental Setup.** For each AU, we trained an SVM classifier using the one-vs-all approach, namely, positive samples were the images where the AU was displayed, and the negative samples were all the other images where the AU was not displayed, including neutral samples. We used a linear $c$-SVM [75] and fixed the $c$ parameter to $c = 10^{-3}$.

We used the MMI Facial Expression [74] database, which contains a total of 329 video sequences with annotations provided for the temporal segments of onset, apex, and offset. In order to increase the number of training samples, we selected multiple frames from the apex segment. Subjects often displayed eye movements or small head movements; therefore, the frames extracted from the apex segment were not identical. Similarly, in order to create negative samples, for $\delta$-appearance and $\delta$-shape representations, we randomly picked pairs of frames with neutral expressions. This resulted in a total of 6349 training samples; however, some AUs (e.g. AU1, AU12) have a relatively small number of samples. We handled the data imbalance issue by limiting the number of negative samples. More explicitly, for each AU, we formed 20% of the training samples from the positive samples, 40% from the negative samples with neutral faces, and 40% from the negative samples with nonneutral faces.

**Results.** We evaluated AU detection performance using five-fold subject-independent cross validation. Table 10.1 presents AU detection results with respect to the four individual features, and their combination via the consensus fusion approach in terms of (a) the alternative forced choice (2AFC) metric [76], (b) the TPR, and (c) the FPR. The 2AFC metric can be defined as the area $A$ underneath the receiver-operator characteristic (ROC) curve, and an upper bound for the uncertainty of the $A$ statistic for $n_p$ positive and $n_n$ negative samples, $s = \sqrt{A(1 - A)/\min\{n_p, n_n\}}$.

**Table 10.1** AU detection performance in terms of (a) the alternative forced choice (2AFC) score, (b) the false positive rate (FPR), and (c) the true positive rate (TPR). Bold text indicates the best (i.e. highest) score

| | AU1 | AU2 | AU4 | AU6 | AU12 | AU25 | AU26 |
|---|---|---|---|---|---|---|---|
| **(a) 2AFC** | | | | | | | |
| Shape | 0.74 | 0.53 | 0.67 | 0.61 | 0.79 | 0.73 | 0.53 |
| Appearance | 0.74 | 0.73 | 0.65 | 0.78 | 0.82 | 0.78 | 0.67 |
| $\delta$-shape | 0.78 | 0.67 | 0.71 | 0.74 | 0.78 | 0.82 | 0.64 |
| $\delta$-appearance | 0.90 | **0.92** | **0.87** | 0.82 | 0.92 | **0.89** | 0.78 |
| Fusion | **0.91** | 0.89 | 0.78 | **0.87** | **0.93** | 0.86 | **0.79** |
| **(b) FPR** | | | | | | | |
| Shape | 0.41 | 0.87 | 0.49 | 0.77 | 0.40 | 0.44 | 0.77 |
| Appearance | 0.45 | 0.46 | 0.50 | 0.35 | 0.31 | 0.32 | 0.58 |
| $\delta$-shape | 0.41 | 0.62 | 0.46 | 0.42 | 0.45 | 0.30 | 0.51 |
| $\delta$-appearance | 0.15 | 0.12 | 0.21 | 0.28 | 0.12 | 0.17 | 0.35 |
| Fusion | **0.02** | **0.03** | **0.04** | **0.12** | **0.06** | **0.02** | **0.11** |
| **(c) TPR** | | | | | | | |
| Shape | 0.89 | 0.93 | 0.82 | 1.00 | 0.98 | 0.90 | 0.83 |
| Appearance | 0.92 | 0.92 | 0.80 | 0.90 | 0.94 | 0.88 | 0.93 |
| $\delta$-shape | 0.98 | 0.96 | 0.87 | 0.90 | 1.00 | 0.93 | 0.79 |
| $\delta$-appearance | 0.96 | 0.96 | 0.95 | 0.91 | 0.96 | 0.95 | 0.91 |
| Fusion | 0.84 | 0.81 | 0.61 | 0.86 | 0.92 | 0.73 | 0.68 |

Looking at the AFC scores (Table 10.1(a)), the best performing individual feature is the $\delta$-appearance feature, and the consensus fusion achieves a higher AFC score than the $\delta$-appearance feature for four AUs (AU1, AU6, AU12, AU26) out of seven AUs. The main advantage of the consensus fusion is the low FPR, as given in Table 10.1(b) (the corresponding TPRs are provided in Table 10.1(c)). We also used the best performing trained models in the real-time demonstration.

**Real-Time Demonstration.** We performed the real-time implementation using C++. For the initial face detection in each session, we used the Viola–Jones face detector [77] and then tracked the face using the SDM method [71]. We redetected the face when tracking failed. The real-time implementation was integrated onto the Nao robot as shown in Fig. 10.1. The computational power of the Nao robot did not allow us to run the AU detection algorithm in real-time. For this reason, we used a pair of external cameras plugged into a laptop (Intel Core i6, 16 GB RAM), and ran the AU detection algorithm on the laptop. As shown in Fig. 10.1, these cameras were attached to Nao's head using custom 3D printed glasses. AU

**Figure 10.1** The robotic platform used during real-time public demonstrations.

detection from the robot's point of view is shown in Fig. 10.2. Vertical and horizontal bars indicate the head pose, and the color green is associated with frontal or nearly frontal head poses that yield more reliable AU detection. The detected AUs are highlighted in blue on the left-hand side of each frame; for example, AU1 and AU2 are detected in Fig. 10.2A.

We demonstrated the real-time AU detection method through face-to-face interactions with the Nao robot in a series of public engagement events. For this purpose, we designed an interactive game where Nao asked participants to help him improve his emotional intelligence by displaying facial expressions of emotion, such as happiness, sadness, etc. The participant could choose to display any AU such as pulling lip corners up (smile), pulling eyebrows up (surprise), dropping the mouth/chin (surprise), lowering the eyebrows (frown), etc. To collect the neutral face that was needed for the $\delta$-appearance and $\delta$-shape representations, we asked the participant at the beginning of the session to stand still and look at the camera. Since the neutral face was collected only for the frontal face, we did not take into account AUs detected in the non–frontal faces.

As illustrated in Fig. 10.2, Nao attempted to recognize each AU displayed by the participant, and inferred the expressed emotion based on the rule based approach, and then asked the participant for feedback in the

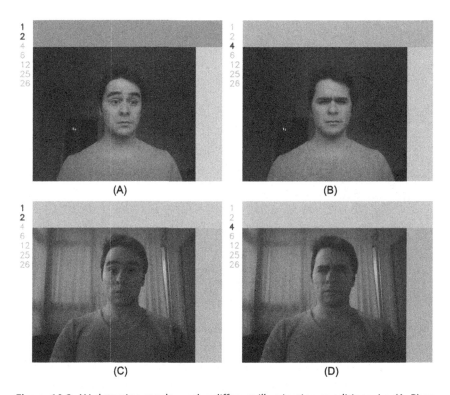

**Figure 10.2** AU detection results under different illumination conditions, i.e. (A–B) vs (C–D). Vertical and horizontal bars indicate the head rotation; the color green is associated with frontal/nearly frontal head poses. The detected AUs in each face image are highlighted in blue: (A, C) AU1 and AU2; (B, D) AU4. (For interpretation of the colors in this figure, the reader is referred to the web version of this chapter.)

form of whether the recognized emotion was correct or not. However, an online learning algorithm was not considered, similarly to [31]. Sample images from the Cambridge Science Festival that took place in Cambridge, United Kingdom, on March 13, 2017,[1] are given in Fig. 10.3. The images illustrate the moment that one of the participants from the public displayed different facial expressions of emotions.

Here, we presented a real-life demonstration of the proposed affect analysis approach in a science communication scenario. However, this approach can be utilized in a health scenario, where, similarly to [32], the robot would provide assistance to children with Autism Spectrum Disorders for improving their facial emotion expression/recognition capability.

[1] https://www.sciencefestival.cam.ac.uk/events/teach-me-emotional-intelligence.

**Figure 10.3** Photos from the public demonstration at the Cambridge Science Festival (Image copyright: University of Cambridge).

## 10.3.2 Automatic Personality Prediction

Several studies have shown that the success of social robots highly depends on assessing and responding to the user's personality (see Section 10.2.2). In this section, we describe how to build an automatic predictor of user personality during human–robot interactions as originally presented in [57]. We also investigate the impact of the participant's personality and the robot's personality on the human–robot interaction.

### 10.3.2.1 Personality Analysis Methodology

**Data Collection and Annotation.** To model the user's personality, we designed an experimental study involving interactions between two human participants and a robot, and collected audio-visual data using a set of first-person vision cameras (also called egocentric cameras) and annotation data by asking participants to complete BFI personality questionnaires [45].

We recruited participants from graduate students and researchers to take part in our experiment. The flow of interaction between the two participants and the robot was structured as follows. The robot was initially seated and situated on the table. The interaction session was initiated by the robot standing up on the table and greeting the participants. The robot initiated the conversation by asking neutrally, "You, on my right, could you please stand up? Thank you! What is your name?" Then the robot continued by asking each of the participants about their occupations, feelings, and so on, by specifying their names at each turn.

We used the Nao robotic platform with the technical details of NaoQi version 2.1, head version 4.0, and body version 25. The robot was controlled remotely in a Wizard of Oz setup during the interaction. To manage the turn taking, an experimenter (i.e. operator), who was seated out of sight behind a sheet of poster board, operated the robot using a computer, the robot's camera, and the other cameras placed in the experimental room. To

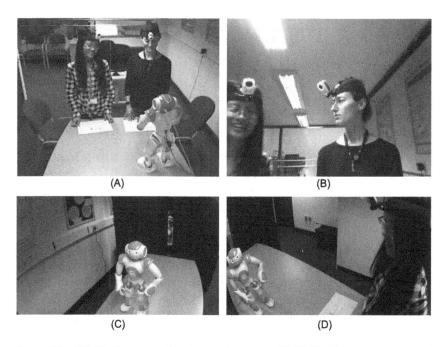

**Figure 10.4** (A) The human–robot interaction setup. (B–D) Simultaneously captured snapshots from the first-person videos: the robot's camera (B) and the ego-centric cameras placed on the foreheads of the participants (C–D).

examine the importance of the robot's personality in the HRI, the robot was made to exhibit either extroverted or introverted personality. Following the previous literature [49,50], we manipulated the robot's behaviors to generate the two types of personality. The extroverted robot displayed hand gestures and talked faster and louder, the introverted robot was hesitant, less energetic, and exhibited no hand gestures in the course of the interaction.

A total of 18 participants (9 female and 9 male) took part in our experiment. Each interaction session lasted from 10 to 15 minutes and was recorded from different camera views. First-person videos were recorded using two Liquid Image ego-centric cameras[2] placed on the forehead of each participant and the robot's camera. The whole scene was also captured using two static Microsoft Kinect depth sensors (version 1)[3] as shown in Fig. 10.4A, resulting in RGB-D recordings. Sound was recorded via the microphones built into the ego-centric cameras.

[2] www.liquidimageco.com/products/model-727-the-ego-1.
[3] en.wikipedia.org/wiki/Kinect.

We recorded 12 interaction sessions and collected approximately 6 hours of multimodal recordings. Each session involved two participants, resulting in 24 individual recordings (some participants took part more than once provided that they had a different partner and were exposed to different robot personalities). The ego-centric recordings and the robot's camera were unsynchronized with the Kinect cameras. For this reason, the experimenter switched the light off and on before each session started. This co-occurred appearance change in the cameras was used to synchronize the multiple videos (i.e. from the two ego-centric cameras, the two Kinect depth sensors, and the robot's camera) in time. Basically, we calculated the amount of appearance change between two successive frames based on gray-level histograms. For further analysis, we segmented each recording into short clips using one question and answer duration. Each clip comprises the robot asking a question to one of the participants and the target participant responding accordingly. This yielded 456 clips where each clip has a duration ranging from 20 to 120 seconds.

In this chapter, we only took into account the recordings from the ego-centric cameras. First-person vision has been shown to be advantageous in analyzing social interactions [78] as it provides the most relevant part of the data. For instance, the people who the camera wearer interacts with tend to be centered in the scene, and are less likely to be occluded when captured from a co-located, first-person perspective rather than from a static, third-person perspective. Fig. 10.4 illustrates simultaneous snapshots from the ego-centric clips.

The participants were asked to complete two different questionnaires, one before the interaction session (pre-study questionnaire) and the other after the interaction session (post-study questionnaire). All measures were on a 10-point Likert scale (from very low to very high). For the pre-study questionnaire, we used the BFI-10 [79] to measure the Big Five personality traits, which is the short version of the Big Five Inventory, and has been used in other studies, e.g. [44]. Each item contributes to the score of a particular trait. The post-study questionnaire consisted of five items (see Table 10.2) that evaluate the participants' engagement with the robot and measure their impressions about the robot's behaviors and abilities.

**Feature Extraction.** We used simple and computationally efficient low-level features to describe motion and changes from the first-person perspective [80]. As mentioned in Section 10.2.2, nonverbal cues conveyed through gaze direction, attention, and head movement carry important information regarding the individual's personality and internal states. These

**Table 10.2** Post-study questionnaire to evaluate the interaction experience with the robot

| Question | Interaction measure |
|---|---|
| I enjoyed the interaction with the robot. | Engagement |
| I thought the robot was being supportive. | Empathy |
| I thought the robot was assertive and social. | Extroversion |
| I thought the robot was being positive. | Positivity |
| I found the robot's behavior realistic. | Realism |

behaviors might lead to significant motion in the first-person videos, which can be characterized by optical flow and motion blur. Attention shifts and rapid scene changes may also cause drastic illumination changes.

Blur values were computed based on the no-reference blur estimation algorithm of [81]. Given a frame, this algorithm yielded two values, vertical (BLUR-Ver) and horizontal blur (BLUR-Hor), ranging from 0 to 1 (the best and worst quality, respectively). We also calculated the maximum blur (BLUR-Max) over the vertical and the horizontal values. For illumination, we simply calculated the mean (ILLU-Mean) and the median (ILLU-Med) of the pixel intensity values per frame.

For optical flow, we used the SIFT flow algorithm proposed in [82]. We computed a dense optical flow estimate for each frame, where we set the grid size to 4. We converted the $x$ and $y$ flow estimate of a pixel into magnitude and angle, and then quantized the angles into eight orientation bins. We calculated the mean (MAG-Mean) and the median (MAG-Med) of the magnitude values per frame. For the angle values, two types of features were computed over a frame: (i) the number of times the angle bin $i$ contained the most motion energy in a frame (ANG-Nrg-$i$) and (ii) the total number of pixels belonging to the angle bin $i$ (ANG-Count-$i$). These features were normalized such that the sum over all eight bins was 1.

Since the frame rate of the ego-centric cameras was high (60 frames per second), all features were extracted from frames sampled every 200 milliseconds instead of at adjacent time instants. A clip was summarized by computing a total of 40 features over the frames. Each feature was computed by performing a series of operations over the blur, illumination, and optical flow features. These operations calculated the mean (Mean), median (Med), and standard deviation (Std) over all frames in a video, calculating the absolute mean (Abs-Mean) over all frames, applying z-score normalization (z) across all frames and taking the first (d1) and the second (d2) temporal derivatives.

**Table 10.3** Significant correlations between the Big Five personality traits of the participants and their interaction experience measures (at a significance level of $p < 0.05$, $^*p < 0.01$). EXT: extroversion, AGR: agreeableness, CON: conscientiousness, NEU: neuroticism, OPE: openness

| Trait | Extroverted robot condition | Introverted robot condition |
|---|---|---|
| EXT | Engagement (0.85*) | – |
|  | Empathy (0.58) |  |
| AGR | Engagement (0.62) | – |
| CON | Positivity (0.71*) | Positivity (0.71) |
| NEU | Realism (0.60) | – |
| OPE | – | Positivity (0.70) |
|  |  | Realism (0.67) |

## 10.3.2.2 Experimental Results

This section presents the correlation analysis between the Big Five personality traits and the interaction experience, and also examines how personality is linked to the automatically extracted first-person vision features. We tested the statistical significance of the correlations (against the null hypothesis of no correlation) using a t-distribution test.

**Relationship Between Personality and Interaction Experience.** We investigated the possible links between the Big Five personality traits of the participants, the *extroversion/introversion* trait of the robot, and the participants' interaction experience with the robot. In Table 10.3, the significant results are given with their respective correlation values in parentheses.

For the extroverted robot condition, the perceived *engagement* with the robot is found to be significantly correlated with participants' *extroversion* trait, which validates the similarity rule [60,42]. We observe that the robot's perceived *empathy* positively correlates with the participants' *extroversion* trait. This might be due to the fact that extroverted people feel more control over their interactions and judge them as more intimate and less incompatible [83,84]. A study of *agreeableness* reported that more agreeable people showed strong self-reported *rapport* when they interacted with a virtual agent [85]. Cuperman and Ickes [41] also indicated that more agreeable people reported having more enjoyable interactions. Similarly, we observe that perceived *engagement* with the robot is highly correlated with the *agreeableness* trait of the participants. A significant relationship is also established between the robot's perceived *realism* and the *neuroticism* trait of the participants. People who score high on *neuroticism* tend to perceive their

**Table 10.4** Selected statistically significant correlations between the participants' personality traits and first-person vision features (at a significance level of $p < 0.01$). BLUR: blur, ILLU: illumination, MAG: optical flow magnitude, ANG: optical flow angle, EXT: extroversion, AGR: agreeableness, CON: conscientiousness, NEU: neuroticism, OPE: openness

| Trait | Extroverted robot condition | Introverted robot condition |
|---|---|---|
| EXT | – | BLUR-Ver-Mean(−0.55); |
|     |   | BLUR-Ratio-Med(−0.49) |
| AGR | BLUR-Ver-Mean(0.36) | BLUR-Max-Med(0.35) |
| CON | BLUR-Ver-Mean(0.34); | BLUR-Ver-Med(−0.53); |
|     | ILLU-Mean-Std(−0.33) | BLUR-Ratio-Med(−0.48); |
|     |   | ANG-Nrg-1(0.35) |
| NEU | BLUR-Ver-Mean(−0.40); | BLUR-Ver-Mean(0.68); |
|     | BLUR-Ratio-Med(−0.36); | BLUR-Max-Std(0.40); |
|     | ILLU-Med-Std(−0.38); | BLUR-Ratio-Med(0.61); |
|     | ANG-Nrg-1(0.41); | MAG-Mean-Mean(0.35); |
|     | ANG-Count-2(−0.42) | MAG-Mean-d1-Abs-Mean(0.38) |
| OPE | BLUR-Max-Mean(0.34); | BLUR-Hor-Mean(0.47); |
|     | ILLU-Med-Std(0.39); | BLUR-Ver-Mean(−0.43); |
|     | ANG-Count-3(0.33) | MAG-Mean-Mean(-0.35); |
|     |   | ANG-Count-1(0.35) |

interactions as being forced and strained [41] and therefore the artificial behaviors of the robot might appear to them as realistic.

For the introverted robot condition, no significant correlations are obtained with participants' *extroversion*, *agreeableness*, and *neuroticism* traits. People who score high on *conscientiousness* tend to interact with others by showing greater attentiveness and responsiveness [41]. This might cause significant correlations with the interaction measure of *positivity* regardless of the robot's personality as the robot always provided feedback to the participant in the course of interaction.

**Relationship Between Personality and First-Person Vision Features.** The goal of this analysis was to study the one-to-one relationships between the Big Five personality traits of the participants and the automatically extracted first-person features. Table 10.4 shows the prominent features and the significant correlations.

In general the introverted robot condition provides a larger number of significant correlations with the extracted features. This can be due to the participants' attention being shifted more when interacting with the introverted robot. For the extroverted robot condition, the *neuroticism* trait of the participants shows significant relationships with all three feature types (blur, illumination, and optical flow), in particular with blur features. No significant correlations are found between participants' *extroversion* trait and the first-person features. For the introverted robot condition, the personality traits of *conscientiousness*, *neuroticism*, and *openness* of the participants show significant relationships with the blur and optical flow features. However, no correlations are found with the illumination features.

In Table 10.4, one significant relationship is seen between *agreeableness* and the vertical blur feature, which can be associated with head nodding and being positive and supportive. We observe that extroverted people tend to enjoy the interaction with the extroverted robot more than the interaction with the introverted robot. Our experimental results further show that *extroversion* is negatively correlated with the blur (motion) features for the introverted robot. This result indicates that less energetic (introverted) people like the introverted robot more, and it is possible to deduce this from the first-person vision features extracted.

For automatic personality prediction, we employed the linear Support Vector Regression method with nested leave-one-subject-out cross-validation. Optical flow-angle features (ANG-Nrg and ANG-Count) yielded the best prediction results in terms of coefficient of determination $(R^2)$ and root-mean-square error $(RMSE)$, where we obtained $\mu_{R^2} = 0.19$ and $\mu_{RMSE} = 1.63$ over all traits. The method successfully modeled the relationship between the first-person vision features and the traits of *agreeableness* $(R^2 = 0.48, RMSE = 1.37)$, *conscientiousness* $(R^2 = 0.27, RMSE = 1.55)$, and *extroversion* $(R^2 = 0.20, RMSE = 1.72)$. Similarly, the study in [53] applied Ridge regression to predict the *extroversion* trait. Although the database, Likert scale, and visual feature set used were completely different, they also obtained the best results with motion-based features $(R^2 = 0.31)$. Referring to this result as a baseline, our results for *agreeableness*, *conscientiousness*, and *extroversion* show that prediction of personality traits from first-person vision in the scope of HRI is a promising research direction.

## 10.4 CONCLUSION AND DISCUSSION

Robotics as a field is continuously evolving to address the ever-changing needs of humans in society. Today the potential of affective and social robotics is enormous, including but not limited to promoting the health and well-being of the elderly living at home [86], improving the quality of life of individuals via physical recovery and rehabilitation [87], assisting the caregivers of children with cognitive and social disabilities [5], assisting children with special medical needs such as diabetes [88], providing personalized education for children [89], and facilitating engagement in group interactions for improving team performance [6]. To deploy social robots in such naturalistic human–robot interaction settings, user modeling and personalization through automatic analysis of expressions, emotions, personality, and engagement is key.

In the light of the survey of the recent research trends and techniques used by social robots, we would like to conclude this chapter by highlighting three open problems in the field, together with a number of pathways that can be used to address these problems.

**Cross-Fertilization Between Affective Computing and Social Robotics Fields.** In recent years significant progress has been achieved in automatic analysis of affective and social signals, particularly of emotions and affective states; even so, computational social robotics has not yet incorporated these latest developments. There is an apparent lack of cross-fertilization between these fields and the field of social robotics. In the fields of affective computing and social signal processing, the current computational techniques integrate multimodal features from visual, audio, and physiological cues over time and utilize models trained with deep learning. However, to date, there has been virtually no effort to integrate these latest trends into social robots and test their viability in the context of human–robot interaction. This is mainly due to the need of real-time processing and to the lack of computational power available on the current robotic platforms. One possible solution to this issue is attaching external cameras onto the robots and performing the real-time processing on an external computer, as described in Section 10.3.1.2. However, this solution does not hold for mobile robots. Another promising direction is cloud robotics, where the captured data is directly streamed to a server via the network for effective and efficient computing (e.g. [90]). This brings additional challenges into play, including the analysis of affect and social signals using live-streamed data that has low spatial and temporal resolution.

**Analysis Under Realistic and Adverse Conditions.** For emotion recognition, most of the successful methods in computer science have focused on facial cues, and have been characterized by multimodal features, in particular combining facial cues with audio cues and bio-signals such as Electrodermal Activity (EDA). Bio-signals are useful when facial cues cannot be observed reliably. However, in real-life applications, it might not be always possible to attach sensors onto the participants to measure their physiological responses. Reducing the cost, the size, and the invasiveness of the physiological sensors that can work robustly under adverse conditions is expected to resolve many of these challenges. Body postures and hand gestures are important sources of information for the analysis of affective and social signals. Therefore, a promising direction is to use deep learning approaches that combine multiple visual cues, such as facial and bodily cues. However, fusing multiple cues in an effective and efficient manner still remains an open challenge in the field. Learning what to fuse and when as suggested in [26,91] will also help deal with missing data, i.e. the cases where one of the cues is not available or is not reliably detected.

**Datasets and Ground Truth.** Most of the available datasets in social robotics have relied on self-reported assessments, in particular, for assessing personality. However, in real-life applications, self-reported assessments might not be available for evaluating the performance of the automatic analyzers. Online crowd-sourcing platforms (e.g. MTurk) have recently gained popularity, due to their efficiency and practicality for collecting responses from crowds for large sets of data within a short period of time. Such efforts have clearly been proven to be efficient at predicting personality [92]. However, exploring novel ways to incorporate annotation disagreements into the analyzers, similarly to [93], is an avenue that needs to be explored further.

In summary, the review provided in this chapter illustrates that the capabilities of current social robots are quite limited. There is a clear need for incorporating the automatic affect analysis and social processing methods into real-life human–robot interaction applications and for improving these techniques to address the challenges of varying environmental lighting, user distance to camera, camera view, and real-time computational requirements. The availability of commercial robotic platforms such as iCat [35], iCub [59], and Nao [40], and developments in collaborative academic research such as the Frontiers Research Topic on *Affective and Social Signals for HRI*[4] provide us with a positive outlook. However, to truly address the existing challenges, researchers from the relevant fields, including but not limited to psychology, nonverbal behavior, vision, social signal processing, affective computing, and HRI, need to constantly interact with one another.

---

[4] http://journal.frontiersin.org/researchtopic/5162/affective-and-social-signals-for-hri.

## ACKNOWLEDGMENTS

This work was funded by the EPSRC under its IDEAS Factory Sandpits call on Digital Personhood (Grant Ref.: EP/L00416X/1).

## REFERENCES

[1] Robot receptionists introduced at hospitals in Belgium, https://www.theguardian.com/technology/2016/jun/14/robot-receptionists-hospitals-belgium-pepper-humanoid (Accessed May 2017).

[2] Japanese bank introduces robot workers to deal with customers in branches, https://www.theguardian.com/world/2015/feb/04/japanese-bank-introduces-robot-workers-to-deal-with-customers-in-branches (Accessed May 2017).

[3] SoftBank's Robot 'Pepper' Flogs Beer and Burgers at Airport, https://www.bloomberg.com/news/articles/2017-02-10/softbank-s-robot-pepper-flogs-beer-and-burgers-at-airport-iyz2t9hb (Accessed May 2017).

[4] K. Wada, T. Shibata, T. Saito, K. Tanie, Psychological and social effects of robot assisted activity to elderly people who stay at a health service facility for the aged, in: IEEE International Conference on Robotics and Automation (Cat. No. 03CH37422), vol. 3, 2003, pp. 3996–4001.

[5] M. Blow, K. Dautenhahn, A. Appleby, C.L. Nehaniv, D. Lee, Perception of robot smiles and dimensions for human–robot interaction design, in: The 15th IEEE International Symposium on Robot and Human Interactive Communication, RO-MAN 2006, Hatfield, Herthfordshire, UK, September 6–8, 2006, 2006, pp. 469–474.

[6] E. Short, K. Sittig-Boyd, M.J. Mataric, Modeling moderation for multi-party socially assistive robotics, in: IEEE International Symposium on Robot and Human Interactive Communication, RO-MAN 2016, IEEE, New York, NY, USA, 2016.

[7] S. Atmatzidou, S. Demetriadis, Advancing students' computational thinking skills through educational robotics: a study on age and gender relevant differences, Robotics and Autonomous Systems 75 (Part B) (2016) 661–670, https://doi.org/10.1016/j.robot.2015.10.008, http://www.sciencedirect.com/science/article/pii/S0921889015002420.

[8] S. Spaulding, G. Gordon, C. Breazeal, Affect-aware student models for robot tutors, in: Proceedings of the 2016 International Conference on Autonomous Agents and Multiagent Systems, AAMAS '16, International Foundation for Autonomous Agents and Multiagent Systems, Richland, SC, USA, ISBN 978-1-4503-4239-1, 2016, pp. 864–872, http://dl.acm.org/citation.cfm?id=2937029.2937050.

[9] D. Leyzberg, S. Spaulding, B. Scassellati, Personalizing robot tutors to individuals' learning differences, in: Proceedings of the 2014 ACM/IEEE International Conference on Human–Robot Interaction, HRI '14, ACM, New York, NY, USA, ISBN 978-1-4503-2658-2, 2014, pp. 423–430, http://doi.acm.org/10.1145/2559636.2559671.

[10] A. Ramachandran, C.M. Huang, B. Scassellati, Give me a break!: personalized timing strategies to promote learning in robot–child tutoring, in: Proceedings of the 2017 ACM/IEEE International Conference on Human–Robot Interaction, HRI '17, ACM, New York, NY, USA, ISBN 978-1-4503-4336-7, 2017, pp. 146–155, http://doi.acm.org/10.1145/2909824.3020209.

[11] T. Schodde, K. Bergmann, S. Kopp, Adaptive robot language tutoring based on Bayesian knowledge tracing and predictive decision-making, in: Proceedings of the 2017 ACM/IEEE International Conference on Human–Robot Interaction, HRI '17, ACM, New York, NY, USA, ISBN 978-1-4503-4336-7, 2017, pp. 128–136, http://doi.acm.org/10.1145/2909824.3020222.

[12] H. Yan, M.H. Ang, A.N. Poo, A survey on perception methods for human–robot interaction in social robots, International Journal of Social Robotics 6 (1) (2014) 85–119.

[13] J.K. Burgoon, L.K. Guerrero, K. Floyd, Nonverbal Communication, Allyn and Bacon, Boston, MA, USA, 2009.

[14] M. Pantic, Automatic analysis of facial expressions, in: Encyclopedia of Biometrics, 2015, pp. 128–134.

[15] E. Sariyanidi, H. Gunes, A. Cavallaro, Automatic analysis of facial affect: a survey of registration, representation, and recognition, IEEE Transactions on Pattern Analysis and Machine Intelligence 37 (6) (2015) 1113–1133.

[16] H. Gunes, C. Shan, S. Chen, Y. Tian, Bodily Expression for Automatic Affect Recognition, John Wiley & Sons, Inc., 2015, pp. 343–377.

[17] P. Ekman, W. Friesen, J. Hager, The Facial Action Coding System, 2nd ed., Weidenfeld and Nicolson, London, UK, 2002.

[18] P. Ekman, J. Campos, R. Davidson, F.D. Waals, Emotions inside out, Annals of the New York Academy of Sciences (2003) 1000.

[19] H. Gunes, B. Schuller, Categorical and dimensional affect analysis in continuous input: current trends and future directions, Image and Vision Computing 31 (2) (2013) 120–136, https://doi.org/10.1016/j.imavis.2012.06.016, http://www.sciencedirect.com/science/article/pii/S0262885612001084.

[20] M.F. Valstar, M. Mehu, B. Jiang, M. Pantic, K. Scherer, Meta-analysis of the first facial expression recognition challenge, IEEE Transactions on Systems, Man and Cybernetics. Part B. Cybernetics 42 (4) (2012) 966–979, https://doi.org/10.1109/TSMCB.2012.2200675.

[21] M.F. Valstar, T. Almaev, J.M. Girard, G. McKeown, M. Mehu, L. Yin, et al., FERA 2015 – second facial expression recognition and analysis challenge, in: 11th IEEE International Conference and Workshops on Automatic Face and Gesture Recognition (FG), vol. 6, 2015, pp. 1–8.

[22] B. Schuller, M. Valstar, R. Cowie, M. Pantic, Avec 2012: the continuous audio/visual emotion challenge – an introduction, in: Proceedings of the 14th ACM International Conference on Multimodal Interaction, ICMI '12, ACM, New York, NY, USA, ISBN 978-1-4503-1467-1, 2012, pp. 361–362, http://doi.acm.org/10.1145/2388676.2388758.

[23] F. De la Torre, W.S. Chu, X. Xiong, F. Vicente, X. Ding, J.F. Cohn, Intraface, in: IEEE International Conference on Automatic Face and Gesture Recognition (FG), 2015.

[24] S. Jaiswal, M. Valstar, Deep learning the dynamic appearance and shape of facial action units, in: IEEE Winter Conference on Application of Computer Vision, 2016.

[25] L. He, D. Jiang, L. Yang, E. Pei, P. Wu, H. Sahli, Multimodal affective dimension prediction using deep bidirectional long short-term memory recurrent neural networks, in: Proceedings of the 5th International Workshop on Audio/Visual Emotion Challenge, AVEC '15, ACM, New York, NY, USA, ISBN 978-1-4503-3743-4, 2015, pp. 73–80, http://doi.acm.org/10.1145/2808196.2811641.

[26] S. Chen, Q. Jin, Multi-modal conditional attention fusion for dimensional emotion prediction, in: Proceedings of the 2016 ACM on Multimedia Conference, MM

'16, ACM, New York, NY, USA, ISBN 978-1-4503-3603-1, 2016, pp. 571–575, http://doi.acm.org/10.1145/2964284.2967286.

[27] F. Ringeval, B. Schuller, M. Valstar, S. Jaiswal, E. Marchi, D. Lalanne, et al., AV+EC 2015: the first affect recognition challenge bridging across audio, video, and physiological data, in: Proceedings of the 5th International Workshop on Audio/Visual Emotion Challenge, AVEC '15, ACM, New York, NY, USA, ISBN 978-1-4503-3743-4, 2015, pp. 3–8, http://doi.acm.org/10.1145/2808196.2811642, 2015.

[28] D. McColl, A. Hong, N. Hatakeyama, G. Nejat, B. Benhabib, A survey of autonomous human affect detection methods for social robots engaged in natural HRI, Journal of Intelligent & Robotic Systems 82 (1) (2016) 101–133.

[29] F. Cid, J.A. Prado, P. Bustos, P. Nunez, A real time and robust facial expression recognition and imitation approach for affective human–robot interaction using Gabor filtering, in: IEEE/RSJ International Conference on Intelligent Robots and Systems, 2013, pp. 2188–2193.

[30] F. Cid, J. Moreno, P. Bustos, P. Núñez, Muecas: a multi-sensor robotic head for affective human robot interaction and imitation, Sensors 14 (5) (2014) 7711–7737.

[31] S. Boucenna, P. Gaussier, P. Andry, L. Hafemeister, A robot learns the facial expressions recognition and face/non-face discrimination through an imitation game, International Journal of Social Robotics 6 (4) (2014) 633–652.

[32] M. Leo, M.D. Coco, P. Carcagnì, C. Distante, M. Bernava, G. Pioggia, G. Palestra, Automatic emotion recognition in robot–children interaction for ASD treatment, in: IEEE International Conference on Computer Vision Workshop (ICCVW), 2015, pp. 537–545.

[33] Robokind Robots Advanced Social Robotics, http://robokind.com/ (Accessed May 2017).

[34] G. Castellano, I. Leite, A. Pereira, C. Martinho, A. Paiva, P. McOwan, Multimodal affect modelling and recognition for empathic robot companions, International Journal of Humanoid Robotics 10 (1) (2013).

[35] A. van Breemen, X. Yan, B. Meerbeek, iCat: an animated user-interface robot with personality, in: Proceedings of the Fourth International Joint Conference on Autonomous Agents and Multiagent Systems, ACM, 2005, pp. 143–144.

[36] D. Schacter, C. Wang, G. Nejat, B. Benhabib, A two-dimensional facial-affect estimation system for human–robot interaction using facial expression parameters, Advanced Robotics 27 (4) (2013) 259–273.

[37] S. Lucey, Y. Wang, M. Cox, S. Sridharan, J.F. Cohn, Efficient constrained local model fitting for non-rigid face alignment, Image and Vision Computing 27 (12) (2009) 1804–1813.

[38] J. Shotton, T. Sharp, A. Kipman, A. Fitzgibbon, M. Finocchio, A. Blake, et al., Real-time human pose recognition in parts from single depth images, Communications of the ACM 56 (1) (2013) 116–124, https://doi.org/10.1145/2398356.2398381, http://doi.acm.org/10.1145/2398356.2398381.

[39] W. Wang, G. Athanasopoulos, G. Patsis, V. Enescu, H. Sahli, Real-Time Emotion Recognition from Natural Bodily Expressions in Child–Robot Interaction, Springer International Publishing, Cham, ISBN 978-3-319-16199-0, 2015, pp. 424–435.

[40] Aldebaran Softbank Group, Who is Nao?, https://www.ald.softbankrobotics.com/en/cool-robots/nao (Accessed 22 May 2017).

[41] R. Cuperman, W. Ickes, Big five predictors of behavior and perceptions in initial dyadic interactions: personality similarity helps extraverts and introverts, but hurts disagreeables, Journal of Personality and Social Psychology 97 (4) (2009) 667–684.

[42] A. Aly, A. Tapus, A model for synthesizing a combined verbal and nonverbal behavior based on personality traits in human–robot interaction, in: Proceedings of the ACM/IEEE International Conference on Human–Robot Interaction, 2013.

[43] P.J. Corr, G. Matthews, The Cambridge Handbook of Personality Psychology, Cambridge University Press, 2009.

[44] A. Vinciarelli, G. Mohammadi, A survey of personality computing, IEEE Transactions on Affective Computing 5 (3) (2014) 273–291.

[45] O.P. John, S. Srivastava, Big five inventory (BFI), in: Handbook of Personality: Theory and Research, vol. 2, 1999, pp. 102–138.

[46] L. Teijeiro-Mosquera, J.I. Biel, J.L. Alba-Castro, D. Gatica-Perez, What your face vlogs about: expressions of emotion and big-five traits impressions in YouTube, IEEE Transactions on Affective Computing 6 (2) (2015) 193–205.

[47] H. Salam, O. Celiktutan, I. Hupont, H. Gunes, M. Chetouani, Fully automatic analysis of engagement and its relationship to personality in human–robot interactions, IEEE Access 5 (2016) 705–721, https://doi.org/10.1109/ACCESS.2016.2614525.

[48] R.J. Larsen, T.K. Shackelford, Gaze avoidance: personality and social judgments of people who avoid direct face-to-face contact, Personality and Individual Differences 21 (6) (1996) 907–917.

[49] R.E. Riggio, H. Friedman, Impression formation: the role of expressive behavior, Journal of Personality and Social Psychology 50 (2) (1986) 421–427.

[50] R. Lippa, The nonverbal display and judgment of extraversion, masculinity, femininity, and gender diagnosticity: a lens model analysis, Journal of Research in Personality 32 (1) (1998) 80–107.

[51] J. Joshi, H. Gunes, R. Goecke, Automatic prediction of perceived traits using visual cues under varied situational context, in: IEEE International Conference on Pattern Recognition (ICPR), 2014, pp. 2855–2860.

[52] M. Schroder, E. Bevacqua, R. Cowie, F. Eyben, H. Gunes, D. Heylen, et al., Building autonomous sensitive artificial listeners, IEEE Transactions on Affective Computing 3 (2) (2012) 165–183, https://doi.org/10.1109/T-AFFC.2011.34.

[53] O. Aran, D. Gatica-Perez, One of a kind: inferring personality impressions in meetings, in: Proceedings of the ACM International Conference on Multimodal Interaction, 2013.

[54] D. Sanchez-Cortes, O. Aran, M.M. Schmid, D. Gatica-Perez, A nonverbal behavior approach to identify emergent leaders in small groups, IEEE Transactions on Multimedia 14 (3) (2012) 816–832.

[55] F. Pianesi, N. Mana, A. Cappelletti, B. Lepri, M. Zancanaro, Multimodal recognition of personality traits in social interactions, in: Proceedings of the 10th International Conference on Multimodal Interfaces, ICMI '08, 2008, pp. 53–60.

[56] O. Celiktutan, H. Gunes, Automatic prediction of impressions in time and across varying context: personality, attractiveness and likeability, IEEE Transactions on Affective Computing 8 (1) (2017) 29–42.

[57] O. Celiktutan, H. Gunes, Computational analysis of human–robot interactions through first-person vision: personality and interaction experience, in: 24th IEEE International Symposium on Robot and Human Interactive Communication (RO-MAN), 2015, pp. 815–820.

[58] F. Rahbar, S.M. Anzalone, G. Varni, E. Zibetti, S. Ivaldi, M. Chetouani, Predicting extraversion from non-verbal features during a face-to-face human–robot interaction, in: Social Robotics, Springer, 2015, pp. 543–553.

[59] L. Natale, F. Nori, G. Metta, M. Fumagalli, S. Ivaldi, U. Pattacini, et al., The iCub platform: a tool for studying intrinsically motivated learning, in: Intrinsically Motivated Learning in Natural and Artificial Systems, Springer, 2013, pp. 433–458.

[60] S. Buisine, J.C. Martin, The influence of user's personality and gender on the processing of virtual agents' multimodal behavior, Advances in Psychology Research 65 (2009) 1–14.

[61] O. Celiktutan, E. Sariyanidi, H. Gunes, Let me tell you about your personality!: real-time personality prediction from nonverbal behavioural cues, in: 11th IEEE International Conference and Workshops on Automatic Face and Gesture Recognition (FG), vol. 1, 2015, IEEE, 2015, p. 1.

[62] C.L. Sidner, M. Dzikovska, Human–robot interaction: engagement between humans and robots for hosting activities, in: Fourth IEEE International Conference on Multimodal Interfaces, 2002, pp. 123–137.

[63] A. Kapoor, R.W. Picard, Y. Ivanov, Probabilistic combination of multiple modalities to detect interest, in: Proceedings of the 17th International Conference on Pattern Recognition, vol. 3, ICPR 2004, 2004, pp. 969–972.

[64] C. Oertel, G. Salvi, A gaze-based method for relating group involvement to individual engagement in multimodal multiparty dialogue, in: Proceedings of the 15th ACM on International Conference on Multimodal Interaction, ICMI '13, ACM, New York, NY, USA, ISBN 978-1-4503-2129-7, 2013, pp. 99–106, http://doi.acm.org/10.1145/2522848.2522865.

[65] C. Peters, S. Asteriadis, K. Karpouzis, Investigating shared attention with a virtual agent using a gaze-based interface, Journal on Multimodal User Interfaces 3 (1) (2010) 119–130.

[66] A. Cerekovic, O. Aran, D. Gatica-Perez, Rapport with virtual agents: what do human social cues and personality explain?, IEEE Transactions on Affective Computing 8 (3) (2017) 382–395, https://doi.org/10.1109/TAFFC.2016.2545650.

[67] J. Sanghvi, G. Castellano, I. Leite, A. Pereira, P.W. McOwan, A. Paiva, Automatic analysis of affective postures and body motion to detect engagement with a game companion, in: Proceedings of the 6th International Conference on Human–Robot Interaction, HRI '11, ACM, New York, NY, USA, ISBN 978-1-4503-0561-7, 2011, pp. 305–312, http://doi.acm.org/10.1145/1957656.1957781.

[68] W. Benkaouar, D. Vaufreydaz, Multi-sensors engagement detection with a robot companion in a home environment, in: Workshop on Assistance and Service Robotics in a Human Environment at IEEE International Conference on Intelligent Robots and Systems (IROS2012), 2012, pp. 45–52.

[69] Kompai Robots, http://kompai.com/ (Accessed May 2017).

[70] H. Salam, M. Chetouani, Engagement detection based on mutli-party cues for human robot interaction, in: International Conference on Affective Computing and Intelligent Interaction (ACII), 2015, IEEE, 2015, pp. 341–347.

[71] X. Xiong, F. De la Torre, Supervised descent method and its applications to face alignment, in: Proceedings of the IEEE Conference on Computer Vision and Pattern Recognition, 2013, pp. 532–539.

[72] E. Sariyanidi, H. Gunes, M. Gökmen, A. Cavallaro, Local Zernike moment representations for facial affect recognition, in: Proceedings of the British Machine Vision Conference, 2013.

[73] E. Sariyanidi, H. Gunes, A. Cavallaro, Biologically-inspired motion encoding for robust global motion estimation, IEEE Transactions on Image Processing 26 (3) (2017) 1521–1535, https://doi.org/10.1109/TIP.2017.2651394.

[74] M. Pantic, M. Valstar, R. Rademaker, L. Maat, Web-based database for facial expression analysis, in: Proceedings of the IEEE International Conference on Multimedia and Expo, 2005, p. 5.

[75] C.C. Chang, C.J. Lin, LIBSVM: a library for support vector machines, ACM Transactions on Intelligent Systems and Technology 2 (2011) 1–27.

[76] G. Littlewort, J. Whitehill, T. Wu, I. Fasel, M. Frank, J. Movellan, M. Bartlett, The computer expression recognition toolbox (CERT), in: Proceedings of the IEEE International Conference on Automatic Face & Gesture Recognition Workshops, 2011, pp. 298–305.

[77] P. Viola, M. Jones, Rapid object detection using a boosted cascade of simple features, in: Proceedings of the IEEE Conference on Computer Vision and Pattern Recognition, vol. 1, 2001, p. I.

[78] A. Fathi, J.K. Hodgins, L.M. Rehg, Social interactions: a first-person perspective, in: IEEE Conference on Computer Vision and Pattern Recognition, 2012.

[79] B. Rammstedt, O.P. John, Measuring personality in one minute or less: a 10-item short version of the Big Five Inventory in English and German, Journal of Research in Personality 41 (1) (2007) 203–212, https://doi.org/10.1016/j.jrp.2006.02.001.

[80] C. Tan, H. Goh, V. Chandrasekhar, L. Liyuan, J. Lim, Understanding the nature of first-person videos: characterization and classification using low-level features, in: Proceedings of the IEEE International Conference on Computer Vision and Pattern Recognition Workshops, 2014.

[81] F. Crete, T. Dolmiere, P. Ladret, M. Nicolas, The blur effect: perception and estimation with a new no-reference perceptual blur metric, Electronic Imaging (2007) 6492.

[82] C. Liu, J. Yuen, A. Torralba, SIFT flow: dense correspondence across scenes and its applications, IEEE Transactions on Pattern Analysis and Machine Intelligence 33 (5) (2011) 978–994.

[83] A.M. von der Putten, N.C. Kramer, J. Gratch, How our personality shapes our interactions with virtual characters – implications for research and development, in: Proceedings of the International Conference on Intelligent Virtual Agents, 2010.

[84] A.W. Heaton, A.W. Kruglanski, Person perception by introverts and extraverts under time pressure: effects of need for closure, Personality & Social Psychology Bulletin 17 (2) (1991) 161–165.

[85] S.H. Kang, J. Gratch, N. Wang, J.H. Watt, Agreeable people like agreeable virtual humans, in: Lecture Notes in Computer Science, 2008, pp. 253–261.

[86] J. Saunders, D.S. Syrdal, K.L. Koay, N. Burke, K. Dautenhahn, Teach me, show me: end-user personalization of a smart home and companion robot, IEEE Transactions on Human-Machine Systems 46 (1) (2016) 27–40, https://doi.org/10.1109/THMS.2015.2445105.

[87] W.G. Louie, S. Mohamed, G. Nejat, Human–Robot Interaction for Rehabilitation Robots: Principles and Practice, Taylor & Francis Group, Boca Raton, FL, USA, 2017, pp. 25–70.

[88] L. Cañamero, M. Lewis, Making new "New AI" friends: designing a social robot for diabetic children from an embodied AI perspective, International Journal of Social Robotics 8 (4) (2016) 523–537, https://doi.org/10.1007/s12369-016-0364-9.

[89] D. Leyzberg, S. Spaulding, B. Scassellati, Personalizing robot tutors to individuals' learning differences, in: Proceedings of the 2014 ACM/IEEE International Conference on Human–Robot Interaction, HRI '14, ACM, New York, NY, USA, ISBN 978-1-4503-2658-2, 2014, pp. 423–430, http://doi.acm.org/10.1145/2559636.2559671.

[90] Y. Yamauchi, Y. Kato, T. Yamashita, H. Fujiyoshil, Cloud robotics based on facial attribute image analysis for human–robot interaction, in: IEEE International Symposium on Robot and Human Interactive Communication, 2016.

[91] F. Li, N. Neverova, C. Wolf, G.W. Taylor, Modout: learning to fuse face and gesture modalities with stochastic regularization, in: International Conference on Automatic Face and Gesture Recognition, 2017.

[92] V. Ponce-López, B. Chen, M. Oliu, C. Corneanu, A. Clapés, I. Guyon, et al., ChaLearn LAP 2016: First Round Challenge on First Impressions – Dataset and Results, Springer International Publishing, Cham, ISBN 978-3-319-49409-8, 2016, pp. 400–418.

[93] V. Sharmanska, D. Hernández-Lobato, J.M. Hernandez-Lobato, N. Quadrianto, Ambiguity helps: classification with disagreements in crowdsourced annotations, in: Proceedings of the IEEE Conference on Computer Vision and Pattern Recognition, 2016, pp. 2194–2202.

# CHAPTER 11

# On Modeling and Analyzing Crowds From Videos

**Nicola Conci\*, Niccoló Bisagno\*, Andrea Cavallaro[†]**
*University of Trento, Trento, Italy
[†]Centre for Intelligent Sensing, Queen Mary University of London, UK

## Contents

## Abstract

The automated analysis of crowds and the identification of crowd behaviors are important for predicting dangerous situations during events, for appropriately designing public spaces, and for the real-time management of people flows. This chapter covers models and algorithms for the analysis of crowds captured in videos for facilitating personal mobility, safety and security, and enabling assistive robotics in public spaces. We discuss the main challenges and solutions for the analysis of collective behaviors in public spaces, which include understanding how people interact and their constantly changing interpersonal relations under clutter and frequent visual occlusions.

## Keywords

Mobility, Behavior analysis, Crowd analysis

**Computer Vision for Assistive Healthcare.**
DOI: https://doi.org/10.1016/B978-0-12-813445-0.00011-3

Figure 11.1 Examples of UCF crowd scenes (A) for segmentation and (B) for people counting.

## 11.1 INTRODUCTION

Crowd analysis supports security and safety (e.g. to prevent dangerous situations), people flow management, and marketing and business intelligence. Understanding and characterizing a crowd (see Fig. 11.1) generally require the extraction of appropriate visual features to highlight regions with motion changes or similarities [1].

Early works on crowd flow analysis focused on segmentation, either globally [2] or at the person level [3]. Other works analyze the optical flow through dense trajectories and model the activities of particles to resemble human presence [4]. Once the parameters of a crowd are learned, anomalies in a crowd can be highlighted [5].

When individuals can be isolated, their motion can be estimated with a feature-based tracker [6] or by using models of social interaction [7,8]. However, because satisfactorily segmenting individuals in a crowd is challenging, most works rely instead on the displacement of image patches that preserve certain appearance properties [9,10]. Moreover, to compensate for occlusions it is possible to fuse data from multiple cameras [11] or to combine visual and nonvisual sensors (e.g. mobile phones [12]).

A crowd can be represented as a collective of agents (e.g. pedestrians), which preserves the fundamental properties of the behavior of an individual, i.e. responding to internal, external, conscious, or unconscious stimuli [13]. When dealing with multiple agents in a crowd the internal and external stimuli may be amplified by the mutual influence that the agents receive and transmit to their neighbors [14]. For example, an action of an agent (e.g. running or shouting) may trigger a reaction in his neighbors, inducing a collective behavior (e.g. a panic reaction). Moreover, the movement of a crowd is influenced not only by the perception and motivations of the individual agents, but also by the environment, including the presence of objects, obstacles, and gates [15].

This chapter provides the reader with an overview on crowd behavior analysis in terms of models and applications. In Section 11.2, we present physics–inspired approaches that model dense crowds as a whole and approaches that model a crowd as a collective of agents.[1] Agent-based models focus on reproducing the collision avoidance behavior of people and include the Distributed Behavioral model [16], the Social Force Model [17], and the Reciprocal Velocity Obstacles [18] model. In Section 11.3, we focus on tools and applications such as motion segmentation, people counting and density estimation [19], and anomaly detection. Inspired by natural and biological processes, methods use coherence to dynamically cluster elements of a crowd that share similar motion properties and indicate the local coordination of individuals. Finally, Section 11.4 concludes the chapter with a summary and a discussion of open issues.

## 11.2 CROWD MODELS

### 11.2.1 The Flow of Human Crowds

Crowd have often been modeled as fluids. The seminal work in this area is the flow of human crowds [20] by Hughes, which describes a dense crowd as a *thinking* fluid. This model is built upon three main hypotheses of the agent's speed, heading direction, and travel time. An agent's *speed* is determined only by (i) its behavioral tendencies, (ii) the density of the population in the neighborhood, and (iii) the terrain. Agents *heading* to the same destination, $G \in \mathbb{R}^2$, are driven by a common tendency or potential. Finally, agents aim to minimize their *travel time*, but also to avoid high densities. To this end a discomfort factor is used as a measure of the agent's willingness to take a longer path in order to avoid a dense area.

If $\rho$ is the density of the crowd, $f(\rho)$ and $d(\rho)$ are respectively the velocity of the pedestrian and the pedestrian's discomfort level, $(x, y, t)$ are the spatio-temporal coordinates, and $T$ is the remaining travel time (potential), then the motion for the flow of an agent is defined by two governing equations,

$$-\frac{\partial \rho}{\partial t} + \frac{\partial}{\partial x}\left(\rho d(\rho) f^2(\rho)\frac{\partial T}{\partial x}\right) + \frac{\partial}{\partial y}\left(\rho d(\rho) f^2(\rho)\frac{\partial T}{\partial y}\right) = 0, \qquad (11.1)$$

---

[1] Agents are autonomous and do not communicate with one other.

and

$$d(\rho)f(\rho) = \frac{1}{\sqrt{(\frac{\partial T}{\partial x}) + (\frac{\partial T}{\partial y})}}, \tag{11.2}$$

where $f(\rho)$ and $d(\rho)$ can simulate different agent behaviors, such as aggressiveness and shyness, and are chosen to allow the above equation to be mappable in space, even if they are time-dependent and nonlinear.

## 11.2.2  Continuum Crowd Model

Inspired by the flow of human crowds, Treuille et al. [21] use a discomfort factor to model *geographical* tendencies of agents and improve the continuum crowd model using hypotheses on the goal area, the speed, and the discomfort factor. The model is a function that aims to minimize the discomfort felt along the path, the traveled distance, and the time to reach the destination.

Each agent tries to reach a geographic goal area, $G \subseteq \mathbb{R}^2$ (which can dynamically change) while moving at the maximum possible speed. The velocity $v_{a_i}(x, y)$ of an agent $a_i$ at location $(x, y)$ is expressed as a maximum speed field $f(.)$, which depends on location $(x, y)$ and moving direction $\theta$:

$$v_{a_i}(x, y) = f(x, y, \theta)n_\theta, \tag{11.3}$$

where $n_\theta$ is a unit vector pointing in the direction $\theta$.

The discomfort factor $d(.)$ models (i) the collision avoidance behavior (e.g. an agent avoids very crowded areas) and (ii) social rules (e.g. a pedestrian crosses a road when reaching a zebra crossing, not before). A discomfort factor $d((x, y)') > d((x, y))$ defines a preference of point $(x, y)$ over $(x, y)'$.

Let $\Pi$ define the set of all paths from $(x, y)$ to the goal. With fixed speed field $f(.)$, discomfort factor $d(.)$, and goal $G$, an agent at point $(x, y)$ will choose the path $P \in \Pi$ minimizing the cost function

$$\alpha \int_P 1\,ds + \beta \int_P 1\,dt + \gamma \int_P d\,dt, \tag{11.4}$$

where the three terms respectively represent the path length, the time to destination, and the discomfort. $ds$ and $dt$, where $ds = f\,dt$, indicate the integration taken with respect to the path length and to the time. The values of $\alpha$, $\beta$, and $\gamma$ take into account the *personality* of each agent.

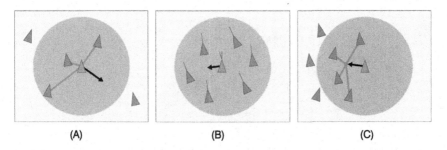

**Figure 11.2** The Boids model. The triangles in the figure represent the agents in the environment and their orientations. The central triangle represents the agent under analysis. Local flock-mates are inside the circle. (A) Separation behavior to avoid collisions. (B) Alignment behavior to steer the direction towards the average of local flock-mates. (C) Cohesion behavior to move towards the average position of other agents in the flock.

## 11.2.3 Distributed Behavioral Model

Instead of setting the path of each agent individually, the distributed behavioral model [16] proposed by Reynolds models the aggregate motion of agents using local rules. This model, known as *Boids*, aims to reproduce the collective motion of a flock of birds using three rules that regulate the interactions between agents: separation, alignment, and cohesion (Fig. 11.2). The *separation* rule provides the collision-avoidance behavior, making the agent steer to avoid crowding local flock-mates. The *alignment* rule influences the agent's velocity and direction such that it matches the average heading of the agent in a fixed neighborhood. Finally, the *cohesion* rule causes the agent to move closer to the average position of other local agents.

The Boids model has generated several lines of research that aim at determining whether the emergent complex behaviors are caused by simple rules and interactions between agents in a crowd, leading to the formulation of new models capable of displaying the desired emergent behaviors such as the Social Force Model, which exploits a Newtonian approach to avoid collisions.

The *Social Force Model* for pedestrian dynamics [17] is frequently used to model the collision avoidance behavior in crowd (Fig. 11.3). The model describes pedestrian motion dynamics taking into account personal goals and environmental constraints. If $f^{att}(a_i)$ is the attractive force that drives the pedestrian towards the destination and $F^{soc}(a_i)$ is the social force that models the interactions of the pedestrian with the environment, then the

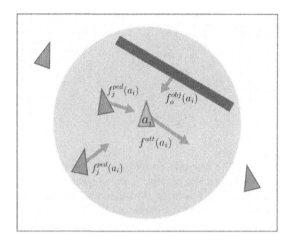

**Figure 11.3** The Social Force Model. The triangles in the figure represent the agents in the environment and their orientations. The central triangle, labeled as $a_i$ represents the agent under analysis. The motion of agent $a_i$ is determined by the attractive force, $f^{att}(a_i)$, toward its destination, repulsive forces with other agents, $f^{ped}(a_i)$, and objects, $f^{obj}(a_i)$, in its neighborhood (indicated with a circle).

net force $F_a$ acting on agent $a_i$ is

$$F_a(a_i) = f^{att}(a_i) + F^{soc}(a_i). \tag{11.5}$$

If $m_i$ is the mass of the agent; $\tau$ is the reaction time; $v_{a_i}$ is the current velocity of the agent; $v_{a_i}^p$ is the preferred velocity of an agent; then the attractive force $f^{att}(a_i)$ is

$$f^{att}(a_i) = m_i \frac{1}{\tau} \left( v_{a_i}^p - v_{a_i} \right). \tag{11.6}$$

Diverse agent behaviors can be included in the framework. For example, the values of the preferred velocity $v_{a_i}^p$ and reaction time $\tau$ can be modified to simulate the *urgency* or the *aggressiveness* of an agent.

The social force $F^{soc}(a_i)$ consists of two factors: a repulsive force $F^{ped}(a_i)$, which reflects the psychological tendency to maintain a social distance between individuals, and an environmental force $F^{obj}(a_i)$, which allows the agent to move without colliding with obstacles such as walls and trees.

Let $A = \{a_1, a_2, \ldots, a_j, \ldots\}$ be the set of agents, $O = \{o_1, o_2, \ldots, o_k, \ldots\}$ the set of obstacles, and $o_k$ a specific obstacle. The social force acting on $a_i$

is

$$F^{soc}(a_i) = F^{ped}(a_i) + F^{obj}(a_i) = \sum_{a_j \in A, j \neq i} f_j^{ped}(a_i) + \sum_{o_k \in O} f_{o_k}^{obj}(a_i), \qquad (11.7)$$

where

$$f_j^{ped}(a_i) = \alpha e^{\frac{-d_{ij}}{\beta}} \hat{n}_{ij}, \qquad (11.8)$$

$$f_{o_k}^{obj}(a_i) = \gamma e^{\frac{-d_{io_k}}{\delta}} \hat{n}_{io_k}, \qquad (11.9)$$

where $\alpha$ is a social scaling constant and $\gamma$ is an obstacle scaling constant; $\beta$ and $\delta$ are space drop-off constants; $d_{ij}$ is the distance between agents $a_i$ and $a_j$; $d_{io_k}$ is the distance between agent $a_i$ and obstacle $o_k$; $\hat{n}_{ij}$ and $\hat{n}_{io_k}$ are the versors in the normal directions between $a_i$ and respectively $a_j$ and $o_k$.

Works such as [14,22] further develop the model by including additional terms in Eq. (11.7) to simulate behaviors such as socially bounded groups of people moving together in the crowd, stationary behavior, queuing behavior, and behavior in case of an emergency.

### 11.2.4 Reciprocal Velocity Obstacles

The Reciprocal Velocity Obstacle (RVO) [18,23] is a framework for collision-free online multiagent navigation. It was first introduced in robotics for collision-free navigation of multiple robots and was then extended to crowd applications [24]. In robotics, a velocity obstacle (VO) is the set of all velocities of a robot that will cause a collision with another robot, assuming that the other robot maintains its current velocity.

The RVO algorithm allows collision-free movement in the environment solving a low-dimensional linear problem. At each step, given the agent's goal, position, and preferred velocity, the algorithm computes a new velocity and direction, which allows smooth movement taking into consideration the position and velocity of each agent in a (fixed) neighborhood.

Let $a_i$ and $a_j$ be two agents moving on a plane, $p_{a_i}$ and $p_{a_j}$ their reference points, $s_{a_i}$ and $s_{a_j}$ generic points of the agent, and $v_{a_i}$ and $v_{a_j}$ their current velocities. The Minkowski sum of two geometric primitives, which represent agents $a_i$ and $a_j$, is

$$a_i \oplus a_j = \{s_{a_i} + s_{a_j} | s_{a_i} \in a_i, s_{a_j} \in a_j\}. \qquad (11.10)$$

Moreover, let $-a_i$ be

$$-a_i = \{-s_{a_i}|s_{a_i} \in a_i\} \tag{11.11}$$

and let $\lambda(p, v)$ define the ray from point $p$ in the direction along $v$:

$$\lambda(p, v) = \{p + tv|t \ge 0\}. \tag{11.12}$$

If the ray originating at $p_{a_i}$ shot in the direction of the relative velocity of $a_i$ and $a_j$ (which is $v_{a_i} - v_{a_j}$) intersects the Minkowski sum of $a_j$ and $-a_i$ centered at $p_{a_j}$, velocity $v_{a_i}$ is in the velocity obstacle of $a_j$. Hence, the VO of $a_j$ to $a_i$ is mathematically defined as

$$VO_{a_j}^{a_i}(v_{a_j}) = \{v_{a_i}|\lambda(p_{a_i}, v_{a_i} - v_{a_j}) \cap a_j \oplus -a_i \ne \emptyset\}. \tag{11.13}$$

If $v_{a_i} \in VO_{a_j}^{a_i}(v_{a_j})$, $a_i$ and $a_j$ will collide at some point. If $v_{a_i} \notin VO_{a_j}^{a_i}(v_{a_j})$, $a_i$ and $a_j$ will not collide and the agent can perform collision-free navigation between moving obstacles. At each time step, the agent chooses a velocity not belonging to any of the velocity obstacles induced by other moving agents. If the agent chooses the velocity closer to the direction toward its preferred goal, it will reach the goal's position without colliding with moving obstacles.

However, if multiple agents use the VO algorithm at the same time considering other agents as moving obstacles, every agent would choose a velocity such that $v_{a_i} \notin VO_{a_j}^{a_i}(v_{a_j})$. This approach would lead to an undesired oscillatory motion. Suppose two agents, $a_i$ and $a_j$, are respectively moving at velocities $v_{a_i} \in VO_{a_j}^{a_i}(v_{a_j})$ and $v_{a_j} \in VO_{a_i}^{a_j}(v_{a_i})$, which means they will collide at some point in time. In the next time step, they autonomously select new velocities such that $v_{a_i}' \notin VO_{a_j}^{a_i}(v_{a_j})$ and $v_{a_j}' \notin VO_{a_i}^{a_j}(v_{a_i})$ in order to avoid colliding with one other. In the next cycle, since $v_{a_i} \notin VO_{a_j}^{a_i}(v_{a_j}')$ and $v_{a_j} \notin VO_{a_i}^{a_j}(v_{a_i}')$, both agents will choose their old velocities $v_{a_i}$ and $v_{a_j}$, which are in the best direction for them to reach their goal. Next time, to avoid collision, they will go back to $v_{a_i}'$ and $v_{a_j}'$ causing the oscillatory behavior.

To avoid these oscillations, the RVO of agent $a_j$ to agent $a_i$ results in a set of velocities that guarantees a *collision-free* and *oscillation-free* trajectory (Fig. 11.4), $RVO_{a_j}^{a_i}(v_{a_j}, v_{a_i})$, and is defined as

$$RVO_{a_j}^{a_i}(v_{a_j}, v_{a_i}) = \{v_{a_i}'|2v_{a_i}' - v_{a_i} \in VO_{a_j}^{a_i}\}, \tag{11.14}$$

which contains all the velocities $v_{a_i}'$ that correspond to the average of the current velocity $v_{a_i}$ and a velocity contained in $VO_{a_j}^{a_i}(v_{a_j})$.

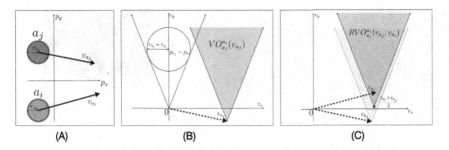

**Figure 11.4** Circular agents $a_i$ and $a_j$ with radius $r_{a_i}$ and $r_{a_j}$ and reference points $p_{a_i}$ and $p_{a_j}$, respectively, are moving at velocities $v_{a_i}$ and $v_{a_j}$, which will cause them to collide (A). The velocity obstacle $VO_{a_j}^{a_i}(v_{a_j})$ induced by agent $a_j$ to agent $a_i$ (B). The reciprocal velocity obstacle $RVO_{a_j}^{a_i}(v_{a_j}, v_{a_i})$ for agent $a_i$ induced by agent $a_j$ (C).

Given the set of velocities $RVO_{a_j}^{a_i}(v_{a_j}, v_{a_i})$, the algorithm selects the one with minimum penalty, $p(.)$, defined as

$$p(v_{a_i}^{cand}) = \omega_{a_i} \frac{1}{T_c(v_{a_i}^{cand})} + \left\| v_{a_i}^{pref} - v_{a_i}^{cand} \right\|, \qquad (11.15)$$

where $v_{a_i}^{cand}$ is the candidate velocity (which deviates from the preferred velocity $v_{a_i}^{pref}$); $T_c(v_{a_i}^{cand})$ is the expected time to collision; and $\omega_{a_i}$ is a parameter that can be defined for every agent, together with the preferred velocity, to simulate different behavior such as aggressiveness and shyness. Finally, given the set of agents $A$, $RVO(a_i, A - a_i)$ selects the optimal collision-free velocity for agent $a_i$ in the next step.[2]

## 11.3 ALGORITHMS AND APPLICATIONS

### 11.3.1 Crowd Motion Segmentation

Motion segmentation is one of the most investigated problems in crowd analysis (see Fig. 11.5). The relevant methods are listed in Table 11.1 and are discussed below.

Ali and Shah [2] segment dense crowds to identify motion patterns using Lagrangian Particle Dynamics and isolate regions with a coherent motion of particles. The method combines dynamical systems and fluid dynamics theories to determine the spatial organization of the flow field through

---

[2] RVO assumes that *all* agents move using the same algorithm.

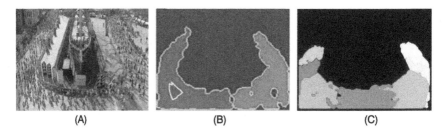

**Figure 11.5** Sample segmentation maps analyzing the test video sequence of a dense crowd (A), obtained from the methods presented in [25] (B), and [26] (C). In these particular examples color labels are used to identify the motion of the flow with similar properties.

**Table 11.1** Comparison of crowd segmentation methods. KEY – Ref.: reference; FTLE: Finite-Time Lyapunov Exponent; LCS: Lagrangian Coherent Structure; CRF: Conditional Random Field; KLT: Kanade–Lucas–Tomasi tracker

| Ref. | Features | Method | Application | Validation |
|------|----------|--------|-------------|------------|
| [2] | Block-based correlation particle advection | FTLE Field and LCS | Detection of areas with coherent motion in high-density crowds | Getty Images Photo-Search Google Video |
| [25] | Optical flow and particle tracking | Streaklines | Segmentation of areas based on motion direction; anomaly detection | Web videos UMN |
| [26] | Optical flow and particle tracking | CRF | Segmentation of areas based on motion direction | UCF |
| [27] | KLT keypoints | Coherent Neighbor Invariance | Detection of coherent motions. Evaluation on people counting | UCF proprietary video |

the Finite-Time Lyapunov Exponent (FTLE) field and the computation of Lagrangian Coherent Structures (LCS). To better cope with spatial and temporal changes, Mehran et al. [25] also use the locations, at a given time, of all the particles that passed through a particular point (the *streaklines*). Motion analysis results can also be used to recognize different crowd behaviors, such as lane, circle, bottleneck, fountainhead, and blocking [7], based on the detection of accumulation points, which are then used to trace back the motion of the crowd. Finally, a Conditional Random Field (CRF) can be used to segment the crowd flow [26]. CRF is a discriminative model

**Figure 11.6** Particles sharing similar motion properties are grouped using Coherent Filtering [27].

used to label sequential data and that specifies the probability of a particular label sequence, given the observation sequence containing orientation features to identify the most probable label in terms of direction.

Unlike the methods presented above that are designed for dense flows only, Zhou et al. [27] present a framework for the detection of coherent motion in the crowd, which characterizes the local spatio-temporal relationships of the elements moving in the scene. Coherent motion patterns are detected based on an online Coherent Neighbor Invariance algorithm, which has been tested in various scenarios with variable crowd density (see Fig. 11.6).

## 11.3.2 Crowd Density Estimation

A crowd can be measured statistically to provide an estimate of its density. Relevant methods for crowd density estimation from video are discussed in this section and summarized in Table 11.2.

Based on the observation that low-density areas are likely to have coarse patterns and high-density areas are instead represented by fine patterns, Marana et al. [28] adopt a self-organizing neural network to classify low- and high-density scenes based on texture intensity. A texture is characterized by features such as contrast, homogeneity, energy, and entropy. The same metrics are also used in [29], jointly with edge and spatial features (area and perimeter). A Gaussian process [30] then regresses feature vectors to estimate the number of people.

An indirect approach for density estimation is presented by Conte et al. [31]. This approach relies on the extraction of SURF features [32] and their motion flow. Points are clustered and passed through an $\epsilon$-SVR regres-

**Table 11.2**  Algorithms for crowd density estimation. KEY – Ref.: reference

| Ref. | Features | Method | Application | Validation |
|------|----------|--------|-------------|------------|
| [28] | Textures | Self-organizing neural network | Density estimation | 300 static images |
| [33] | Foreground; edge map | Neural net with one hidden layer | View-invariant density estimation | n/a |
| [35] | Foreground map; edges; textures | Multivariate ridge regression | Density estimation with sparse data | UCSD Mall |

sor (a variation of the Support Vector Machine) that estimates the number of people in the crowd. This method can cope with parallax and perspective deformations by computing the distance of the moving points from the camera through an Inverse Perspective Mapping (IPM). In fact, moving farther away from the camera implies a reduced visibility of features and amount of motion, thus reducing the accuracy of the estimate. Similarly, a viewpoint-invariant solution is presented by Kong et al. [33] using the histogram features of blobs obtained after background subtraction [34]. To achieve invariance to perspective distortions, features are normalized with respect to relative density and orientation scale.

### 11.3.3  People Counting

While density estimation provides holistic quantitative information about a crowd, other methods have focused on estimating the actual number of people in the crowd. Table 11.3 summarizes these methods, which are then described in this section.

An evaluation of regression-based algorithms for crowd counting is presented in [36]. While regression is effective when dealing with high-dimensional data, the sparsity and imbalance of training data make the problem more challenging. A solution is proposed by Chen et al. [35] with a cumulative attribute concept for learning a regression model for sparse and imbalanced data. Each attribute is discriminative and cumulative, and separates the training images into two groups based on whether the attribute is positive or negative.

The authors in [37] propose a multisource and multiscale approach that combines low-confidence features like head detection together with repetition of texture elements and frequency-domain analysis to estimate counts. The method is tested on high-density crowd images, with several hundreds of annotated heads.

**Table 11.3** Algorithms for people counting in a crowd. KEY – Ref.: reference; SVR: Support Vector Regressor; SURF: Speeded-Up Robust Feature; MRF: Markov Random Field; CNN: Convolutional Neural Network

| Ref. | Features | Method | Application | Validation |
|---|---|---|---|---|
| [29] | Shape and size of moving regions; edge features | Gaussian Process Regression | People counting | UCSD Crowd counting |
| [31] | SURF | $\epsilon$-SVR regressor | People counting | PETS 2009 |
| [37] | Head detection textures | MRF | Multisource high-density people counting | Annotated Flickr data |
| [38] | Image patches from training set | CNN | Cross scene crowd counting | 108 crowd scenes 200k head annotations |
| [39] | Image slices | CNN Regression CNN Classification | People counting at a line of interest | 31 days of videos in 6 locations |
| [40] | Crowd images | Multicolumn CNN | Crowd counting and transfer learning | Shanghaitech |

Zhang et al. [38] propose a method for crowd density estimation and counting based on a deep learning framework that can generalize across scenes (changes in perspective, scale, density) with a nonparametric fine-tuning of the network. Given a target video from unseen scenes, samples with similar properties from the training scenes are retrieved and added to training data to fine-tune the crowd CNN model. Alternatively, Cao et al. [39] rely on a CNN learning framework for crowd counting in very dense scenes using a virtual line of interest to count people who cross it. The data for learning consist of temporal image-slices extracted around this line of interest and the corresponding temporal slices containing the optical flow.

Finally, the authors of [40] propose a method for single-image crowd counting that copes with different camera perspectives and crowd densities. The model adopts a multicolumn convolutional neural network [41] that copes with the different scales of objects through automatic learning of the visual features.

**Table 11.4** Algorithms for group and anomaly detection. KEY – Ref.: reference; ETIN: Evolving Tracklet Interaction Network; PCA: Principal Component Analysis; SVM: Support Vector Machine; HoG: Histogram of Oriented Gradients; RVO: Reciprocal Velocity Obstacles; EnKF: Ensemble Kalman Filter; LDA: Latent Dirichlet Allocation

| Ref. | Features | Method | Application | Validation |
|------|----------|--------|-------------|------------|
| [44] | Optical flow; Motion tracklets | ETIN | Discovery of social groups | CAVIAR; PETS2009; UNIV |
| [42] | HoG; Motion trajectories Group shape | Gaussian Process; PCA; SVM | Group action | NUS-HGA BeHave |
| [43] | Position; velocity | Linear Trajectory Avoidance | Multitarget tracking | ETH |
| [45] | Optical flow; Social Force Model | LDA | Anomaly detection | UMN web data |
| [46] | Motion trajectories | RVO; EnKF | Anomaly detection | PETS 2016 UCSD |
| [47] | Raw Video | Spatio-temporal CNN | Crowd attribute recognition | WWW Crowd |

## 11.3.4 Detecting Groups

A group is a particularly interesting social configuration that represents an intermediate structure between the individuals and the crowd as a whole. The contribution of each group member is usually distinguishable and the interactions among the entities in the group is the discriminating feature to detect the class of behavior [42] or to improve the detection and tracking [43]. For example, Feng and Bhanu [44] use a hierarchical approach to dynamically merge tracklets in order to construct an interaction graph that defines the existence of a group. The above-mentioned methods are summarized in Table 11.4.

## 11.3.5 Anomaly Detection

The characterization of a crowd in terms of motion properties and density can help determine the presence of anomalies or threats (see Table 11.4). The anomaly detection literature has mostly focused on the deviations in the motion properties of the crowd from previous observations [45]. For example, the interaction forces among particles, treated as human instances, are used to define a *force flow* that models the regular movement of the crowd. Latent Dirichlet Allocation [48] is then trained to detect anomalies. Moreover, Bera et al. [46] propose a real-time framework to detect anomalies in low- to medium-density crowd scenes. The global motion properties of the crowd are derived using a tracker and the RVO motion model. The local features of the moving agents are then analyzed to highlight an anomaly when the value of the feature exceeds a threshold. Finally, Shao et al. [47] extend the use of CNNs to the temporal domain to learn temporal attributes of the crowd.

## 11.4 CONCLUDING REMARKS

This chapter presented an overview of crowd analysis models and various applications, including motion segmentation and anomaly detection. An important limitation for further progress in crowd analysis algorithms is the lack of video data of crowds, and their annotation. Simulation frameworks can address this limitation and help validate the algorithms in a variety of scenarios, environmental conditions, crowd density, types of behavior, and events. Furthermore, this virtualization could serve as a common ground for validation, guaranteeing replicability. The development of appropriate simulation tools and the exploitation of novel deep-learning architectures [49] are therefore promising research lines for an effective modeling and analysis of crowds in videos.

## REFERENCES

[1] J.C.S.J. Junior, S.R. Musse, C.R. Jung, Crowd analysis using computer vision techniques, IEEE Signal Processing Magazine 27 (2010) 66–77.
[2] S. Ali, M. Shah, A Lagrangian particle dynamics approach for crowd flow segmentation and stability analysis, in: IEEE Conference on Computer Vision and Pattern Recognition, 2007, pp. 1–6.
[3] P. Tu, T. Sebastian, G. Doretto, N. Krahnstoever, J. Rittscher, T. Yu, Unified crowd segmentation, in: European Conference on Computer Vision, 2008, pp. 691–704.
[4] J. Shao, C.C. Loy, X. Wang, Learning scene-independent group descriptors for crowd understanding, IEEE Transactions on Circuits and Systems for Video Technology (2016).
[5] T. Li, H. Chang, M. Wang, B. Ni, R. Hong, S. Yan, Crowded scene analysis: a survey, IEEE Transactions on Circuits and Systems for Video Technology 25 (2015) 367–386.
[6] S. Saxena, F. Brémond, M. Thonnat, R. Ma, Crowd behavior recognition for video surveillance, in: International Conference on Advanced Concepts for Intelligent Vision Systems, 2008, pp. 970–981.

[7] B. Solmaz, B.E. Moore, M. Shah, Identifying behaviors in crowd scenes using stability analysis for dynamical systems, IEEE Transactions on Pattern Analysis and Machine Intelligence 34 (2012) 2064–2070.

[8] D. Sugimura, K.M. Kitani, T. Okabe, Y. Sato, A. Sugimoto, Using individuality to track individuals: clustering individual trajectories in crowds using local appearance and frequency trait, in: International Conference on Computer Vision, 2009, pp. 1467–1474.

[9] S. Ali, M. Shah, Floor fields for tracking in high density crowd scenes, in: European Conference on Computer Vision, 2008, pp. 1–14.

[10] M. Rodriguez, J. Sivic, I. Laptev, J.Y. Audibert, Data-driven crowd analysis in videos, in: International Conference on Computer Vision, 2011, pp. 1235–1242.

[11] R. Eshel, Y. Moses, Tracking in a dense crowd using multiple cameras, International Journal of Computer Vision 88 (2010) 129–143.

[12] L. Tokarchuk, M. Irfan, L. Marcenaro, Crowd analysis using visual and non-visual sensors, a survey, in: IEEE Global Conference on Signal and Information Processing, 2016.

[13] A.H. Maslow, R. Frager, J. Fadiman, C. McReynolds, R. Cox, Motivation and Personality, Harper & Row, New York, 1954.

[14] M. Moussaïd, N. Perozo, S. Garnier, D. Helbing, G. Theraulaz, The walking behaviour of pedestrian social groups and its impact on crowd dynamics, PLoS ONE 5 (2010) e10047.

[15] P.V.K. Borges, N. Conci, A. Cavallaro, Video-based human behavior understanding: a survey, IEEE Transactions on Circuits and Systems for Video Technology 23 (2013) 1993–2008.

[16] C.W. Reynolds, Flocks, herds, and schools: a distributed behavioral model, ACM SIG-GRAPH Computer Graphics 21 (1987) 25–34.

[17] D. Helbing, P. Molnar, Social force model for pedestrian dynamics, Physical Review E 51 (1995) 4282.

[18] J. Van den Berg, M. Lin, D. Manocha, Reciprocal velocity obstacles for real-time multi-agent navigation, in: IEEE International Conference on Robotics and Automation, 2008, pp. 1928–1935.

[19] S.A.M. Saleh, S.A. Suandi, H. Ibrahim, Recent survey on crowd density estimation and counting for visual surveillance, Engineering Applications of Artificial Intelligence 41 (2015) 103–114.

[20] R.L. Hughes, The flow of human crowds, Annual Review of Fluid Mechanics 35 (2003) 169–182.

[21] A. Treuille, S. Cooper, Z. Popović, Continuum crowds, ACM Transactions on Graphics 25 (2006) 1160–1168.

[22] D. Helbing, I. Farkas, T. Vicsek, Simulating dynamical features of escape panic, preprint, arXiv:cond-mat/0009448, 2000.

[23] J. Van Den Berg, S.J. Guy, M. Lin, D. Manocha, Reciprocal n-body collision avoidance, Robotics Research 70 (2011) 3–19.

[24] J. van den Berg, S. Patil, J. Sewall, D. Manocha, M. Lin, Interactive navigation of multiple agents in crowded environments, in: Symposium on Interactive 3D Graphics and Games, 2008, pp. 139–147.

[25] R. Mehran, B. Moore, M. Shah, A streakline representation of flow in crowded scenes, in: European Conference on Computer Vision, 2010, pp. 439–452.

[26] H. Ullah, N. Conci, Structured learning for crowd motion segmentation, in: IEEE International Conference on Image Processing, 2013, pp. 824–828.

[27] B. Zhou, X. Tang, X. Wang, Coherent filtering: detecting coherent motions from crowd clutters, in: European Conference on Computer Vision, 2012, pp. 857–871.

[28] A.N. Marana, S.A. Velastin, L.F. Costa, R.A. Lotufo, Estimation of crowd density using image processing, in: IEE Colloquium on Image Processing for Security Applications (Digest No. 1997/074), vol. 11, 1997, pp. 1–8.

[29] A.B. Chan, Z.S.J. Liang, N. Vasconcelos, Privacy preserving crowd monitoring: counting people without people models or tracking, in: IEEE Conference on Computer Vision and Pattern Recognition, 2008, pp. 1–7.

[30] C.E. Rasmussen, C.K. Williams, Gaussian Processes for Machine Learning, vol. 1, MIT Press, Cambridge, 2006.

[31] D. Conte, P. Foggia, G. Percannella, F. Tufano, M. Vento, A method for counting people in crowded scenes, in: IEEE International Conference on Advanced Video and Signal Based Surveillance, 2010, pp. 225–232.

[32] H. Bay, A. Ess, T. Tuytelaars, L. Van Gool, Speeded-up robust features (SURF), Computer Vision and Image Understanding 110 (2008) 346–359.

[33] D. Kong, D. Gray, H. Tao, A viewpoint invariant approach for crowd counting, in: International Conference on Pattern Recognition, vol. 3, 2006, pp. 1187–1190.

[34] C. Stauffer, W.E.L. Grimson, Adaptive background mixture models for real-time tracking, in: IEEE Conference on Computer Vision and Pattern Recognition, vol. 2, 1999, pp. 246–252.

[35] K. Chen, S. Gong, T. Xiang, C. Change Loy, Cumulative attribute space for age and crowd density estimation, in: IEEE Conference on Computer Vision and Pattern Recognition, 2013, pp. 2467–2474.

[36] C.C. Loy, K. Chen, S. Gong, T. Xiang, Crowd counting and profiling: methodology and evaluation, in: Modeling, Simulation and Visual Analysis of Crowds, Springer, 2013, pp. 347–382.

[37] H. Idrees, I. Saleemi, C. Seibert, M. Shah, Multi-source multi-scale counting in extremely dense crowd images, in: IEEE Conference on Computer Vision and Pattern Recognition, 2013, pp. 2547–2554.

[38] C. Zhang, H. Li, X. Wang, X. Yang, Cross-scene crowd counting via deep convolutional neural networks, in: EEE Conference on Computer Vision and Pattern Recognition, 2015, pp. 833–841.

[39] L. Cao, X. Zhang, W. Ren, K. Huang, Large scale crowd analysis based on convolutional neural network, Pattern Recognition 48 (2015) 3016–3024.

[40] Y. Zhang, D. Zhou, S. Chen, S. Gao, Y. Ma, Single-image crowd counting via multi-column convolutional neural network, in: Conference on Computer Vision and Pattern Recognition, 2016, pp. 589–597.

[41] D. Ciregan, U. Meier, J. Schmidhuber, Multi-column deep neural networks for image classification, in: IEEE Conference on Computer Vision and Pattern Recognition, 2012, pp. 3642–3649.

[42] Z. Cheng, L. Qin, Q. Huang, S. Yan, Q. Tian, Recognizing human group action by layered model with multiple cues, Neurocomputing 136 (2014) 124–135.

[43] S. Pellegrini, A. Ess, K. Schindler, L. Van Gool, You'll never walk alone: modeling social behavior for multi-target tracking, in: International Conference on Computer Vision, 2009, pp. 261–268.

[44] L. Feng, B. Bhanu, Understanding dynamic social grouping behaviors of pedestrians, IEEE Journal of Selected Topics in Signal Processing 9 (2015) 317–329.

[45] R. Mehran, A. Oyama, M. Shah, Abnormal crowd behavior detection using social force model, in: IEEE Conference on Computer Vision and Pattern Recognition, 2009, pp. 935–942.

[46] A. Bera, S. Kim, D. Manocha, Realtime anomaly detection using trajectory-level crowd behavior learning, in: EEE Conference on Computer Vision and Pattern Recognition Workshops, 2016, pp. 50–57.

[47] J. Shao, C.C. Loy, K. Kang, X. Wang, Slicing convolutional neural network for crowd video understanding, in: IEEE Conference on Computer Vision and Pattern Recognition, 2016, pp. 5620–5628.

[48] D.M. Blei, A.Y. Ng, M.I. Jordan, Latent Dirichlet allocation, Journal of Machine Learning Research 3 (2003) 993–1022.

[49] Y. LeCun, Y. Bengio, G. Hinton, Deep learning, Nature 521 (2015) 436–444.

# CHAPTER 12

# Designing Assistive Tools for the Market

**Manuela Chessa[#], Nicoletta Noceti[#], Chiara Martini, Fabio Solari, Francesca Odone**
DIBRIS, University of Genoa, Italy

## Contents

## Abstract

In this chapter we consider assistive environments and discuss the possible benefits for an aging population. As a study case we summarize the current state of a research project taking place in collaboration with Galliera Hospital (Genoa, Italy). A protected discharge model is being developed to assist elderly users after they have been dismissed from the hospital and before they are ready to go back home, with the perspective of coaching them towards a healthy lifestyle. We focus in particular on the vision-based modules designed to automatically estimate a frailty index of the patient that would allow physicians to assess the patient's health status and state of mind. We contribute to this frailty index by supplying a set of motility features obtained by continuously analyzing the motion patterns of the patient. We show very promising results which have received positive feedback from the geriatricians participating in the project.

---

[#] M. Chessa and N. Noceti contributed equally to the chapter.

**Computer Vision for Assistive Healthcare.**
DOI: https://doi.org/10.1016/B978-0-12-813445-0.00012-5

## Keywords

Assistive environments, Vision-based setups, Motion analysis, Elderly monitoring, Motility index estimation, RGB-D sensors

## 12.1 INTRODUCTION

The design of smart environments, or Ambient Assisted Living (AAL) is a key application for assistive computer vision. AAL may provide an enhanced environment where a user can "feel at home" while receiving a special care or help in basic everyday activities [1]. It is well known that assistive technologies can have positive impacts on different dimensions of health and quality of life for disabled and vulnerable citizens.

One of the motivations for AAL is the significant worldwide increase in the aging population. According to the World Health Organization (WHO), the world's elderly population (defined as people aged 60 and older) has increased drastically in the past decades and will reach about 2 billion in 2050. In Europe, the percentage of the European Union population above 65 years of age is foreseen to rise to 30% by 2060 [2]. Smart home technologies can help to enhance quality of life, prolong independent living, and reduce caregivers' necessary time and health care costs in general without losing the sense of security that a continuous and unobtrusive monitoring provides. Thus, the benefits of these technologies are not only for the elderly, but also for their families, caregivers, and society in general.

Hospitalization is the first cause of functional decline in the elderly: 30% to 60% of elderly patients lose some independence in basic Activities of Daily Living (ADL) in the course of a hospital stay [3]. In addition to the disabling effect of the acute event, hospitalization itself might represent an additional stressor in terms of environmental hazard, reduced caloric intake, low physical activity or prolonged bed rest, depressed mood, and social isolation [4,5]. Prolonged hospital stays may increase the risk of infections and other iatrogenic complications, worsen the patient's quality of life (especially in the elderly), and imply a waste of economic and human resources. Moreover, more than half of patients spend extra time in hospital after recovery from an acute condition waiting for the arrangement of either special assistance at home or the admission to long-term care facilities [6].

Starting from these observations, we carried out a feasibility study for a short-stay post-hospitalization protected residence, built in the vicinity of the hospital and designed as a smart environment with all the comforts of a private home. Here the patient can be monitored by a system of sensors, while physicians and nurses have the opportunity to check on the patient remotely. This model has the potential of a positive economic impact coming from the reduction in personnel costs. It would also have a medical impact since the patients will experience a sense of comfort and independence as if they were at home, while being constantly monitored.

Some of the core functionalities of the system are based on computer vision, in particular the automatic estimation of a motility index, contributing to the assessment of the so-called *frailty index* [7,8], used by physicians to monitor the overall well-being of a patient. Our current evaluation takes into account primarily the estimation of motility indices, for instance the amount of time the patient spends at rest, the length of walking sessions, the number of stand-to-sit transitions, the average walk speed. Our approach is based primarily on the analysis of 3D motion tracks obtained by RGB-D sensors (i.e. sensors acquiring both images -RGB- and depth -D- data streams), focusing in particular on 3D positions of the head over time.

The chapter is organized as follows. Section 12.2 discuss related works on assistive and smart environments, with a focus on vision-based solutions; in Section 12.3 we discuss in detail the post-hospitalization facility, describing the sensors, the acquisition and storage of the acquired data, and the analysis towards the estimation of the motility quantities, useful to devise a global motility index. The experimental analysis is also presented. Section 12.4 presents a final discussion, and highlights future lines of investigation and open issues.

## 12.2 THE STATE OF THE ART

Smart homes are purposely designed living spaces that provide interactive technologies and unobtrusive support systems that enable people to improve their level of independence, activity, participation, or well-being [9]. Often, smart technologies can permit people to live in their own homes rather than in hospitals or institutions [10,11]. The challenge with smart home technologies is to create a home environment that is safe and secure in order to reduce falls, disability, stress, fear, or social isolation [12]; the usual aim is for the technologies to perform their functions without disturbing the user and without causing any pain, inconvenience, or movement restrictions.

Different smart homes may exploit different types of sensors, depending on the specific requirements. An assistive environment may include bed sensors to detect the patient's heart rate, respiration, and restlessness; motion sensors to detect motion within the home or apartment and to provide data regarding the patient's location over time; kitchen safety sensors used in combination with motion sensors to detect activities of daily living and to raise an alarm if necessary; fall detection sensors to passively monitor if a patient falls; cameras and RGB-D sensors to detect activity and to provide motion and activity analysis.

Most smart home technologies passively collect and share elder well-being information with the family members and primary care providers. These devices collect multiple types of data, including physiological, location, or movement data. Algorithms transform the raw data into activity patterns that can be used for early detection and intervention by healthcare providers or families [13].

Closely related to the concept of smart homes, are techniques used to address human activity detection. Most work in this area uses the concept of "activity" as a building block for health monitoring and assisted living. The process of identifying a specific activity encompasses the selection of the appropriate set of sensors, the correct preprocessing of their provided raw data, and the learning/reasoning that uses this information [2].

There have been several studies on smart homes for the elderly; in this section we review the ones that are most closely related to our work. The SWEET-HOME project aims to provide audio-based interaction technology that allows the user to have full control over the home environment, to detect distress situations, and to ease the social inclusion of the elderly and the frail [14]. The USEFIL [15] project tries to advance the independent living of the elderly, developing services that contribute to prolonging the seniors' quality of life and support them during their daily life activities, bearing in mind the special needs of the seniors and their acceptance of the new technologies invading their daily lives. An unobtrusive sensor network such as a Wearable Wrist Unit, a camera strategically placed behind a mirror coupled with a microphone, and a Kinect sensor provides in-home monitoring. Heterogeneous data are fused on-site to provide a more accurate description of the context, which is then sent to the server for trend analysis and recognition of early warning signs of health deterioration. The project "At home despite the age" [16] is a model for senior housing that consist of a dwelling that fosters an active and independent life style with smart technologies. ICT4LIFE [17] introduces an innovative approach for the real-time monitoring of patients, with the aim of preventing falls and social isolation, and to promote patients independence. The common denominator of all these solutions is the fact that they leverage the availability of long-term observations of the monitored environment.

In the literature, there are several studies that investigate the problem of daily activity recognition from data fusion of heterogeneous sensors. Monitoring the daily activities of the elderly is addressed in [18]; the authors have proposed a description-based approach where the human body context (e.g. sitting, standing, walking) is provided from a set of cameras, while the environmental context is obtained from accelerometers attached to objects of daily living. A rule-based reasoning engine is used to process and combine the two context types. Similarly, in [19] the authors have evaluated a video monitoring system for the identification of activities of daily living of older people in a model apartment where the home appliances are equipped with pressure and contact sensors. People's movements and posture are tracked over time by using RGB-D cameras. Multisensor environments, in particular those with heterogeneous sensors, open the need to investigate robust data fusion techniques; for example, in [20] the authors have demonstrated the fusion of data from inertial sensors worn at the waist, chest, thigh, and side of a person body for activity recognition. They propose a multisensor fusion framework, which consists of a sensor selection module and a hierarchical classifier. Other approaches (see e.g. [21]) have also used a classifier and fusion approach for inertial sensors of activity recognition, though these approaches present some disadvantages, for example the presence of motion noise, and the need for intersensor calibration.

In the reminder of the section, we focus on vision-based approaches, leveraging the use of RGB or RGB-D data streams. The use of vision sensors raises privacy issues, as well as considerations on the level of acceptance and perception of elder patients [9]. Smart technologies appear to be accepted by older adults, their family members, healthcare professionals, and caregivers. Nevertheless, older adults' perception of privacy can inhibit the acceptance and subsequent adoption of cameras [22]. For this reason, it is necessary to adopt vision sensors that minimize the intrusion into a patient's life, for example using only depth sensors in the bedrooms, and performing the processing online without the need of storing any image data.

As previously mentioned, in [23] the authors present a cognitive vision approach to recognizing a set of interesting ADLs for the elderly at home. This approach is composed of a video analysis component and an activity recognition component. The video analysis component includes person detection, person tracking, and human posture recognition. The activity recognition component contains a set of video event models and a dedicated video event recognition algorithm. In the study, the authors collaborate with medical experts (gerontologists from a hospital in Nice, France) to define and model a set of scenarios related to elderly activities of interest. Some of these activities require detecting a fine description of the human body such as postures. For this purpose, they propose ten 3D key human postures useful for recognizing a set of interesting human activities regardless of the environment.

Among the tools available in the computer vision field, of particular interest for application in the assistive living domain, is human posture recognition. Human posture can be determined and classified by considering several explicit or statistical models, in 2D or 3D workspaces. In [24], the authors propose an automatic video monitoring system, which takes as input video streams on which they perform detection, classification and tracking processes, and a priori knowledge for 3D scene modeling and recognition of events. They extract attributes (e.g. duration, walking speed) of automatically recognized physical task events in order to identify differences between groups of Alzheimer patients and healthy participants. A prototype fall detection system to be employed in a University hospital in Minnesota is proposed in [25]; the system uses features extracted from 3D point clouds created from Kinect depth images. It is worth noting that the choice of relying uniquely on depth images is due to privacy reasons. Finally, in [26] a video monitoring system for event recognition of older people using a hierarchical description-based approach and a 2D-RGB camera as input is presented. The authors addressed the issues of recognizing both physical tasks and instrumental activities of daily living during a clinical protocol for an Alzheimer disease study.

## 12.3  A STUDY CASE

In this section, we present a use case consisting of a model of protected discharge, where a patient, after being discharged from the hospital, is hosted for a few days (up to one week) inside an apartment. The facility is located in a hospital[1] and is equipped as a comfortable apartment where the patient

---

[1] The Galliera Hospital in Genova, Italy, http://www.galliera.it/.

can be monitored by a system of unobtrusive sensors, while physicians and nurses have the opportunity of monitoring the patient remotely.

A key element of the model is to acquire a multitude of heterogeneous data and carry out an offline data analysis with the purpose of automatically assessing the health and mental status of the patient. This automatic evaluation will complement the evaluation carried out by the physicians. More specifically, we are currently estimating a *motility index* that evaluates the overall health status of the patient, and that should represent a more objective, continuous, and quantitative counterpart of the generally adopted methodology to assess the mental and physical status of the patients, i.e. the Multidimensional Prognostic Index (MPI), which we will briefly introduce in Section 12.3.2.

In the following, we first describe the apartment, then we give an overview of the sensors installed in the facility, with specific reference to the vision devices. Then, we describe the processing stages and the algorithms that allow us to compute an estimation of the patients motility index starting from the acquired raw visual data.

## 12.3.1 The Sensorized Apartment

Fig. 12.1 shows the floor plan of the apartment. It consists of a common area (Fig. 12.2, right) with kitchenette and living room, a gym, and two bedrooms (Fig. 12.2, left). To make the atmosphere homey, an architectural study was conducted to choose appropriate colors, layout, and furniture, leading to an environment more similar to a regular apartment rather than to a hospital room. The common area hosts the majority of the ambient, vision, and pressure sensors. Health monitoring devices are located on a console table in the same area. In our system, for privacy purposes, cameras and RGB-D sensors are not located in the bedrooms and bathrooms, but thanks to position tags and presence sensors, it is possible to perform an unobtrusive monitoring of the patients in these areas. In the following, we first describe the vision sensors present in the apartment; to follow, we provide a brief overview of the other sensors.

### Vision-Based Components

In this section we focus on the vision components of the system and discuss their configuration. Fig. 12.3 shows the arrangement of the visual sensors in the living room of the apartment, highlighting in particular the fields of view and their overlapping areas.

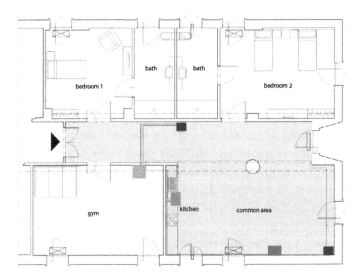

**Figure 12.1** Floor plan of the apartment. The blue and red rectangles represent the RGB-D sensors, the green squares correspond to the cameras. They are all wired to the workstation (orange square). (For interpretation of the colors in this figure, the reader is referred to the web version of this chapter.)

**Figure 12.2** Left panel: one of the two bedrooms; Right panel: the common area.

The RGB-D sensors are Asus Xtion Pro, with only the depth channel available. This choice comes from the possibility of also using them in the bedrooms, thus avoiding privacy issues [25]. Each sensor acquires a depth stream with VGA resolution ($640 \times 480$ pixels, at 30 fps). They cover a field of view of about 58 deg horizontal, 45 deg vertical, and 70 deg diagonal, with a range of operation between 0.8 m and 3.5 m. The first RGB-D sensor ($RGBD_1$) is located over the kitchen sink. Its field of view (FOV) is

**Figure 12.3** A sketch of the visual sensors, their fields of view, and their overlap. (For interpretation of the colors in this figure, the reader is referred to the web version of this chapter.)

highlighted in blue in Fig. 12.3, right, and it covers the entire kitchen and table area, i.e. where patients are supposed to have breakfast, lunch, and dinner. The second sensor ($RGBD_2$) is located near the TV in front of the sofa; its field of view is highlighted in red in Fig. 12.3, and it covers the living room, i.e. the sofa, the armchair, the library, and the area of the vital monitoring devices. It is worth noting that in order to accomplish some design constraints without affecting the appearance of the room, $RGBD_1$ and $RGBD_2$ are located at different heights from the floor. Each RGB-D sensor is connected via USB to a small PC–home theater system, again chosen to be as unobtrusive and as quiet as possible, running Ubuntu Operating System and the acquisition software we developed. The cameras are high resolution mini-dome IP cameras acquiring $1920 \times 1080$ frames at 25 fps. They are located in the two opposite corners of the room, indicated in red in Fig. 12.3. The mutual position of RGB-D sensors and cameras is intended to provide a partial overlap of the fields of view while covering complementary areas. The two home theater PCs and the cameras are connected to a local ethernet network.

## Other Sensors

Although this chapter focuses on vision-based sensors, we provide a very brief overview of all nonvision sensors currently installed in the facility. The physicians identified a minimal set of vital parameters to be monitored, including weight, blood pressure, heart rate, oxygen saturation (SpO2 level), glucose. They recommended that these parameters be measured twice a day directly by the patient. Based on these requirements, we identified a set of wearable and noninvasive devices, selected to guarantee the patient's complete freedom of movement (no cables, data are transmitted via wireless communication) and to allow automatic analysis of the acquired data (the data are transmitted to a remote workstation running an appropriate management software). The collection of all measurements is done via bluetooth by an LG G3 smartphone, which sends the data to the Ri-Healthy[2] platform, which stores the data and also allows physicians and nurses to remotely monitor the parameters.

The apartment is also equipped with other environment sensors, including an indoor localization system, presence sensors, cabinet door sensors, and chair occupancy sensors. Each patient is also required to wear an LG G Watch R5 equipped with a triaxial accelerometer. The feedback provided by this sensor is used primarily to classify a set of predefined activities.

## 12.3.2 Evaluating the Motility of Patients

Physicians, in particular geriatricians, use questionnaires and evaluation scales to assess the mental and physical status of their patients. Based on the results of these evaluations they estimate the Multidimensional Prognostic Index (MPI), a meaningful index of the patient's overall status which has also been shown to correlate with the likelihood of short-term and mid-term mortality in elderly people [8]. The MPI is based on a number of heterogeneous evaluations, usually taking into account eight different domains: Activities of Daily Living (ADL); Instrumental Activities of Daily Living (IADL); Short Portable Mental Status Questionnaire (SPMSQ); Mini Nutritional Assessments (MNA); Exton-Smith scale (ESS); Comorbidity Index Rating Scale (CIRS); Medical adherence; Living condition. Each of these domains consists of a series of items that are evaluated in a qualitative way by medical personnel through questions asked to patients. The obtained index (frailty index) has a value between 0.0 (low risk) and 1.0 (high risk of mortality). If the index has a value between 0.0 and 0.33 the prognostic risk of mortality is low; if the value is between 0.34 and 0.66 the prognostic risk of mortality is moderate; if the value is between 0.67 and 1.00 the prognostic risk is severe.

In our work, we refer in particular to the well-known Exton-Smith scale [27], a five-item questionnaire that allows doctors to determine physical condition, mental condition, activity, motility, and incontinence. For each item, a score from 1 to 4 is assigned, and consequently a risk scale

[2] http://www.rihealthy.com/.

is determined. We analyze this scale in terms of motility, activity distribution, and postural transfers. In particular, we compute the following *motility quantities*, which have been identified with the help of geriatricians:

- number of postural changes, i.e. from sitting to standing and vice versa: this is done by looking at the variation in heights of the detected skeletons (through RGB-Ds) in the scene;
- the total time spent moving and being still (in seconds and in percentage): this is done by checking the variation in the distribution of the velocity modulus;
- number of instances of walking, i.e. how many times in a given observation period people start walking;
- average distance and velocity of a single walk;
- total distance walked;
- longest distance covered in a single walk;
- longest walking time.

These measures can provide a *motility index* to be used in order to build up the final MPI. Our approach to obtain their estimates is discussed in the remainder of the chapter.

## 12.3.3 Our Monitoring System

Fig. 12.4 shows the pipeline of our system, from the acquisition and processing stages, to the computation of the *motility quantities*, and finally to the estimation of the *motility index*. In the following, we discuss how we acquire and store the data to prepare them for future analysis. Later, we provide details on the estimation of the motility quantities.

### 12.3.3.1 Data Acquisition, Storage, and Preprocessing

The software for the acquisition from each RGB-D sensor runs continuously on the home theater PCs. In the presence of people in the field of view, they acquire and store the 3D position of the skeleton joints for each user, which is assigned a unique identifier. The association with the timestamp is also stored for future reference. The two RGB-D sensors produce independent yet temporally related streams of acquisitions—each with respect to its own reference frame—and depending on the position the user may be associated with one or two simultaneous joint observations. We store the observations—in terms of tuples of the form

$$(RGBD_{id}, TimeStamp, USER_{id}, JP_1, \ldots, JP_N),$$

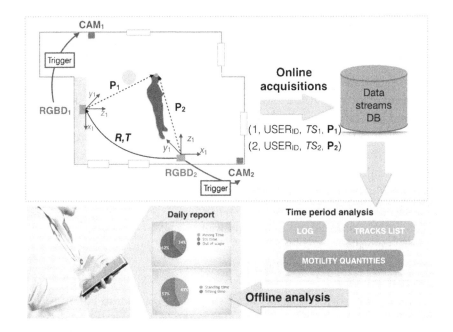

**Figure 12.4** A sketch of the pipeline of our system.

where $RGBD_{id} = \{1, 2\}$, while $JP_i = (X_i, Y_i, Z_i)$ is the 3D position of the $i$th joint—in files covering 30 minutes of acquisitions. Although we acquire all the skeleton joints, in the analysis we only take into account the head joint since it meets the requirements of stability and precision for our reference application [28,29].

In order to be able to merge observations obtained from the two sensors we need to adjust their space-time mutual relationship:

- The home theater PCs are temporally synchronized so that the time for the two views correspond;
- The reference systems of the two views are related by a global rigid transformation, a roto-translation. We estimated this transformation by collecting simultaneous observations from the small intersection between the two fields of view and applying a Direct Linear Transformation (DLT) algorithm with RANdom SAmple Consensus (RANSAC) to cope with the presence of outliers.

Given a temporal period of analysis $\mathcal{T}$, we preprocess the observations acquired via the following steps:

  i  We apply the roto-translation to all observations obtained by $RGBD_2$ to bring them in the reference frame of $RGBD_1$;

ii  We merge the observations from the two views by temporally ordering all the tuples; then, for all observations occurring within a second, we compute the average position. Also, we apply a smoothing on the transitions between the two views, and discard the oscillations;

iii  In the presence of multiple users, we solve data association using the output of the localization sensors installed in the facility.

When detecting a human in the scene, the RGB-D sensor also triggers the acquisition from the corresponding camera. Under normal conditions of the system, the video streams are stored, but are accessible only when necessary (e.g. for the safety of the user) in order to guarantee the privacy of the patients. In our investigation phase, videos are mainly adopted for a visual inspection allowing the validation of our results. Although they do not play a leading role in our current investigation, RGB cameras can be exploited for further analysis of ADL (see the discussion on open issues in Section 12.4).

The results of the data preprocessing can be formalized as follows. Given the period of analysis $\mathcal{T} = [t_0, t_\mathcal{T}]$ and an observed user $uid$, we first identify sets of joint observations consecutive in time (it is not guaranteed that the user remains visible by the sensors for the entire period of analysis). Formally, this amounts to identifying a series of time instants

$$T_m = [t_0 = t_{m0}, t_{m1}, \ldots, t_{mk-1}, t_{mk} = t_\mathcal{T}] \tag{12.1}$$

such that a pair $(t_{mi}, t_{mi+1})$ delimits a continuous interval of observations.

We then produce a temporal series of measures in the form

$$\mathcal{M}_\mathcal{T}^{uid} = \bigcup_{t_{mi}} \{\mathcal{O}_{t_k}^{uid}\}_{t_k=t_{mi}}^{t_{mi+1}} \tag{12.2}$$

with

$$\mathcal{O}_t^{uid} = \left(RGBD_{id}, TimeStamp_t, \mathbf{P}_t^{head}\right), \tag{12.3}$$

where $t_0$ and $t_\mathcal{T}$ are the instants temporally delimitating the window of analysis $\mathcal{T}$, $\mathbf{P}_t^{head} = (X_t^{head}, Y_t^{head}, Z_t^{head})$ is the 3D position of the head joint and time $t$, and $TimeStamp_t$ is the timestamp of time $t$.

For readability, hereafter we omit the $uid$ and $head$ references. Thus, unless otherwise stated and without losing in generality, our discussion will refer to the analysis of the head joint of a single user.

### 12.3.3.2 Motility Quantities Estimation

In this section we address the problem of evaluating motility quantities used later to estimate the *motility index* of a user. Motility quantities are based on statistics on the joint observations included in the set $\mathcal{M}_{\mathcal{T}}$, and they can be computed at a global level on the time period $\mathcal{T}$ or on subintervals called *tracks*. In the following, we detail the computation of these statistics and discuss the experimental results.

#### Global Motility Quantities

Referring to the list presented in Section 12.3.2, from a global point of view we estimate the following:

- **Number of postural changes** ($TR_{2stand}$, $TR_{2sit}$). The estimate is based on the analysis of the $Y$ (height) coordinate of the head joint and its temporal variations. More specifically, we fix a threshold $\tau_y$ and an interval around it defining the range of values that identify a transition between sitting and standing posture or vice versa. It is worth noting that this value strongly depends not only on the height of a person, but also on the environmental elements and from case to case more than one value may be needed. For instance, in our facility we have to consider the presence of chairs and a sofa for which the sitting postures are characterized by two different values of $Y$ (see Fig. 12.5A).
  The number of sit-to-stand transitions $TR_{2stand}$ can be determined as the cardinality of the set collecting the time instants $t_i$, $t_0 < t_i < t_{\mathcal{T}}$, such that a $k > 0$ exists for which $Y_{t_i-k} < \tau_y$ while $Y_{t_i+k} > \tau_y$. The identification of the stand-to-sit transitions is straightforward.

- **Number of instances of walks** ($W$). We identify here the time instants in which the magnitude of the velocity of the joint goes beyond a certain threshold $\tau_v$ (see Fig. 12.5B). To compute the velocity module we first approximate the instantaneous velocity components as differences between consecutive joints 3D positions, i.e. $\mathbf{V}_t = \mathbf{P}_t - \mathbf{P}_{t-1} = (V_{xt}, V_{yt}, V_{zt})$, and then compute the module as $V_t = \sqrt{V_{xt}^2 + V_{yt}^2 + V_{zt}^2}$. The number of instances of walk $W$ is computed as the cardinality of the set including the time instants $t_i | t_0 < t_i < t_{\mathcal{T}}$ for which a $k > 0$ exists such that $V_{t_i-k} < \tau_v$ and $V_{t_i+k} > \tau_v$.

- **Number of instances of stops** ($S$). Similarly to the previous point, this quantity is estimated as the cardinality of the set including the time instants $t_i | t_0 < t_i < t_{\mathcal{T}}$ for which a $k > 0$ exists such that $V_{t_i-k} > \tau_v$ and $V_{t_i+k} < \tau_v$.

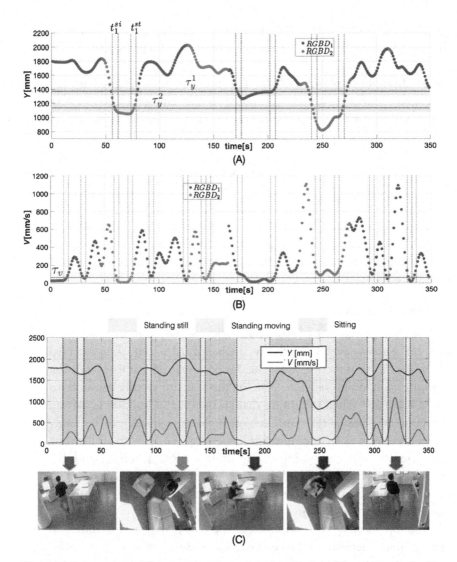

**Figure 12.5** An example of temporal analysis of the head height (A) and its velocity (B). Posture and velocity changes are detected and are used to identify tracks (C).

- **Total time spent moving (*TM*) and still (*TS*).** Since $\mathcal{M}_{\mathcal{T}}$ includes observations averaged at a temporal resolution of one second, this statistic amounts to counting the number of observations in $\mathcal{M}_{\mathcal{T}}$ whose corresponding $V$ is above or below the threshold $\tau_v$, for moving and still instances respectively. Note that only observations in which the subject is standing are considered for the estimation of this quantity.

- **Total time spent sitting ($T_{sit}$).** This is estimated as the number of observations for which the $Y$ coordinate of the head is below the threshold $\tau_y$.
- **Total distance walked ($DW$).** The total distance spanned by the observations in $\mathcal{M}_T$ can be computed as the sum of all consecutive displacements between joint positions. In our setting, this displacement corresponds to the velocity magnitude, thus

$$DW(\mathcal{T}) = \sum_{t_{mi}} \sum_{t_k=t_{mi}+1}^{t_{mi+1}} ||\mathbf{V}_{t_k}||. \qquad (12.4)$$

Note that $TT = TM + TS + T_{sit} = \#\mathcal{M}_T$, which amounts to the number of actual observations in the time period $\mathcal{T}$. In general, $TT$ is lower than the length of $\mathcal{T}$ since the subjects may spend some time out of the field of view of the sensors.

### Tracks Detection and Analysis

A track is a continuous set of temporally adjacent observations where the *dynamic and postural state* of the user is unaltered. A new track is detected when the user is (re-)entering the scene, his posture is subject to a transition from sitting to standing or vice versa, or the velocity rapidly grows from zero to a reference value, indicating that an instance of walking is starting or stopping. It is straightforward to notice that the tracks can be easily identified by appropriately combining $T_m$ (see Eq. (12.1)) with all the time instants collected for computing the quantities $TR_{2stand}$, $TR_{2sit}$, and $W$, ending up with a sequence of ordered time instants $T_{tk} = [t_0, \ldots, t_h, \ldots, t_T]$, where $t_h$ is labeled according to one of the sets mentioned above. With this in mind, we can consider three different classes of tracks, corresponding the user *sitting*, *standing still*, and *moving*. The plot in Fig. 12.5C shows an example of tracks detected and color-coded according to the reference class. Let us denote with $TK_T = [Tk_1, \ldots, Tk_M]$ the set of $M$ tracks associated with the time period $\mathcal{T}$, where the $j$th track is a series of $X$ and observations consecutive in time, $Tk_j = [O_{t_j}, \ldots, O_{t_{j+1}}]$, with $t_j \in T_{tk}$. Only on tracks labeled as moving do we compute a set of statistics—with straightforward mathematical definitions summarized in Table 12.1—as follows:

- **Average distance ($AD$) and average velocity ($AV$) of moving tracks**, see Eqs. (12.6) and (12.8);
- **Longest distance ($LD$) covered in a single walk**, see Eq. (12.9);
- **Longest walking time ($LT$)**, see Eq. (12.10).

**Table 12.1**  A summary of the mathematical formulation for the tracks analysis

| | |
|---|---|
| Distance covered in a single track | $$DW(Tk_j) = \sum_{t=t_j}^{t_{j+1}} \|\mathbf{V}_t\| \qquad (12.5)$$ |
| Average distance | $$AD(\mathcal{T}) = \frac{1}{M}\sum_{j=1}^{M} DW(Tk_j) \quad (12.6)$$ |
| Track length | $$L(Tk_j) = \#Tk_j \qquad (12.7)$$ |
| Average velocity | $$AV(\mathcal{T}) = \frac{1}{M}\sum_{j=1}^{M} \frac{DW(Tk_j)}{L(Tk_j)} \quad (12.8)$$ |
| Longest distance | $$LD(\mathcal{T}) = \max_{j=1}^{M} DW(Tk_j) \quad (12.9)$$ |
| Longest track | $$LT(\mathcal{T}) = \max_{j=1}^{M} L(Tk_j) \quad (12.10)$$ |

## Motility Index Estimation

Now that we have evaluated the motility quantities we can combine them into a single value called the Motility Index ($MI$) that summarizes the overall level of dynamism of the observed subjects. More formally, the index is estimated as

$$MI(\mathcal{T}) = (1-\alpha)\left(\frac{T_{sit}+TS}{TT}\right) + \alpha\left[C\left(1 - \frac{TR_{2sit}+TR_{2stand}+W+S}{TT}\right)\right],$$
(12.11)

where the first term quantifies the percentage of inactivity time, while the second determines the relative amount of postural and dynamic transitions with respect to the entire time period. The parameter $\alpha$ is a value chosen to weight the importance of the two terms of the equation, while $C$ is a factor to make the second term numerically comparable with the first one. Similarly to the MPI, the motility index approaches 1 when the motility of the subject is not satisfactory.

## Experimental Analysis

In this section, we discuss the experimental evaluation we performed to assess the methodologies described above on a set of manually annotated

**Table 12.2** Description of the annotated sequences recorded in the monitored apartment

| Sequence | Age | Time interval [min] | $T^*_{sit}$ [s] | $TS^*$ [s] | $TM^*$ [s] |
|---|---|---|---|---|---|
| Seq. #1 | 23 | 90 | 478 | 79 | 296 |
| Seq. #2 | 22 | 150 | 5260 | 755 | 1165 |
| Seq. #3 | 24 | 120 | 224 | 174 | 181 |
| Seq. #4 | 36 | 30 | 126 | 79 | 384 |
| Seq. #5 | 40 | 30 | 99 | 77 | 196 |

data. The annotation was performed using the video sequences acquired by $CAM_1$ and $CAM_2$ in the same temporal span of the joint acquisitions considered here. Note that only the joint data are used in the following evaluation.

In Table 12.2, we summarize the main characteristics of the five sequences we considered. In each sequence, a single subject (volunteer) is observed while moving without constraints in the apartment, and spontaneously performing common daily life activities (e.g. walking, sitting at the table, standing in front of the stove). For these reasons, the acquired data represent a suitable testbed for the evaluation of our methods.

For each sequence, we report the age of the volunteer, the total time in which he/she has been monitored, and the ground truth for a subset of the statistics mentioned in Section 12.3.3.2, namely the total time spent sitting ($T^*_{sit}$), standing still ($TS^*$), and moving ($TM^*$). All quantities are expressed in seconds. It is worth noting that within the temporal period of observation the subjects spent a certain amount of time out of the fields of view of the vision sensors (e.g. in the bedroom).

A qualitative overview of the space occupancy and of the trajectories commonly produced in the environment is shown in Fig. 12.6. In Fig. 12.6A, we show the projections of the head joint positions on the image plane representing a visual sketch of the environment. Projections were obtained by estimating the homography between the $X$ and $Z$ coordinates of the head joints and their corresponding location on the image. Different colors refer to different RGB-D sensors. In Fig. 12.6B, an alternative visualization highlights the space occupancy. From a visual inspection of the two representations, it is easy to identify common patterns of movements and to get an idea of the scene portions where people spend most of their time.

We now discuss in greater detail the results of the analysis we performed according to Section 12.3.3.2. We start by comparing our estimates with

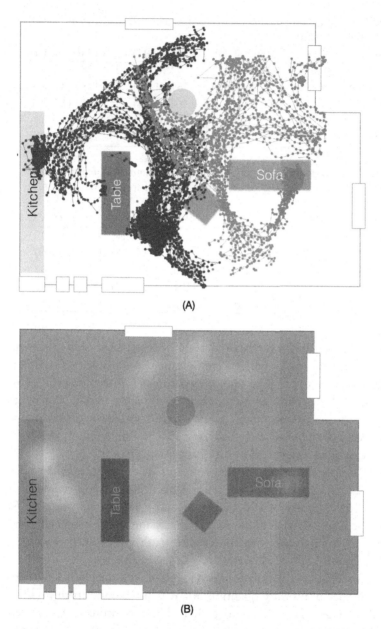

(A)

(B)

**Figure 12.6** An example of distribution of head joint positions over time (for Seq. #2). (A) The trajectories of head positions projected onto an image plane representing the environment (blue and red refer to $RGBD_1$ and $RGBD_2$, respectively). (B) A heat map obtained from the distribution of points shown above. (For interpretation of the colors in this figure, the reader is referred to the web version of this chapter.)

**Table 12.3** Estimates of the global statistics and associated relative errors with respect to the ground truth

| Sequence | $TT$ [s] | $T_{sit}$ (Err) [s] (%) | $TS$ (Err) [s] (%) | $TM$ (Err) [s] (%) |
|---|---|---|---|---|
| Seq. #1 | 853 | 496 (3.8%) | 74 (6.3%) | 283 (4.4%) |
| Seq. #2 | 7180 | 5239 (0.4%) | 752 (0.4%) | 1189 (2.1%) |
| Seq. #3 | 579 | 202 (9.8%) | 164 (5.7%) | 213 (17.7%) |
| Seq. #4 | 589 | 128 (1.6%) | 84 (6.3%) | 377 (1.8%) |
| Seq. #5 | 372 | 92 (7.1%) | 81 (5.2%) | 167 (14.8%) |

**Table 12.4** Accuracies on transitions of posture or dynamics for which we have a ground truth available. The corresponding ground truth value is given in parentheses

| Sequence | $TR_{2stand}$ | $TR_{2sit}$ | $W$ | $S$ | #$TRACKS$ |
|---|---|---|---|---|---|
| Seq. #1 | 10 (8) | 8 (8) | 6 (5) | 13 (14) | 30 (21) |
| Seq. #2 | 20 (23) | 21 (23) | 37 (38) | 57 (48) | 115 (109) |
| Seq. #3 | 5 (5) | 5 (5) | 11 (13) | 10 (12) | 22 (25) |
| Seq. #4 | 9 (10) | 8 (10) | 9 (9) | 18 (14) | 36 (30) |
| Seq. #5 | 3 (3) | 3 (3) | 7 (7) | 7 (7) | 15 (15) |

ground truth values, when available. Table 12.3 shows, for each sequence, the actual amount of time in which a measurement is available ($TT$). Also, it reports the estimates of the total time spent sitting ($T_{sit}$), standing still ($TS$), and moving ($TM$), together with the relative error computed in percentage with respect to the ground truth. It can be seen that the performance is in general very accurate, also considering that the annotation, which is manually performed, is affected by an intrinsic uncertainty.

Table 12.4 lists the results of the estimates of the posture and velocity transitions, i.e. the number of sit-to-stand ($TR_{2stand}$) and stand-to-sit ($TR_{2sit}$) and the number of still-to-moving ($W$) and moving-to-still ($S$) transitions, respectively, that allow us to identify the tracks composing the trajectories. A comparison with the ground truth values (in parentheses) shows the reliability of our estimates.

In the protocol of evaluation designed together with the physicians there are also a number of motility quantities for which a ground truth is either not available or is very difficult to gather. These measures are reported in Table 12.5; the columns refer to the total distance walked ($DW$), the average distance covered in the tracks ($AD$), the average velocity of moving tracks ($AV$), the longest distance covered in a single track ($LD$), and the longest track ($LT$). The results were positively evaluated with a qualitative comparison with the corresponding video sequences.

**Table 12.5**  Estimates of statistics for which a ground truth is not available

| Sequence | $DW$ [m] | $AD$ [m] | $AV$ [m/s] | $LD$ [m] | $LT$ [s] |
|----------|----------|----------|------------|----------|----------|
| Seq. #1 | 189.1 | 3.7 | 0.2 | 38.3 | 174 |
| Seq. #2 | 1123.6 | 7.2 | 0.2 | 157.6 | 853 |
| Seq. #3 | 277.2 | 9.5 | 0.4 | 85.5 | 130 |
| Seq. #4 | 149.8 | 2.8 | 0.2 | 22.0 | 56 |
| Seq. #5 | 99.4 | 3.8 | 0.2 | 15.4 | 33 |

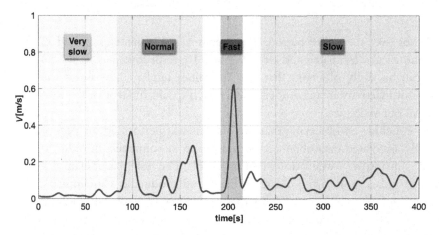

**Figure 12.7** The velocity estimated over time on a sequence where a subject walks at different speeds.

It is worth noting that the capability of appropriately quantifying the level of dynamism of a subject is of high relevance in this application domain, also considering that elderly people are characterized by a range of walking speeds that is in general lower than for adults. We thus provide a simple example showing the possibility of further analyzing the data considering more subtle dynamic properties. In particular, Fig. 12.7 shows the trend of velocity magnitude in a specific sequence where the volunteer was asked to move at different velocities in the apartment. The plot we show encodes these changes. Note that the velocity values for normal walking are comparable to the estimates on the average velocity given in Table 12.5.

We finally provide in Table 12.6 the motility index estimations for the five subjects according to Eq. (12.11). In addition to the final $MI$, in the table we also list the values of the two main terms of the equation, namely the inactivity time (the percentage of time spent sitting or standing still) and an estimation of inactivity from the number of postural or dynamic

**Table 12.6** The estimation of the motility index (*MI*) for the five subjects (see text for details)

| Sequence | % Inactivity time | Inactivity from transitions | *MI* |
|---|---|---|---|
| Seq. #1 | 0.67 | 0.57 | 0.65 |
| Seq. #2 | 0.83 | 0.81 | 0.83 |
| Seq. #3 | 0.63 | 0.46 | 0.60 |
| Seq. #4 | 0.36 | 0.25 | 0.34 |
| Seq. #5 | 0.47 | 0.46 | 0.46 |

transitions over the entire period. The values of the parameters $\alpha$ and $C$ were empirically estimated and set to 0.2 and 10, respectively.

It can be easily observed that the *MI* values tend to increase as the amount of inactivity increases (Seq. #2), which speaks in favor of the reliability of our strategy.

We conclude by observing that this analysis was performed on a set of manually annotated data obtained using a group of volunteers in order to provide a quantitative evaluation of the approach. The system is now up and running, and is monitoring elderly patients and producing a preliminary stream of estimates. The quality of these estimates is comparable to the outcome of the analysis discussed here. Moreover, the use of these measures for the automatic estimation of a more informative and multidimensional frailty index is currently under investigation.

## 12.4 DISCUSSION

### The Objectives We Reached

In this chapter we presented a model for a protected discharge facility which has been designed, implemented, and validated within the Galliera Hospital (Genova, Italy). Here the patient, after being discharged from the hospital, is hosted for about one week and can be monitored by a system of sensors, while physicians and nurses have the opportunity of monitoring him or her remotely. As a general goal of the project, first we wanted to study how advanced technology enables elderly patients to improve their lifestyle and habits in order to reduce the need for assistance from medical staff and thus the costs for both the public healthcare sector and patients. Second, from the data analysis view point, our main challenge was to evaluate the status of the patient, based on relatively short observations (while the literature usually exploits long observations of about 6–12 months [30]). For this reason, in close collaboration with geriatricians and domain experts, we identified a set of meaningful measurements that allow us to derive an estimate of a *motility index* to be associated with the patient's overall health status. Today, the model hosts a variety of sensors and runs a set of algorithms to automatically evaluate motility measurements. In this chapter we reported a qualitative and quantitative experimental assessment based on a set of data we acquired with the help of five volunteers who spent a few days in the apartment. The corresponding acquisition thus represents a suitable testbed for the analysis of daily life activities. We now discuss a set of noticeable aspects:

- **Usability and generality.** The model can be seen as a proof of concept of a wider variety of environments and situations. It does not rely on any specificity of the installation, and could be easily adapted to private homes or protected residences. The hardware setup is low-cost, easily configurable, and does not need to obey specific constraints. The only issue to be considered is the problem of occlusions; therefore, it is advisable to avoid excess furniture.

- **Accessibility.** As an assistive tool, the proposed model is not used directly by the patient. The end-users of the technology are physicians, and the proposed model is highly accessible to them as it produces reports that have been designed in close collaboration with doctors and are based on the type of questionnaires they normally use to assess a patient's status. Instead, for the patients, there is no accessibility issue to consider as they should be virtually unaware of the monitoring installation.

- **Robustness.** The whole model of protected discharge is undergoing a live test that started nearly a year ago. The various components, including the vision-based project discussed in this chapter, have been continuously evaluated and quantitative tests are carried out each time a new computational model is inserted or updated. Quantitative tests requiring a manual annotation of actions performed by the subjects are only performed with the help of volunteers. Current tests did not highlight any specific issue; instead, the system appears to be robust to illumination changes and to scene variations.

- **Usefulness.** Considering the steady growth of the elderly population, the proposed model has a potential impact on society that goes beyond the specific problem we have considered.

## Open Problems

We conclude with an account of *open problems and research opportunities*. An important aspect of our model is its modularity and the fact it will easily accommodate new data analysis algorithms. With regard to computer vision, we will consider the general problem of action or activity analysis, addressing other aspects of MPI and with the aim of enriching our automatic frailty index. Following the geriatricians' suggestions, working jointly on video streams and RGB–D streams, we will analyze gait and posture, evaluate nutrition habits (based on recognizing cooking and eating actions), socialization cues, and the patient's overall independence (ability to perform tasks without the help of a caregiver). Closely related with this goal is personal care independence. For this issue, we will consider the use of the sole depth sensors to protect the privacy of patients.

## Perspective

The model of protected discharge we are contributing to has a very innovative structure inspired by the concept of a *living lab*, where the design of new functionalities can proceed in parallel with live testing. In principle, this concept could allow us to validate new sensors and to design, implement, and test new methods. Importance should be given to the integration of different observations and to the need of optimizing the considerable size of the dataset we are dealing with on a daily basis. In the longer term, we are proposing a paradigm shift in assistance, where intelligent environments act as personalized, socially aware, and evolving *cognitive prostheses*, adaptively integrating their cognitive capabilities. We aim to perform a continuous comprehensive geriatric assessment of the assisted person's frailty index, and to coach them in keeping an active and healthy lifestyle, informing medical staff in case medical assistance is required.

## ACKNOWLEDGMENTS

The authors would like to thank Andrea Capriccioli for developing the initial modules of motion analysis, Imavis Srl for providing and maintaining the cameras currently installed in the facility, Annalisa Barla and Alessandro Verri for fruitful discussions. The authors are grateful to Galliera Hospital, and in particular to Alberto Pilotto, Alberto Cella, and Gian Andrea Rollandi for the opportunity of participating in the MoDiPro Project. A final thanks to the volunteers who agreed to spend time in the apartment during the various stages of the work.

## REFERENCES

[1] Q. Ni, A.B. García Hernando, I.P. de la Cruz, The elderly's independent living in smart homes: a characterization of activities and sensing infrastructure survey to facilitate services development, Sensors 15 (5) (2015) 11312–11362.

[2] K. Giannakouris, Ageing characterises the demographic perspectives of the European societies, Statistics in Focus 72 (2008) 2008.

[3] C. Lafont, S. Gérard, T. Voisin, M. Pahor, B. Vellas, et al., Reducing "iatrogenic disability" in the hospitalized frail elderly, The Journal of Nutrition, Health & Aging 15 (8) (2011) 645–660.

[4] S. Volpato, G. Onder, M. Cavalieri, G. Guerra, F. Sioulis, C. Maraldi, et al., Characteristics of nondisabled older patients developing new disability associated with medical illnesses and hospitalization, Journal of General Internal Medicine 22 (5) (2007) 668–674.

[5] S. Volpato, J. Daragjati, M. Simonato, A. Fontana, L. Ferrucci, A. Pilotto, Change in the Multidimensional Prognostic Index Score during hospitalization in older patients, Rejuvenation Research 19 (3) (2016) 244–251.

[6] J. Lenzi, M. Mongardi, P. Rucci, E. Di Ruscio, M. Vizioli, C. Randazzo, et al., Sociodemographic, clinical and organisational factors associated with delayed hospital discharges: a cross-sectional study, BMC Health Services Research 14 (128) (2014) 1–8.

[7] R.V. Azzopardi, S. Vermeiren, E. Gorus, A.K. Habbig, M. Petrovic, N. Van Den Noortgate, et al., Linking frailty instruments to the international classification of functioning, disability, and health: a systematic review, Journal of the American Medical Directors Association 17 (11) (2016) 1066–e1.

[8] A. Pilotto, L. Ferrucci, M. Franceschi, L.P. D'Ambrosio, C. Scarcelli, L. Cascavilla, et al., Development and validation of a multidimensional prognostic index for one-year mortality from comprehensive geriatric assessment in hospitalized older patients, Rejuvenation Research 11 (1) (2008) 151–161.

[9] B. Adair, K. Miller, E. Ozanne, R. Hansen, A.J. Pearce, N. Santamaria, et al., Smart-home technologies to assist older people to live well at home, Journal of Aging Science (2013).

[10] P. Menschner, A. Prinz, P. Koene, F. Köbler, M. Altmann, H. Krcmar, J.M. Leimester, Reaching into patients' homes – participatory designed AAL services, Electronic Markets 21 (1) (2011) 63–76.

[11] C. Chen, P. Dawadi, CASASviz: Web-based visualization of behavior patterns in smart environments, in: IEEE International Conference on Pervasive Computing and Communications Workshops (PERCOM Workshops), 2011, IEEE, 2011, pp. 301–303.

[12] J. Barlow, T. Venables, Will technological innovation create the true lifetime home?, Housing Studies 19 (5) (2004) 795–810.

[13] K.L. Courtney, G. Demeris, M. Rantz, M. Skubic, Needing Smart Home Technologies: The Perspectives of Older Adults in Continuing Care Retirement Communities, Radcliffe Medical Press, 2008.

[14] M. Vacher, D. Istrate, F. Portet, T. Joubert, T. Chevalier, S. Smidtas, et al., The sweet-home project: audio technology in smart homes to improve well-being and reliance, in: Annual International Conference of the IEEE, Engineering in Medicine and Biology Society (EMBC), 2011, pp. 5291–5294.

[15] https://www.usefil.eu/.

[16] http://mimowieku.pl/.

[17] http://www.ict4life.eu.

[18] Y. Cao, L. Tao, G. Xu, An event-driven context model in elderly health monitoring, in: Symposia and Workshops on Ubiquitous, Autonomic and Trusted Computing, 2009, UIC-ATC'09, IEEE, 2009, pp. 120–124.

[19] N. Zouba, F. Bremond, M. Thonnat, An activity monitoring system for real elderly at home: validation study, in: Seventh IEEE International Conference on Advanced Video and Signal Based Surveillance (AVSS), 2010, IEEE, 2010, pp. 278–285.

[20] L. Gao, A.K. Bourke, J. Nelson, Activity recognition using dynamic multiple sensor fusion in body sensor networks, in: Annual International Conference of the IEEE, Engineering in Medicine and Biology Society (EMBC), 2012, pp. 1077–1080.

[21] R. Liu, M. Liu, Recognizing human activities based on multi-sensors fusion, in: 4th International Conference on Bioinformatics and Biomedical Engineering (iCBBE), 2010, IEEE, 2010, pp. 1–4.

[22] K.L. Courtney, Privacy and Senior Willingness to Adopt Smart Home Information Technology in Residential Care Facilities, Schattauer, 2008.

[23] N. Zouba, B. Boulay, F. Bremond, M. Thonnat, Monitoring activities of daily living (ADLs) of elderly based on 3D key human postures, in: Cognitive Vision, Springer, 2008, pp. 37–50.

[24] V. Joumier, R. Romdhane, F. Bremond, M. Thonnat, E. Mulin, P. Robert, et al., Video activity recognition framework for assessing motor behavioural disorders in Alzheimer disease patients, in: International Workshop on Behaviour Analysis and Video Understanding (ICVS 2011), 2011, p. 9.

[25] T. Banerjee, M. Rantz, M. Li, M. Popescu, E. Stone, M. Skubic, S. Scott, Monitoring Hospital Rooms for Safety Using Depth Images, AI for Gerontechnology, Arlington, VA, USA, 2012.

[26] C.F. Crispim, V. Bathrinarayanan, B. Fosty, A. Konig, R. Romdhane, M. Thonnat, F. Bremond, Evaluation of a monitoring system for event recognition of older people, in: 10th IEEE International Conference on Advanced Video and Signal Based Surveillance (AVSS), 2013, pp. 165–170.

[27] M. Bliss, R. McLaren, A. Exton-Smith, Mattresses for preventing pressure sores in geriatric patients, Monthly Bulletin of the Ministry of Health and Public Health Laboratory Service 25 (1966) 238–268.

[28] R.A. Clark, Y.H. Pua, K. Fortin, C. Ritchie, K.E. Webster, L. Denehy, A.L. Bryant, Validity of the Microsoft Kinect for assessment of postural control, Gait & Posture 36 (3) (2012) 372–377.

[29] A. Canessa, M. Chessa, A. Gibaldi, S.P. Sabatini, F. Solari, Calibrated depth and color cameras for accurate 3D interaction in a stereoscopic augmented reality environment, Journal of Visual Communication and Image Representation 25 (1) (2014) 227–237.

[30] C.N. Scanaill, S. Carew, P. Barralon, N. Noury, D. Lyons, G.M. Lyons, A review of approaches to mobility telemonitoring of the elderly in their living environment, Annals of Biomedical Engineering 34 (4) (2006) 547–563.

# INDEX

Printed in the United States
By Bookmasters